LABORATORY MANUAL FOR

Human Anatomy
for Use with Models
and Prosected Cadavers

WITH ILLUSTRATIONS
PREPARED ESPECIALLY FOR COLORING

SECOND EDITION

Victor P. Eroschenko, Ph.D.

PROFESSOR OF HUMAN ANATOMY

DEPARTMENT OF BIOLOGICAL SCIENCES

AND

WAMI Medical Program

UNIVERSITY OF IDAHO, MOSCOW, IDAHO

An imprint of Addison Wesley Longman, Inc.

Menlo Park, California • Reading, Massachusetts • New York • Harlow, England
Don Mills, Ontario • Sydney • Mexico City • Madrid • Amster'

To:

Sharon, Shannon, Sarah, Shaun,

Tatiana, Kathryn, Diane, and Elke

Executive Editor: Bonnie Roesch
Developmental Editor: Cyndy Taylor
Project Coordination and Text Design: Electronic Publishing Services Inc.
Cover Designer: Kay Petronio
Cover Photograph: Mehau Kulyk & Victor De Schwanberg/Science Photo Library; Hilda Muinos; Mark Nielsen
Art Studio: R. R. Donnelley & Sons-Barbados
Electronic Production Manager: Mike Kemper
Manufacturing Manager: Helene G. Landers
Electronic Page Makeup: Electronic Publishing Services Inc.
Text Printer: Malloy Lithographing, Inc.
Cover Printer: Malloy Lithographing, Inc.

Library of Congress Cataloging in Publication Data

Eroschenko, Victor P.
    Laboratory manual for human anatomy for use with models and
prosected cadavers : with illustrations prepared especially for
coloring / Victor P. Eroschenko and WAMI Medical Program. — 2nd ed.
      p. cm.
    Rev. ed. of : Laboratory manual for human anatomy using cadavers.
c1990.
    Originally published : New York : HarperCollins College Publishers,
c1996.
    ISBN 0-673-99558-5 (paper, spiral bound)
    1. Human anatomy—Laboratory manuals.   I. Eroschenko, Victor P.
Laboratory manual for human anatomy using cadavers.   II. WAMI
Medical Program.   III. Title.
    [DNLM: 1. Anatomy—laboratory manuals. 2. Cadaver—laboratory
manuals. Not Acquired / QS 25 E71L 1996a]
QM34.E76 1997
611'.0078—dc21
DNLM/DLC
for Library of Congress   95-17910
                                                                CIP

ISBN 0-673-99558-5

13 14 15 -OPM- 07 06 05

CONTENTS

PREFACE

Introduction

The basic concepts for studying human anatomy remain the same in this second edition of the *Laboratory Manual for Human Anatomy for Use with Models and Prosected Cadavers* as they were in the first edition. The manual is especially prepared as a complete guide for use in laboratory courses where models of human body and/or human cadavers are utilized to study anatomy. In anatomy laboratories where human cadavers are not available to beginning students, accurate models of the human body may substitute. The text and anatomical illustrations have been prepared with this possibility in mind. Using the manual, the students will learn the correct anatomical terminology, description, location, and appearance of different organs in the human body.

The intended audience for *Laboratory Manual for Human Anatomy for Use with Models and Prosected Cadavers* are students preparing for such allied health professions as nursing, physical therapy, sports sciences, physical education, recreation, laboratory technology, and related fields in which knowledge of human anatomy is important. This manual contains sufficient information on human anatomy to satisfy the demands of most introductory courses.

The distinctive features of *Laboratory Manual for Human Anatomy for Use with Models and Prosected Cadavers* include a thorough description of anatomical terms and structures, and numerous black and white illustrations designed for coloring by students. By reading the text and examining the illustrations, the student should be able to identify and locate the described anatomical structure on a model or cadaver with minimal assistance.

Like the first edition of *Laboratory Manual for Human Anatomy for Use with Models and Prosected Cadavers,* the present edition uses the traditional systemic approach to study anatomy. Each chapter starts with a list of laboratory **Objectives**, or goals. This is followed by a thorough **Description** of the anatomical terms, structures, and/or concepts, with reference to specific illustrations. The **Identification** section asks the student to identify the described structures on a model and/or prepared human cadaver and then color the parts in the illustrations. Each chapter ends with a detailed **Summary and Checklist** of major topics covered and specific **Laboratory Exercises** that test the student's knowledge of the chapter.

New to the Second Edition

The *Laboratory Manual for Human Anatomy for Use with Models and Prosected Cadavers* has undergone major revision. The descriptive text has been carefully revised, edited, and reduced. Chapters on the nervous system, in particular, have been extensively revised. In other chapters, new tables summarize information, further reducing the written material. The number of labels in most illustrations has been reduced to emphasize particular structures described in the corresponding section or chapter. Some illustrations have been eliminated because they duplicated existing figures or were no longer needed. Other illustrations have been enlarged or improved.

A major addition to the *Laboratory Manual for Human Anatomy for Use with Models and Prosected Cadavers* is the interactive laboratory exercises at the end of each chapter, to reinforce the material presented. The exercises are presented in three different formats: fill-in questions, matching questions, and labeling of anatomical structures.

To further assist the students and to reinforce what has been seen and learned in the laboratory, a mini-atlas of selected photographs of dissected human cadavers has been added at the end of the manual to supplement the line illustrations.

Also new to the second edition is an Instructor's Manual, which includes: helpful suggestions and directions; suggested structures to be demonstrated in the laboratory; list of useful materials for each laboratory exercise; answers to all laboratory exercises from the Laboratory Manual.

Study Tips

- Familiarize yourself with the names of all the labeled structures in the illustrations.

- Use the blank squares next to the labels on the illustration to color-code the structures. Then color each anatomical structure in the line drawings with the same color you used to color-code it.

- Review the chapter summaries.

- Perform the laboratory exercises at the end of each chapter.

Acknowledgments

I am very happy to acknowledge those professionals whose criticism, suggestions, and comments helped me to improve the manuscript. All suggestions and advice were incorporated into the manual, wherever possible. As a result, the second edition of *Laboratory Manual for Human Anatomy for Use with Models and Prosected Cadavers* is greatly improved. I thank the following individuals: Eydie Kendall-Wassmuth of the University of Idaho at Moscow, Idaho, Professors Ann Repka of the University of Colorado at Boulder, John Moore of Parkland College, Gary Iwamoto of the University of Illinois at Urbana, John Stencel of Olney Central College, and Russ Cagle of Willamette University.

Finally, the efforts, enthusiasm, advice, and friendship of the Executive Editor Bonnie Roesch of HarperCollins Publishers and the assistance of Cyndy Taylor, Development Editor, are also greatly appreciated. This manuscript would not be complete without the editorial assistance of Ms. Maggie Schwarz. Her meticulous attention to detail and her skillful editing are acknowledged and appreciated.

Victor P. Eroschenko, Ph.D.

Part One

Organization of the Human Body

CHAPTER 1
Anatomical Terminology

Chapter 1

Anatomical Terminology

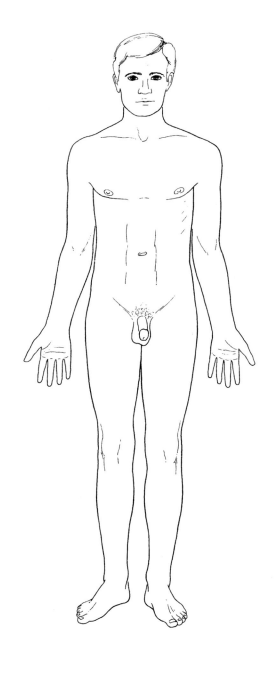

Objective

The objective of Chapter 1, "Anatomical Terminology," is to familiarize you with:

1. **The anatomical positions**
2. **Body regions and planes**
3. **Body cavities and their subdivisions**

The Anatomical Position *[Figure 1.1]*

For this terminology to apply, the body must be in the reference or universally accepted position. The reference position is the **anatomical position**. The body is upright or erect and the eyes face forward. The feet are close together and the upper limbs lie at the sides. The palms of the hands face forward, the fingers are extended, and the thumbs face away from the body. All descriptions of external and internal body structures are based on the anatomical position.

Directional Terminology *[Figure 1.1]*

Directional terminology is used to describe structures and/or surfaces located in different regions of the body. The terms, definitions, and examples are listed below. (Anatomical terms that apply to animals are in parentheses).

Term	*Definition*	*Example*
Anterior (ventral)	Toward or nearer front of body	Sternum is anterior to heart
Posterior (dorsal)	Toward or nearer back of body	Esophagus is posterior to heart
Superior (cranial)	Toward head or upper body	Head is superior to heart
Inferior (caudal)	Nearer lower part of body	Abdomen is inferior to thorax
Medial	Closer to midline of body	Heart is medial to lungs
Lateral	Farther from midline	Lungs are lateral to heart
Proximal	Closer to point of origin	Shoulder is proximal to heart
Distal	Farther from body mass or origin	Wrist is distal to upper arm
Superficial	Closer to surface of body	Skin is superficial to muscle
Deep	Farther from surface of body	Muscles are deep to skin
Ipsilateral	On same side of body	Right lung and right kidney are ipsilateral
Contralateral	On opposite sides of body	Right kidney is contralateral to left kidney

Terminology of Movement

The direction of movement is stated in terms of the anatomical position and anatomical terminology. Terminology of movement is presented later in the manual, in Chapter 6 on the Muscular System.

Body Planes and Sections [Figure 1.1]

A plane is an imaginary line that passes through the body. The body planes that lie at right angles to each other and are most frequently used to illustrate internal arrangements are:

Sagittal Plane: A vertical plane that divides the body or organ(s) into right and left parts. The **midsagittal** or **median plane** divides the body or its organ(s) into symmetrical and equal halves. The vertical planes that pass parallel to the midsagittal (median) plane are called **parasagittal planes**. These planes divide the body or the organ(s) into unequal right and left parts.

Frontal (Coronal) Plane: Any **vertical** plane that passes at a right angle to the midsagittal plane and divides the body or its organ(s) into anterior and posterior parts.

Horizontal (Transverse) Plane: Any **horizontal** plane that passes at a right angle to the sagittal and frontal planes and divides the body or its organ(s) into **superior** and **inferior** parts.

Body Regions [Figure 1.1]

The body can be subdivided into the following general regions:

Axial Region: The region that consists of the **head, neck,** and **trunk.**

Appendicular Region: The region that consists of the **limbs.**

Trunk or **Torso:** The region that consists of the **thorax, abdomen,** and **pelvis.**

Body Cavities [Figure 1.2]

The human body contains two main **cavities:** the **posterior (dorsal) body cavity** and the **anterior (ventral) body cavity.** These cavities house, protect, compartmentalize, and allow for free movement of the internal organs.

Posterior (Dorsal) Body Cavity [Figure 1.2a]

This body cavity is subdivided into the **cranial cavity** and the **vertebral,** or **spinal, cavity.** The cranial cavity houses the **brain.** The vertebral cavity contains the **spinal cord.** These two cavities communicate with each other through a large opening at the base of the skull called the **foramen magnum.**

Anterior (Ventral) Body Cavity [Figure 1.2b]

This cavity is larger than the dorsal cavity. It contains most internal organs of the chest and abdomen. The **ventral,** or **anterior, body cavity** is subdivided into two

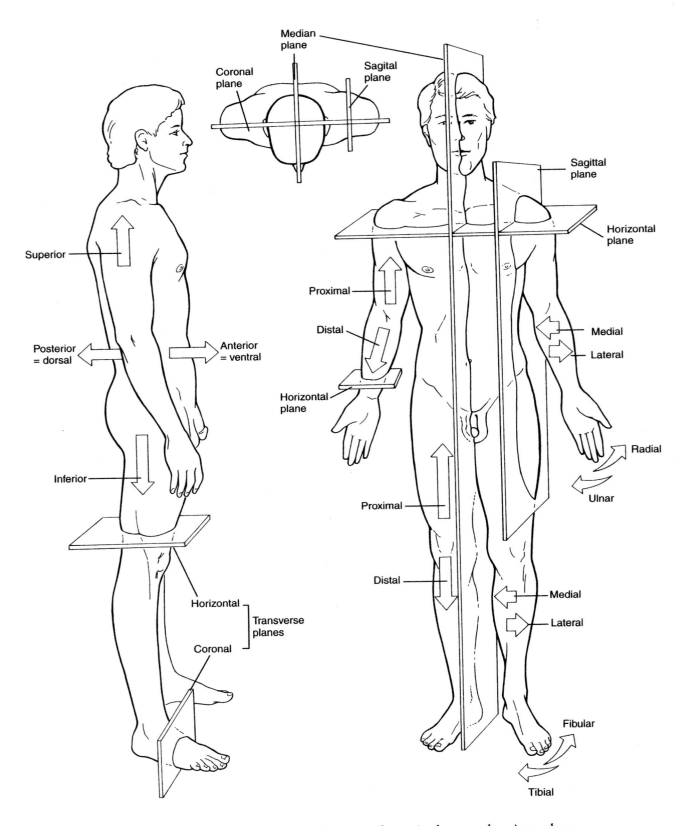

Figure 1.1 The anatomical position, directional terminology, and various planes of section.

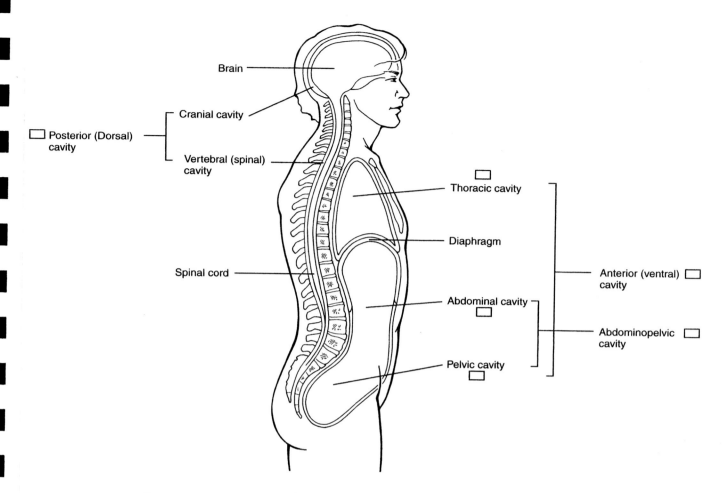

Posterior (Dorsal) cavity
Cranial cavity
Vertebral (spinal) cavity
Brain
Spinal cord
Thoracic cavity
Diaphragm
Abdominal cavity
Pelvic cavity
Anterior (ventral) cavity
Abdominopelvic cavity

Figure 1.2a A midsagittal (median) section showing the body cavities.

main cavities. The **ventral cavity** is separated into a superior **thoracic cavity** and an inferior **abdominopelvic cavity** by the muscular diaphragm. The organs in the ventral body cavity are the **viscera** or **visceral organs**.

The **thoracic cavity** is divided into the **pericardial cavity**, which surrounds the heart, and the right and left **pleural cavities**, which surround the lungs. The **mediastinum**, a **median** partition between the lungs, extends from the sternum to the vertebral column. Passing through or located within the mediastinum are the heart, thymus, trachea, esophagus, and several major blood vessels, such as the aorta.

The **abdominopelvic cavity** is divided into a **superiorly** located **abdominal cavity** and an **inferiorly** located **pelvic cavity**. The abdominal and pelvic regions are not separated by a membranous or a muscular wall.

Surface Landmarks *[Figures 1.3 and 1.4]*

The following are anatomical terms frequently used to describe surface landmarks on the human body and their corresponding lay terms. All surface landmarks are based on the person being in the anatomical position.

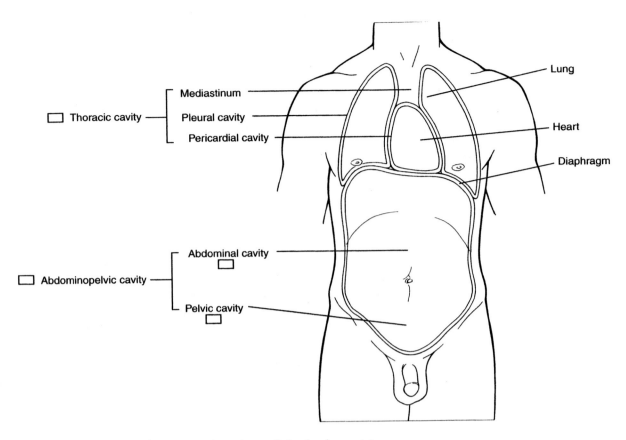

Figure 1.2b An anterior view of the body cavities.

Anatomical Term	Lay Term
Cephalic	**Head Region**
Cranial	Skull
Facial	Face
Cervical	Neck
Frontal	Forehead
Orbital	Eye
Buccal	Cheek
Nasal	Nose
Oral	Mouth
Mental	Chin
Trunk	**Anterior Body Region**
Thorax	Chest
Abdomen	Belly
Pubic	Pubis
Mammary	Breast
Umbilical	Navel
Inguinal	Groin
Back	**Posterior Body Region**
Olecranal	Elbow

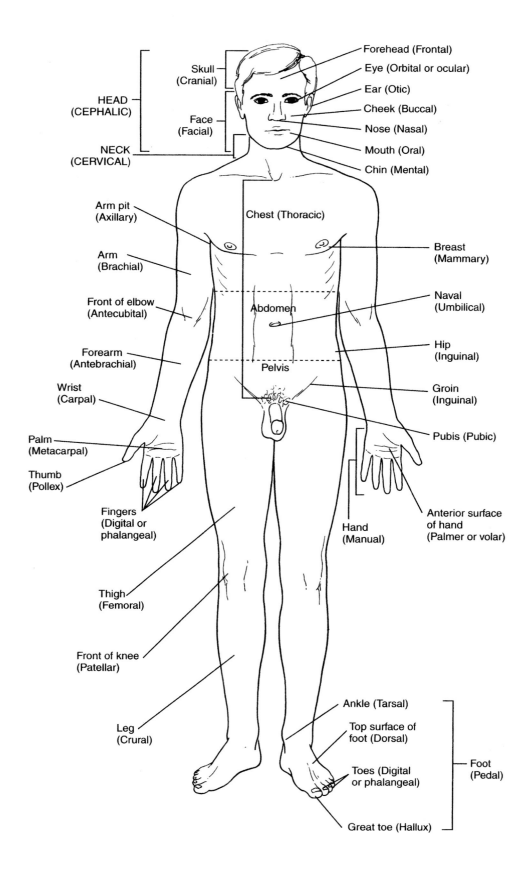

Figure 1.3 Anterior surface landmarks in anatomical position.

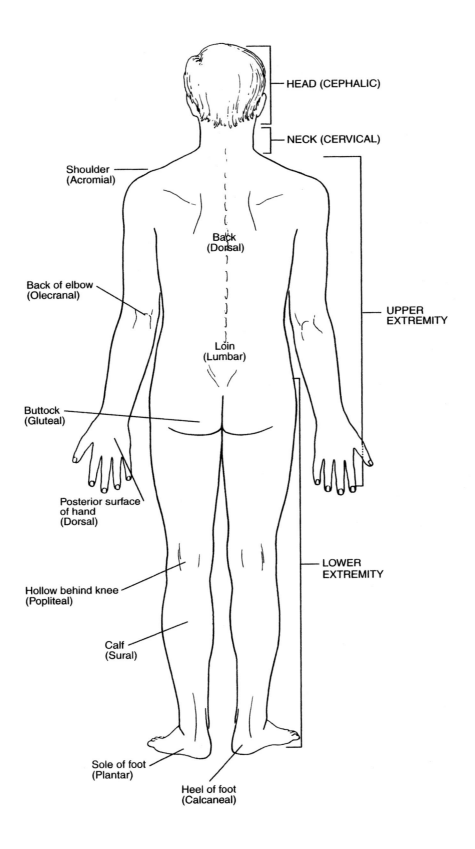

HEAD (CEPHALIC)

NECK (CERVICAL)

Shoulder
(Acromial)

Back
(Dorsal)

Back of elbow
(Olecranal)

UPPER
EXTREMITY

Loin
(Lumbar)

Buttock
(Gluteal)

Posterior surface
of hand
(Dorsal)

Hollow behind knee
(Popliteal)

LOWER
EXTREMITY

Calf
(Sural)

Sole of foot
(Plantar)

Heel of foot
(Calcaneal)

Figure 1.4 **Posterior surface landmarks in anatomical position.**

Lumbar	Loin
Gluteal	Buttock
Popliteal	Hollow behind knee
Sural	Calf of leg
Calcaneal	Heel of foot

Upper Extremity **Arms and Shoulders**

Acromial	Shoulder
Axillary (Axilla)	Armpit
Brachial	Arm
Cubital	Hollow of elbow
Antebrachial	Forearm
Carpal	Wrist region
Palmar	Front of hand
Metacarpal	Palm region
Pollex	Thumb
Digits or Phalanges	Fingers

Lower Extremity **Legs and Hips**

Coxal	Hip joint
Patellar	Front of knee
Tarsal	Ankle
Digital or Phalanges	Toes
Hallux	Great toe
Pedal	Foot
Plantar	Sole of foot

CHAPTER SUMMARY AND CHECKLIST

I. THE ANATOMICAL POSITION

A. **Reference Position**
1. Body erect, feet together, upper limbs at side, palms facing anteriorly, thumbs at right angles to body.

B. **Directional Terminology**
1. Anterior: near front of body
2. Posterior: near back of body
3. Superior: toward upper region of body
4. Inferior: toward lower part of body
5. Medial: closer to midline of body
6. Lateral: closer to side of body
7. Proximal: closer to the point of origin
8. Distal: farther away from point of origin
9. Superficial: near surface of body
10. Deep: farther from surface of body
11. Ipsilateral: on same side
12. Contralateral: on opposite side

C. Body Planes and Sections
1. Midsagittal: vertical plane that divides the body into equal halves
2. Parasagittal: vertical plane that divides the body into unequal parts
3. Frontal: divides body into anterior and posterior parts
4. Horizontal: divides body into superior and inferior parts

D. Body Regions
1. Axial: body region consisting of head, neck, and trunk
2. Appendicular: body region consisting of limbs
3. Trunk: body region consisting of thorax, abdomen, and pelvis

E. Body Cavities
1. Posterior (Dorsal): cranial and vertebral cavities
2. Anterior (Ventral): subdivided into thoracic and abdominopelvic
   a. thoracic cavity: pericardial and pleural cavities
   b. abdominopelvic: abdominal and pelvic cavities

F. Surface Landmarks in Anatomical Position
1. Body parts seen externally
2. Anatomical terminology and common terms

Laboratory Exercises 1

LABORATORY EXERCISE 1.1

Part I

Anatomical Landmarks

Using the listed terms, supply the correct term for each description.

anterior   inferior
axial   lateral
contralateral   medial
deep   superficial
forward   superior

1. In anatomical position, the palms face _____

2. Location of sternum to heart _____

3. Location of wrist to elbow _____

4. Location of esophagus to heart_____

5. Location of skin to muscle _____

6. Location of lungs to heart_____

7. Region that consists of the head, neck, and trunk _____

8. Location of right lung to left lung _____

9. Location of head to heart _____

10. Location of muscle to skin _____

Part II

Anatomical Terminology

Select the most appropriate match.

Head_____   A. Frontal
Skull_____   B. Cervical
Armpit_____   C. Axillary
Neck_____   D. Sagittal
Chest_____   E. Transverse
Buttock_____   F. Gluteal
Thumb_____   G. Pollex
Big toe_____   H. Hallux
Lower back_____   I. Lumbar
Divides body into right and left_____   J. Cranial
Divides body into anterior and posterior_____   K. Thorax
Divides body into superior and inferior_____   L. Cephalic

Part III

Anatomical Structures

Using the listed terms, label the structures and then color them.

Abdominal cavity   Pelvic cavity
Cranial cavity   Thoracic cavity
Diaphragm   Spinal cord

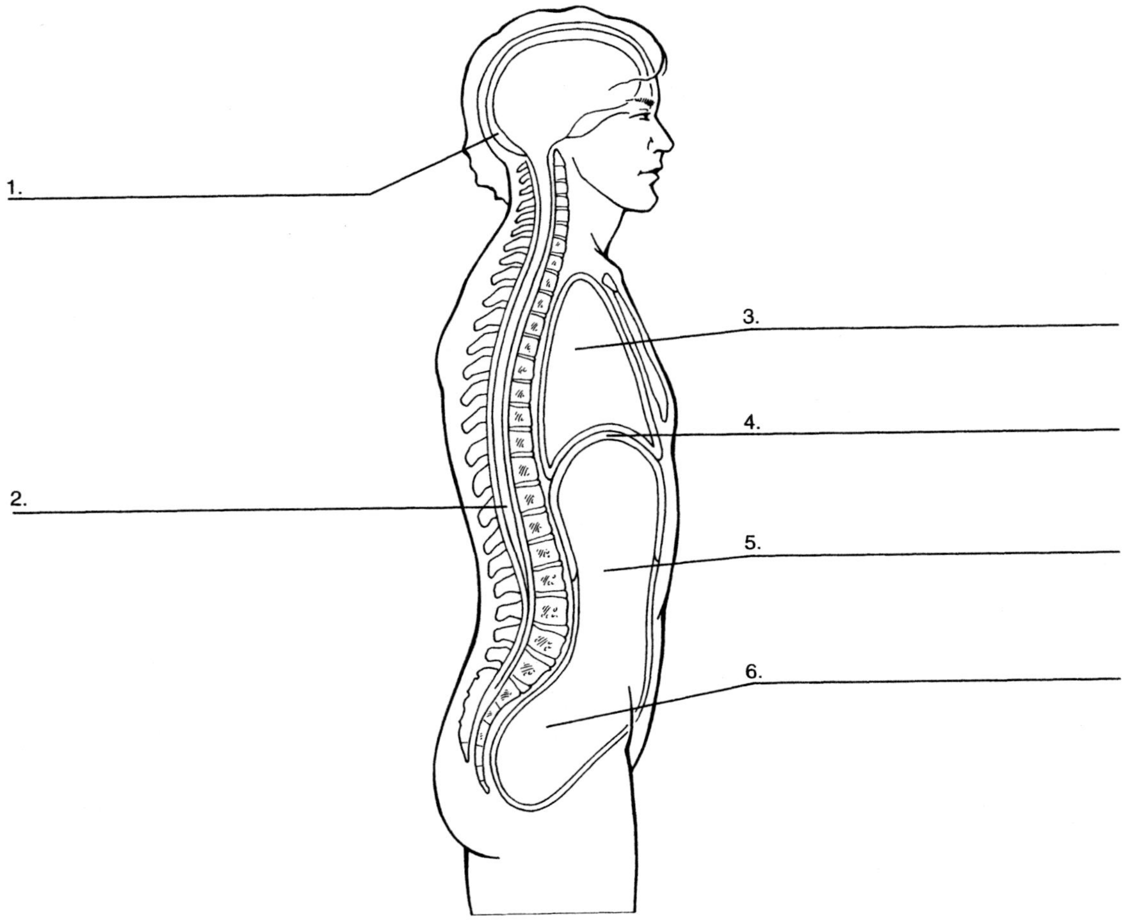

Figure 1.5 Midsaggital (medial) section showing body cavities.

Part Two
The Skeletal System

Chapter 2

Introduction to Bones and the Axial Skeleton

THE SKULL

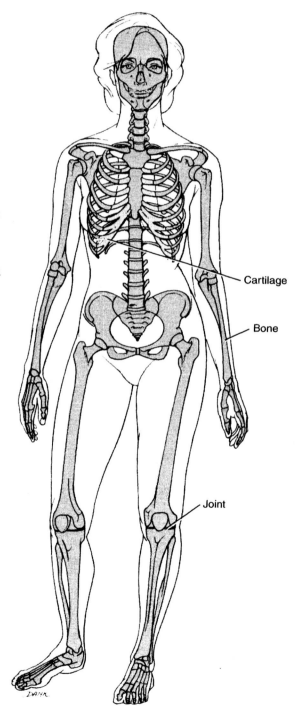

Cartilage

Bone

Joint

Objective

The objective of Chapter 2, "Introduction to Bones and the Axial Skeleton" is to present the:

1. **General characteristics of bones**
2. **Anatomy of a typical bone**
3. **Cranial and facial bones of the skull**

General Characteristics of Bones

Classification of Bone Types [Figure 2.1]

Most bones in the human body are classified into four main categories according to their shape.

Bone Type	Description	Example
Long bones	Bones that are longer than wide	Bones of the extremities
Short bones	Bones without long axis	Bones of wrist and ankles
Flat bones	Thin, flat, somewhat curved bones	Bones of cranium, sternum, ribs, and scapula
Irregular bones	Bones that do not fit any of the above categories	Vertebrae and certain facial bones
Sesame bones	Bones that develop inside tendons	Sesamoid bones

Surface Markings on Bones

The external surfaces of bones exhibit numerous protrusions, depressions, bulges, foramina, and other anatomical structures. These surface markings serve as attachment sites for muscles, tendons, and ligaments; as points of articulation with other bones; and as passageways for blood vessels and nerves.

Depressions and Openings in Bones

Bone Markings	Description	Example
Fissure	Narrow, cleftlike opening or passageway for vessels or nerves	Superior orbital fissure
Foramen	Round or oval opening	Carotid foramen
Meatus	A canal through bone	External auditory meatus

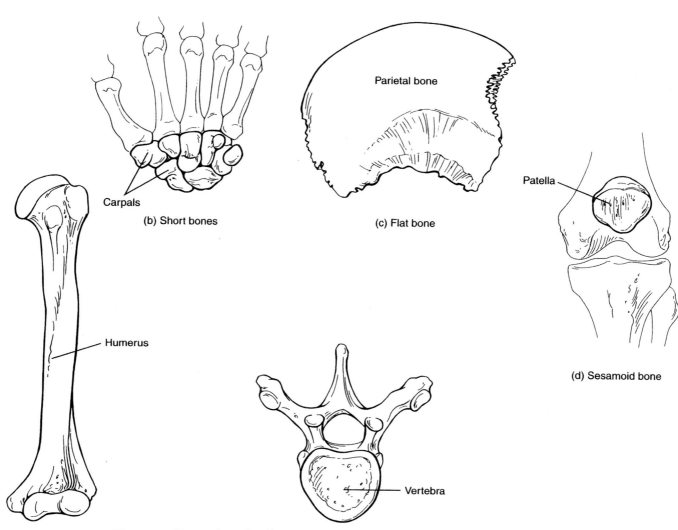

Figure 2.1 Examples of different bone shapes: (a) the humerus, an example of long bone; (b) short bones of the wrist; (c) flat bone of the skull; (d) an example of sesamoid bone, the patella; and (e) vertebra, which is an irregular bone.

Bone Markings	Description	Example
Sinus	Cavity within bone filled with air and lined with epithelium	Frontal sinus
Groove	Shallow depression or furrow in bone for vessels or nerves	Radial groove
Fossa	Shallow depression in bone	Suprascapular fossa

Bony Processes for Articulation in Joints

Processes	Description	Example
Head	Rounded, articulating end of bone	Head of femur
Facet	Smooth, nearly flat articular surface	Articular facet of vertebra

Processes	Description	Example
Condyle	Bony projection with smooth articulating surface	Condyle of femur
Neck	Narrowing of bone between head and shaft of long bone	Neck of femur
Ramus	Extension or branch of bone	Ramus of mandible

Bony Processes for Connective Tissue Attachments (Ligaments and Tendons)

Processes	Description	Example
Tuberosity	Large, rounded process with rough surface	Radial tuberosity
Tubercle	Small rounded process	Greater tubercle of humerus
Trochanter	Large, blunt, irregular projection on femur	Greater trochanter of femur
Crest	Prominent narrow ridge on bone	Iliac crest
Line	Narrow ridge on bone, less prominent than crest	Linea aspera of femur
Spine	Slender, sharp projection on bone	Spinous process of vertebra
Epicondyle	Projection or ridge on or above condyle	Epicondyle of humerus

Anatomy of a Typical Bone—
Description [Figures 2.2 and 2.3]

The bones of the human skeleton normally exhibit **compact bone** externally and **spongy (cancellous) bone** internally. Compact bone surrounds the medullary (marrow) cavity. The typical long bone has a tubular shaft called the **diaphysis.** Its collar is composed of compact bone, which is dense and homogeneous with few spaces. Internal to the compact bone is the spongy bone. This bone contains many spaces and interconnecting small plates of bones, called **trabeculae.** In living bones these spaces are filled with bone marrow. On each end or extremity of the long bone are the **epiphyses,** which have a thin external layer of compact bone and a spongy interior.

The outer bony surface of the long bone is lined by a connective tissue covering called **periosteum.** Where a bone forms a joint with another bone, a thin layer of hyaline cartilage lines the epiphyses. This is the **articular cartilage.** In growing children, a layer of cartilage called the **epiphyseal plate** separates diaphysis from epiphyses. In adults, the epiphyseal cartilage is replaced by bone and is visible externally as a thin **epiphyseal line.**

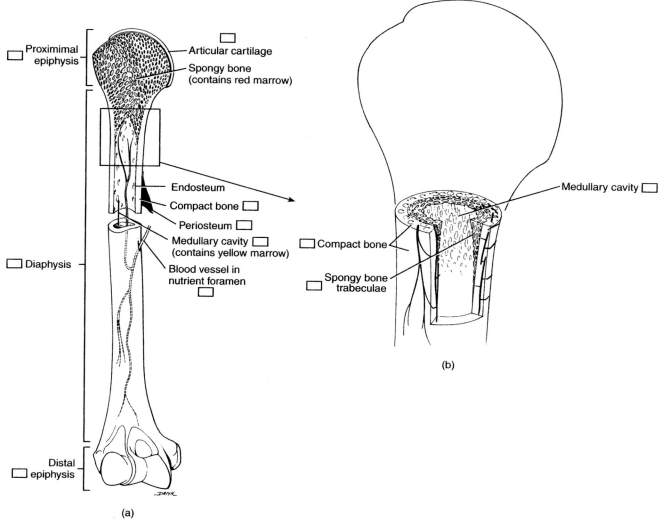

Figure 2.2 Osseous tissue. (a) Macroscopic appearance of a long bone that has been partially sectioned. (b) Histological structure of bone.

During the **ossification process**, blood vessels perforate the developing bone from the periosteum and supply it with blood. On the dried adult bone, the entrance of such vessels is seen as the **nutrient foramen**.

Flat bones are thin and do not contain medullary cavities. Flat bones contain two surface layers of compact bone that enclose a layer of spongy bone called **diploe** (Figure 2.3).

Anatomy of a Typical Bone— Identification *[Figures 2.2 and 2.3]*

Examine the long bone and a flat bone on the illustration and color their parts (Figures 2.2 and 2.3). Then identify the following structures on a model or on human bones.

1. **compact bone**
2. **spongy bone**

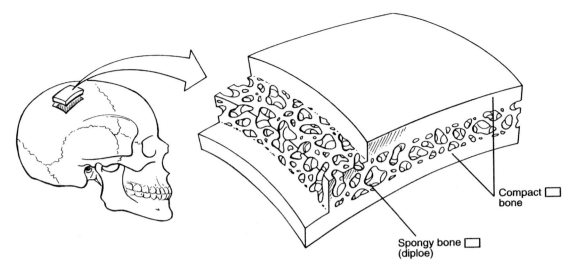

Figure 2.3 Cross-section of a flat bone.

3. **medullary cavity**
4. **diaphysis**
5. **epiphyses**
6. **epiphyseal line**
7. **nutrient foramen**

Organization of the Skeletal System— Description

The bones of the human skeleton are separated into the **axial skeleton** and the **appendicular skeleton.** The axial skeleton consists of bones that form the long axis of the human body. It includes the cranial bones, bones of the thorax, and the vertebral column. The bones of the axial skeleton are described below and in Chapter 3, and the appendicular skeleton in Chapters 4 and 5.

The Bones of the Adult Skull—Description

[Figures 2.4 to 2.12]

Most of the 22 bones in the skull are flat or irregular (Figures 2.4 to 2.7). They are divided into 8 **cranial** and 14 **facial** bones.

The **parietal** and **temporal** cranial bones are paired. The unpaired cranial bones are the **frontal**, **occipital**, **sphenoid**, and **ethmoid**. The occipital, parietal, and frontal bones form the **calvarium**, or the skull cap.

All bones of the adult skull, except the mandible, form immovable, interlocking joints called **sutures**. The most prominent skull sutures are:

The **coronal suture**, located between the frontal and the two parietal bones (Figure 2.5).
The **sagittal suture**, located between the two parietal bones (Figure 2.6).
The **lambdoidal suture**, located between the parietal bones and the occipital bone (Figure 2.6).

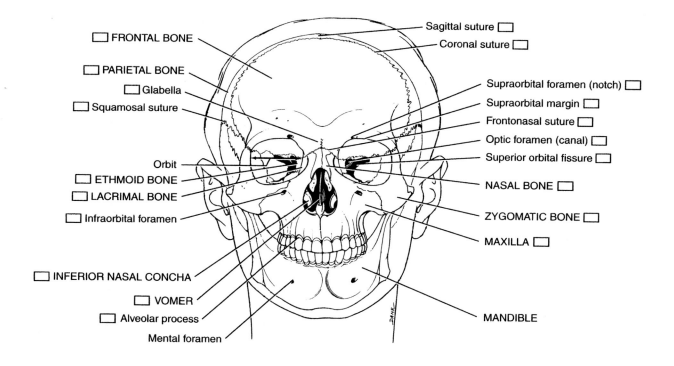

FRONTAL BONE □

PARIETAL BONE □

□ Glabella

□ Squamosal suture

Orbit

ETHMOID BONE □

LACRIMAL BONE □

□ Infraorbital foramen

□ INFERIOR NASAL CONCHA

□ VOMER

□ Alveolar process

Mental foramen

Sagittal suture □

Coronal suture □

Supraorbital foramen (notch) □

Supraorbital margin □

Frontonasal suture □

Optic foramen (canal) □

Superior orbital fissure □

NASAL BONE □

ZYGOMATIC BONE □

MAXILLA □

MANDIBLE

Figure 2.4 Frontal view of the skull.

The **squamous suture**, located between the parietal bones and temporal bones (Figure 2.5).

The Cranial Bones of the Skull

THE FRONTAL BONE [Figure 2.4]

The single **frontal bone** forms the anterior superior portion of the skull (forehead), the roof over the orbits (eye sockets), and the most anterior portion of the cranial (skull) floor, or underside of the calvarium. The main features of the frontal bone are:

1. **Glabella**, a smooth area between the orbits.
2. **Frontonasal suture**, a junction of the frontal bone and the nasal bones.
3. **Supraorbital margin**, a prominent bony ridge above each orbit.
4. **Supraorbital foramen**, a perforation above each bony ridge that allows blood vessels and nerves to reach the superficial structures above the eyelids.
5. **Coronal suture**, a posterior junction of the frontal bone with the two parietal bones (Figure 2.5).

THE PARIETAL BONES [Figures 2.5 and 2.6]

The paired **parietal bones** are located posterior to the **coronal suture** and form most of the roof and sides of the skull. These bones meet in the dorsal midline of the skull to form the **sagittal suture**. Posteriorly, the parietal bones merge

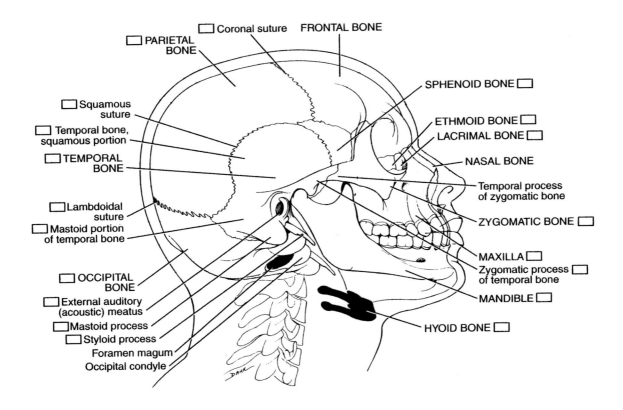

Figure 2.5 Lateral view of the skull.

with the occipital bone to form the **lambdoidal suture**. On each side of the skull, the parietal bone and temporal bone join to form the **squamous suture**.

THE TEMPORAL BONES [Figures 2.5, 2.6, 2.7 and 2.8]

The paired **temporal bones** are situated inferior to the parietal bones and form the inferolateral sides of the skull and a portion of the cranial floor. Each temporal bone has **squamous, mastoid, tympanic, zygomatic,** and **petrous** regions. Features of the **squamous** region of the temporal bone:

1. Are flattened superior region of the temporal bone that connects with the parietal bone at the squamous suture.
2. Contains the **mandibular fossa**, a site of articulation with the mandible (jawbone) (Figure 2.7).

Features of the mastoid region of the temporal bone are:

1. Located just posterior and inferior to the **external auditory meatus,** or **external ear canal**.
2. Forms the most inferior and posterior region of the temporal bone.
3. Contains a **mastoid process** that projects downward immediately posterior to the external auditory meatus. (Figure 2.7).

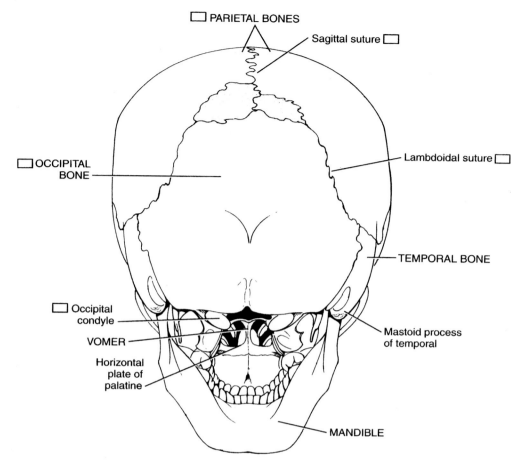

PARIETAL BONES □

Sagittal suture □

Lambdoidal suture □

OCCIPITAL □
BONE

TEMPORAL BONE

Occipital □
condyle

VOMER

Horizontal
plate of
palatine

Mastoid process
of temporal

MANDIBLE

Figure 2.6 Posterior-inferior view of the skull.

Features of the **zygomatic process** of the temporal bone are:

1. Projects anteriorly from the inferior region of the temporal bone.
2. Joins with the **zygomatic bone** of the face.
3. Forms the **zygomatic arch**, or the cheekbone (Figures 2.5 and 2.7).

Features of the **tympanic** portion of the temporal bone are:

1. Contains the external auditory meatus.
2. Bears a thin, pointed **styloid process** that projects inferiorly (Figure 2.7).

Features of the **petrous portion** of the temporal bone (Figure 2.8a) are:

1. Lines part of the floor of the cranial cavity.
2. Houses organs responsible for hearing and balance.
3. Contains the following three foramina:
   a. **Carotid foramen**, or **carotid canal**, which accommodates the carotid artery.
   b. **Jugular foramen**, an opening for the internal jugular vein and three cranial nerves: glossopharyngeal (cranial nerve IX), vagus (cranial nerve X), and spinal accessory (cranial nerve XI).
   c. **Stylomastoid foramen**, a passageway for the facial (cranial nerve VII) nerve and stylomastoid artery (Figure 2.7).

THE OCCIPITAL BONE [Figures 2.5, 2.6, 2.7, and 2.8]

The unpaired **occipital bone** forms the most posterior wall and base of the skull. Its features are:

1. Joins the parietal bone at the **lambdoidal suture**.
2. Contains a large opening at its base called the **foramen magnum**.
3. Located on each side of the foramen magnum are oval processes called the **occipital condyles**.
4. Site of articulation of occipital condyles with the first vertebra (atlas) of the vertebral column, which allows us to nod our head.
5. Bears a prominent bony ridge, the **external occipital protuberance** (Figure 2.7), which can be felt as a bump.
6. Contains the foramen magnum, the passageway through which medulla connects with the spinal cord.
7. Contains the **hypoglossal canal** at the base and on the inner side of the occipital condyles. This canal allows for the passage of the hypoglossal nerve (cranial nerve XII).

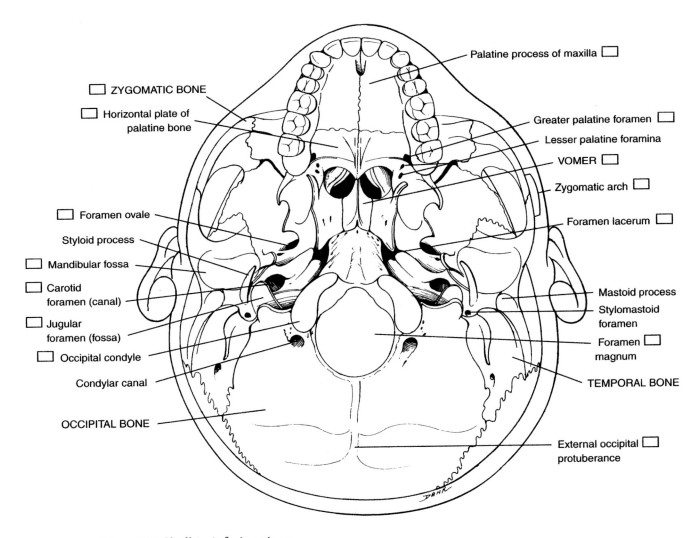

Figure 2.7 Skull in inferior view.

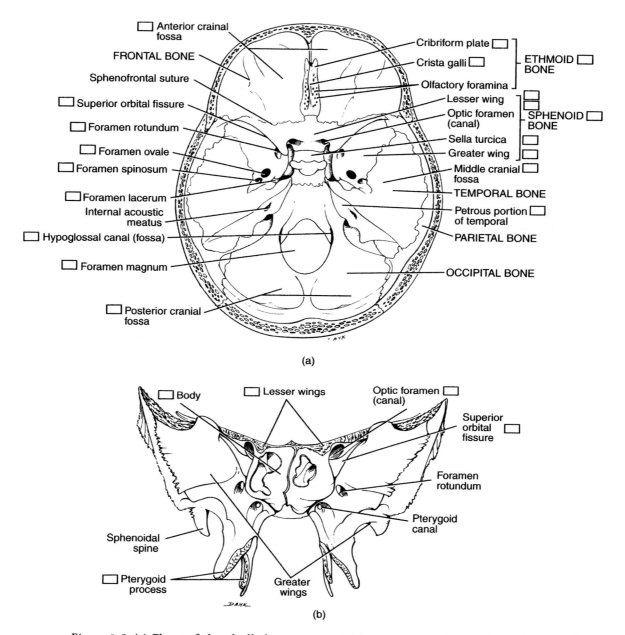

Anterior crainal fossa

FRONTAL BONE

Sphenofrontal suture

Superior orbital fissure

Foramen rotundum

Foramen ovale

Foramen spinosum

Foramen lacerum

Internal acoustic meatus

Hypoglossal canal (fossa)

Foramen magnum

Posterior cranial fossa

Cribriform plate

Crista galli

Olfactory foramina

ETHMOID BONE

Lesser wing

Optic foramen (canal)

SPHENOID BONE

Sella turcica

Greater wing

Middle cranial fossa

TEMPORAL BONE

Petrous portion of temporal

PARIETAL BONE

OCCIPITAL BONE

(a)

Body

Lesser wings

Optic foramen (canal)

Superior orbital fissure

Foramen rotundum

Pterygoid canal

Sphenoidal spine

Pterygoid process

Greater wings

(b)

Figure 2.8 (a) Floor of the skull showing cranial fossae. The sphenoid bone is viewed in the floor of the cranium from above. (b) Anterior view of the sphenoid bone.

THE SPHENOID BONE [Figures 2.5 and 2.8]

The single **sphenoid bone** in the cranial floor is shaped like a butterfly or bat. It makes contact with all the cranial bones mentioned above. Features of the sphenoid bone are:

1. Contains a central **body**, two pairs of anterior **lesser wings**, and posterior **greater wings**.
2. A central depression in the body called the **sella turcica** houses the pituitary gland (hypophysis).

3. Houses the sphenoid sinuses.
4. A pair of **pterygoid processes** on the inferior side form part of the posterior walls of the nasal cavities (Figure 2.8b).

Several important foramina pass through the sphenoid bone (Figures 2.4, 2.5, and 2.8a, b). These foramina are:

1. **Optic foramina,** in the bases of the lesser wings, provide a passageway for the optic nerves (cranial nerve II) from the eyes to the base of the brain.
2. **Superior orbital fissure** is lateral to the optic foramen. This long fissure is the passageway for the oculomotor nerve (cranial nerve III), trochlear nerve (cranial nerve IV), ophthalmic branch of the trigeminal nerve (cranial nerve V), and abducens nerve (cranial nerve VI).
3. **Foramen rotundum,** posterior to the superior orbital fissure, is a canal for the maxillary branch of the trigeminal nerve.
4. **Foramen ovale** is posterior to the foramen rotundum. This large oval foramen houses the mandibular branch of the trigeminal nerve (cranial nerve V).
5. **Foramen lacerum,** between the sphenoid bone and the petrous portion of the temporal bone.

THE ETHMOID BONE [Figures 2.8 and 2.9]

The unpaired **ethmoid bone** lies anterior to the sphenoid bone between the orbits, where it forms the roof of the nasal cavity. The bone consists of four main parts: the **cribriform plate** superiorly, a middle **perpendicular plate**, and two **lateral masses** that project inferiorly from the cribriform plate.

The cribriform plate forms the roof of the nasal cavity. It is perforated by numerous small **olfactory foramina** through which the olfactory nerves (cranial nerve I) extend from the nasal olfactory epithelium. A small triangular process, the **crista galli**, projects superiorly from the middle of the cribriform plate. This process is the attachment site for a connective tissue sheet that secures the brain in the skull (falx cerebri).

The perpendicular plate forms the superior portion of the nasal septum and divides the nasal cavity into the right and left compartments. On the medial surfaces of the lateral masses are the **superior** and **middle nasal conchae (turbinates)** that project into the nasal compartment from the lateral walls.

The Cranial Fossae [Figure 2.8]

The floor, or the interior side of the skull, contains three bony areas: the **anterior, middle,** and **posterior cranial fossae**.

The frontal lobes of the brain (cerebral hemispheres) are located on the **anterior cranial fossa**, while the olfactory bulbs of the olfactory nerves lie on the cribriform plate.

Below the anterior cranial fossa is the **middle cranial fossa,** which houses the temporal lobes of the brain. The body of the sphenoid bone is located in the midline of this fossa.

The **posterior cranial fossa** is largely formed by the occipital bone and houses the cerebellum, pons, and medulla.

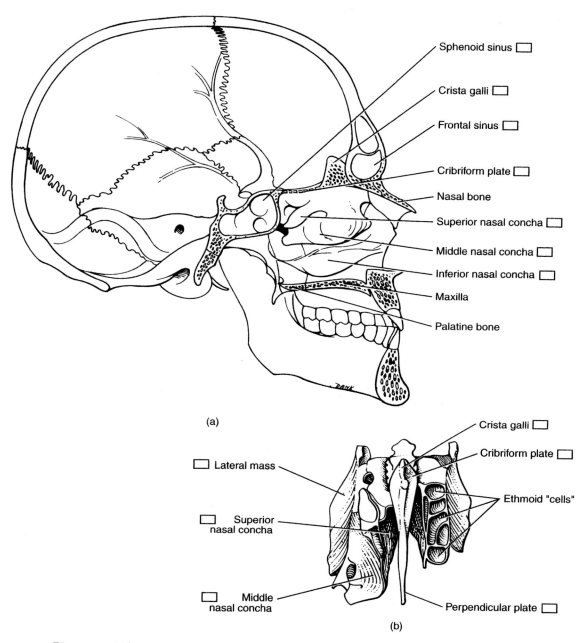

Sphenoid sinus ☐

Crista galli ☐

Frontal sinus ☐

Cribriform plate ☐

Nasal bone

Superior nasal concha ☐

Middle nasal concha ☐

Inferior nasal concha ☐

Maxilla

Palatine bone

(a)

Crista galli ☐

Cribriform plate ☐

☐ Lateral mass

Ethmoid "cells"

☐ Superior
nasal concha

☐ Middle
nasal concha

Perpendicular plate ☐

(b)

Figure 2.9 Ethmoid bone. (a) Median view of ethmoid bone in the inside of skull. (b) Anterior view of ethmoid bone and its parts.

The Facial Bones [Figure 2.4]

There are 14 facial bones. The **mandible** and **vomer** are the only unpaired bones of the facial skeleton. The remaining facial bones are paired: the **maxillary** (maxillae), **zygomatic, nasal, lacrimal, palatine,** and **inferior conchae.**

THE NASAL BONES [Figures 2.4 and 2.10]

The **nasal bones** are small and merge in the midline to form the bony bridge of the nose. These bones are situated between the frontal and the maxillary bones.

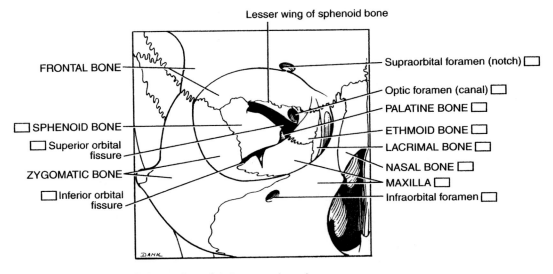

Lesser wing of sphenoid bone

FRONTAL BONE

SPHENOID BONE □

□ Superior orbital
fissure

ZYGOMATIC BONE

□ Inferior orbital
fissure

DANK

Supraorbital foramen (notch) □

Optic foramen (canal) □
PALATINE BONE □
ETHMOID BONE □
LACRIMAL BONE □
NASAL BONE □
MAXILLA □
Infraorbital foramen □

Figure 2.10 Detail of the right orbit in anterior view.

THE MAXILLARY BONES, OR MAXILLAE
[Figures 2.4, 2.5, 2.7, and 2.9]

The **maxillary bones,** or **maxillae,** join in the midline to form the upper jaw, most of the roof of the mouth, and the floor and lateral walls of the nasal cavities. Features of the maxillary bones are:

1. They form part of the floor of the orbits.
2. Inferior to each orbit, the **infraorbital foramen** (Figures 2.4 and 2.10) is the passageway for the infraorbital nerve and artery.
3. The horizontal **palatine process** of the maxillary bones forms most of the hard palate, or roof, of the mouth.
4. Hollow **alveolar processes** contain the **alveoli** (teeth sockets), which house the upper teeth.
5. The **inferior orbital fissure** in the orbit contains the maxillary branch of the trigeminal nerve (cranial nerve V), the zygomatic nerve, and blood vessels.
6. Contain the **maxillary sinuses,** the largest sinuses in the head.

THE ZYGOMATIC BONES [Figures 2.4, 2.5, and 2.7]

The **zygomatic bones** form the cheekbones and much of the wall and floor of the orbits. A posterior-extending temporal process of the zygomatic bone joins the zygomatic process of the temporal bone to form the **zygomatic arch.**

THE MANDIBLE [Figures 2.4, 2.5, and 2.11]

The **mandible** bone forms the lower jaw and chin. It consists of a horizontal **body** and a vertical **ramus,** which joins with the body of the bone at the **mandibular angle.** The **alveolar processes** in the mandible contain the **alveoli.** These house the lower teeth. Each ramus contains two processes at the superior end. The **condylar process** articulates with the mandibular fossa of the temporal bone. The **coronoid process** attaches with the temporalis muscle, which moves the jaw. The depression between these two elevated processes is the **mandibular notch.**

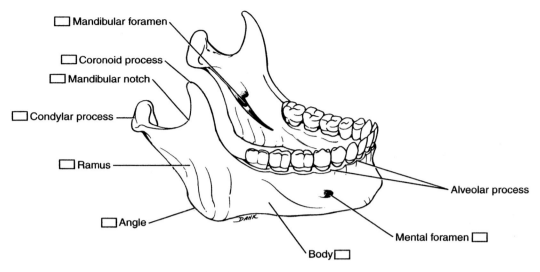

Figure 2.11 Lateral view of the mandible.

Two pairs of foramina penetrate the mandible. The **mental foramen** is on the anterior-lateral side of the mandibular body, and the **mandibular foramen** is on the medial side of each ramus.

THE LACRIMAL BONES [Figures 2.4, 2.5, and 2.10]

The **lacrimal bones** are the smallest facial bones. These thin bones form the anterior part of the medial wall of each orbit.

THE PALATINE BONES [Figures 2.7 and 2.12]

The **palatine bones** contribute to the posterior third of the hard palate. (The anterior two-thirds are formed by the **horizontal plates** of the maxillary bones.) The hard palate of each palatine bone contains the **greater palatine foramen** and the **lesser palatine foramen**. Nerves and blood vessels pass through these openings (Figure 2.7).

THE INFERIOR NASAL CONCHAE [Figures 2.4 and 2.9]

The lateral wall of each nasal cavity exhibits three lateral bony shelves or extensions, the **nasal conchae**. The superior and middle conchae are lateral projections of the ethmoid bone. The **inferior nasal conchae** are the largest conchae.

THE VOMER [Figures 2.7 and 2.12]

The **vomer** is a slender bone that forms the inferior and posterior portion of the nasal septum. Its superior border connects to the perpendicular plate of the ethmoid bone.

The Paranasal Sinuses [Figures 2.9 and 2.12]

The frontal, sphenoid, ethmoid, and paired maxillary bones contain air-filled, mucosa-lined cavities or chambers called the **paranasal sinuses**.

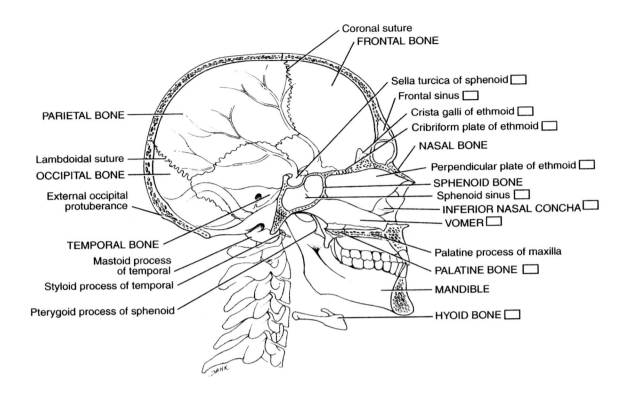

Coronal suture
FRONTAL BONE
Sella turcica of sphenoid ☐
Frontal sinus ☐
Crista galli of ethmoid ☐
Cribriform plate of ethmoid ☐
NASAL BONE
Perpendicular plate of ethmoid ☐
SPHENOID BONE
Sphenoid sinus ☐
INFERIOR NASAL CONCHA ☐
VOMER ☐
Palatine process of maxilla
PALATINE BONE ☐
MANDIBLE
HYOID BONE ☐

PARIETAL BONE
Lambdoidal suture
OCCIPITAL BONE
External occipital protuberance
TEMPORAL BONE
Mastoid process of temporal
Styloid process of temporal
Pterygoid process of sphenoid

Figure 2.12 Median view of the skull.

Bones of the Skull—Identification

Frontal-Anterior Views of the Skull

[Figures 2.4, 2.5, and 2.11]

For all of the skull bones listed below, examine the diagrams of the bones and color their parts. Then identify the bones and structures on the model or human skull.

1. **Frontal Bone**
   a. **Supraorbital foramen**
   b. **Supraorbital margin**
   c. **Glabella**
2. **Maxillary Bone**
   a. **Infraorbital foramen**
   b. **Alveolar processes**
   c. **Vomer**
3. **Mandible**
   a. **Alveolar processes**
   b. **Mental foramen**
4. **Zygomatic Bone**

5. **Ethmoid Bone**
   a. **Perpendicular plate**
   b. **Middle nasal conchae**
   c. **Inferior nasal conchae**
6. **Sutures**
   a. **Frontozygomatic suture**
   b. **Frontonasal suture**
   c. **Sagittal suture**
7. **Fissures**
   a. **Superior orbital fissure**
   b. **Inferior orbital fissure**

Posterior-Lateral Views of the Skull

[Figures 2.5, 2.6, and 2.11]

1. **Parietal Bone**
2. **Temporal Bone**
   a. **Mastoid process**
   b. **Squamous portion**
   c. **External auditory meatus**
   d. **Styloid process**
   e. **Zygomatic process**
3. **Zygomatic Bone**
   a. **Temporal process**
4. **Occipital Bone**
   a. **External occipital protuberance**
5. **Mandible**
   a. **Body**
   b. **Condylar process**
   c. **Coronoid process**
   d. **Mandibular notch**
   e. **Ramus**
   f. **Mandibular angle**
   g. **Mandibular foramen**
6. **Sutures**
   a. **Coronal suture**
   b. **Squamous suture**
   c. **Lambdoidal suture**
   d. **Sphenofrontal suture**

Inferior View of the Skull [Figure 2.7]

1. **Temporal Bone**
   a. **Mandibular fossa**
   b. **Carotid foramen**
   c. **Jugular foramen**
   d. **Stylomastoid foramen**
2. **Maxilla**
   a. **Palatine process**
   b. **Greater and lesser palatine foramina**
3. **Occipital Bone**
   a. **Foramen magnum**
   b. **Occipital condyles**

The Floor of the Cranial Cavity [Figures 2.8 and 2.12]

1. **Sphenoid Bone**
   a. **Greater wing**
   b. **Lesser wing**
   c. **Sella turcica**
   d. **Optic foramen**
   e. **Foramen ovale**
   f. **Foramen spinosum**
   g. **Foramen lacerum**
2. **Temporal Bone**
   a. **Middle cranial fossa**
3. **Ethmoid Bone**
   a. **Crista galli**
   b. **Cribriform plate**
   c. **Olfactory foramina**
4. **Frontal Bone**
   a. **Anterior cranial fossa**
5. **Occipital Bone**
   a. **Hypoglossal canal**
   b. **Posterior cranial fossa**

Median View of the Interior of the Skull [Figures 2.9 and 2.12]

1. **Frontal sinus**
2. **Sphenoid sinus**
3. **Superior nasal conchae**
4. **Middle nasal conchae**
5. **Inferior nasal conchae**

The Hyoid Bone [Figures 2.5, 2.12 and 2.13]

The hyoid bone is a single, U-shaped structure located in the upper neck between the mandible and larynx (voice box). It does not articulate with any other bones and is suspended in the neck by ligaments and muscles through which it attaches to the styloid processes of the temporal bone.

The hyoid bone consists of a horizontal body attached to paired projections called the **cornu**, or horns. The **lesser cornu** are the small projections at the lateral ends of the body, and the **greater cornu** extend posteriorly from the body. Muscles and ligaments attach to the cornu.

Identification [Figure 2.13]

Examine and color the illustrations of the hyoid bone. Then identify the following parts of the hyoid bone on a human skeleton or model.

1. **Body**
2. **Greater cornu**
3. **Lesser cornu**

CHAPTER SUMMARY AND CHECKLIST

I. **GENERAL CHARACTERISTICS OF BONES**

A. **Classification of Bone Types in the Body**
   1. Long bones
   2. Short bones
   3. Flat bones
   4. Irregular bones
   5. Sesamoid bones

B. **Surface Markings on Bones**
   1. Depressions and openings in the bones
      a. fissure: cleftlike opening
      b. foramen: round or oval opening
      c. meatus: canal through bone
      d. sinus: cavity within bone
      e. groove: shallow depression or furrow in bone

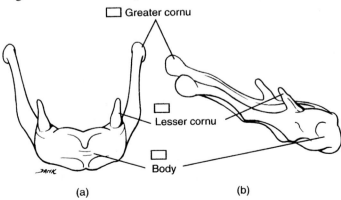

☐ Greater cornu

☐ Lesser cornu

☐ Body

(a)   (b)

Figure 2.13 Hyoid bone. (a) Anterior view. (b) Right lateral view.

f. fossa: shallow depression in bone
2. Bony processes for articulation in joints
   a. head: round articulating end of bone
   b. facet: flat articulating end of bone
   c. condyle: rounded bony projection
   d. neck: narrowing between head and shaft of bone
3. Bone processes for connective tissue attachments
   a. tuberosity: large process with rough surface
   b. tubercle: small, round process
   c. trochanter: large irregular projection on femur
   d. crest: prominent narrow ridge on bone
   e. line: narrow ridge on bone
   f. spine: pointed projection on bone
   g. epicondyle: small projection near condyle

II. ANATOMY OF A TYPICAL BONE

A. Characteristic Features
1. Consists of both compact and spongy bone
2. External compact bone; internal spongy bone
3. Living bone contains bone marrow

B. Long Bone
1. Tubular shaft is diaphysis; collar is compact bone
2. Internal spongy bone surrounds medullary cavity
3. Terminal ends are epiphyses
4. Outer connective membrane is periosteum
5. Articular (hyaline) cartilage lines epiphyses
6. Epiphyseal line is remnant of epiphyseal plate
7. Nutrient foramen is passageway for blood vessel

C. Flat Bone
1. No medullary cavities
2. Two surfaces of compact bone
3. Middle layer in spongy bone is called diploe

III. ORGANIZATION OF THE SKELETAL SYSTEM

A. The Axial Skeleton
1. Bones that form the long axis of the body

B. Bones of Adult Skull
1. There are 8 cranial and 14 facial bones
2. All bones except the mandible are immovable
3. Four major sutures in the skull:
   a. coronal
   b. sagittal
   c. lambdoidal
   d. squamous

IV. CRANIAL BONES OF THE SKULL

A. Frontal Bone
1. Unpaired bone in front

2. Bone forms forehead of skull and roof of orbits
3. Bone forms anterior cranial floor in skull
4. Glabella: smooth anterior area
5. Supraorbital margin: ridge above orbit
6. Margin perforated by supraorbital foramen
7. Bone contains frontal sinus

B. Parietal Bones
1. Paired bones posterior to frontal bone
2. Form sagittal suture in midline
3. Form lambdoidal suture posteriorly

C. Temporal Bones
1. Paired bones
2. Squamous, tympanic, mastoid, and petrous parts
3. Squamous part is near parietal bones
4. Petrous part contains hearing and balance organs
5. Mastoid process extends downward from mastoid part
6. Contain carotid, jugular, and stylomastoid foramina

D. Occipital Bone
1. Unpaired, most posterior bone, contains foramen magnum
2. Joins parietal bones at lambdoidal suture
3. Occipital condyles articulate with atlas
4. Prominent external occipital protuberance

E. Sphenoid Bone
1. Situated in cranial floor
2. Lesser and greater wings extend laterally
3. Sella turcica: central depression
4. Contains optic foramina and superior orbital fissures
5. Contains rotundum, ovale, and lacerum foramina

F. Ethmoid Bone
1. Located anteriorly to sphenoid bone
2. Cribriform plate forms roof of nasal cavity
3. Olfactory foramina in cribriform plate
4. Crista galli project from cribriform plate
5. Superior and middle nasal conchae project laterally

G. Cranial Fossae
1. Anterior cranial fossa
2. Middle cranial fossa
3. Posterior cranial fossa

V. FACIAL BONES

A. Nasal Bones
1. Paired, small bones in midline form bridge of nose

B. Maxillary Bones
1. Paired bones join in midline to form upper jaw
2. Upper teeth housed in the bone
3. Contain infraorbital foramina

4. Borders inferior orbital fissure
5. Contain maxillary sinus
6. Form most of hard palate

C. Zygomatic Bones
   1. Paired bones, form cheekbones and floor of orbits

D. Mandible Bone
   1. Single bone of lower jaw and only movable bone
   2. Has body, ramus, and mandibular angle
   3. Houses lower teeth
   4. Condyloid process articulates with temporal bone
   5. Coronoid process attaches to temporalis muscle
   6. Mental and mandibular foramina in bone

E. Lacrimal Bones
   1. Paired, thin bones in anterior wall of orbit
   2. The smallest facial bones

F. Palatine Bones
   1. Paired bones contribute to hard palate

G. Inferior Nasal Conchae
   1. Separate facial bones, largest nasal conchae

H. Vomer
   1. Plow-shaped bone that forms part of nasal septum

I. Paranasal Sinuses
   1. In frontal, sphenoid, ethmoid, and maxillary bones
   2. Mucosa-lined cavities or chambers
   3. Lighten skull and improve sound

VI. THE HYOID BONE

A. Characteristic Features
   1. Located between mandible and larynx
   2. Does not articulate with other bones
   3. Important site for muscle attachment
   4. Bone has lesser and greater cornu (horns)

Laboratory Exercises 2

NAME _____

LAB SECTION _____ DATE _____

LABORATORY EXERCISE 2.1

Part I

The Bone

Complete the sentences with one of the listed words.

compact   spongy
diaphysis   epiphysis
epiphyseal line   groove
facet   condyle
foramen   sinus
periosteum   articular cartilage

1. On the external surface, bone is _____. Internally, bone is

   _____.

2. The tubular shaft in bone is the _____.

3. The extremity of the bone is the _____. It is covered by a

   thin layer of _____.

4. Bone is covered by a connective tissue called _____.

5. Where cartilage is replaced by bone is the _____.

6. A shallow depression in bone is called a(n) _____.

7. A round or oval-shaped opening in bone is a _____.

8. An air-filled cavity within a bone is a(n) _____.

9. The smooth, articular surface of bone is the _____.

10. A prominent bone projection is called a(n) _____.

Part II

Types of Bones and Skull

Using the listed terms, label the types of bones and then color them.

Long bone   Irregular bone
Sesame bone   Short bone
Flat bone

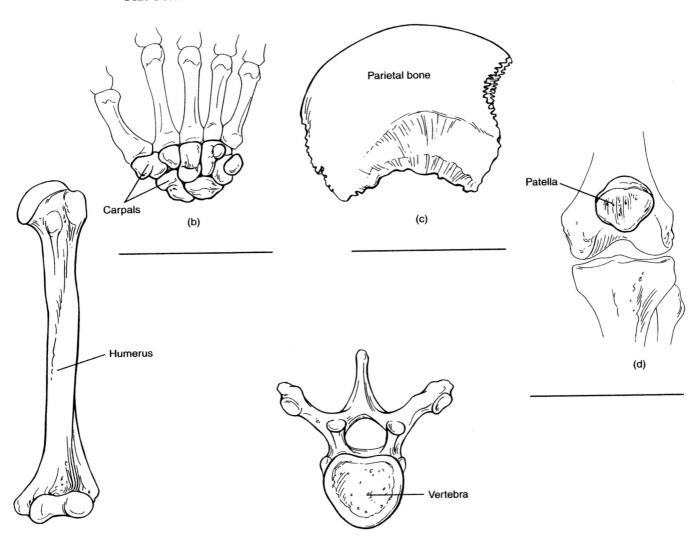

Figure 2.14 Examples of different bone shapes.

LABORATORY EXERCISE 2.2

Using the listed terms, label the bones and markings on the skull and then color them.

Alveolar process	Mandible	Squamous suture
Ethmoid bone	Maxilla	Superior orbital fissure
Frontal bone	Zygomatic bone	Supraorbital foramen
Frontonasal suture	Nasal bone	Supraorbital margin
Glabella	Optic foramen	Vomer
Inferior nasal concha	Orbit	Lacrimal bone
Inferior orbital fissure	Parietal bone	Middle nasal concha
Infraorbital foramen	Perpendicular plate	

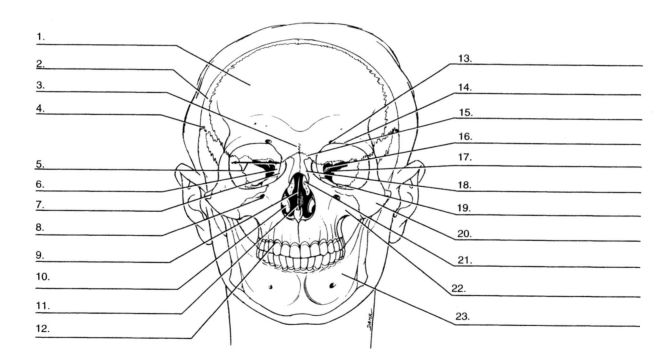

Figure 2.15 Frontal-anterior view of the skull.

LABORATORY EXERCISE 2.3

Using the listed terms, label the bones and markings on the skull and then color them.

Coronal suture
Ethmoid bone
Mandible
Occipital bone
Occipital condyle
Parietal bone
Squamous suture
Styloid process
Temporal process of zygomatic bone
Zygomatic process of temporal bone
Frontal bone

Temporal bone
Sphenoid bone
Zygomatic bone
External auditory meatus
Hyoid bone
Lambdoidal suture
Mastoid process
Foramen magnum
Lacrimal bone
Nasal bone
Maxilla

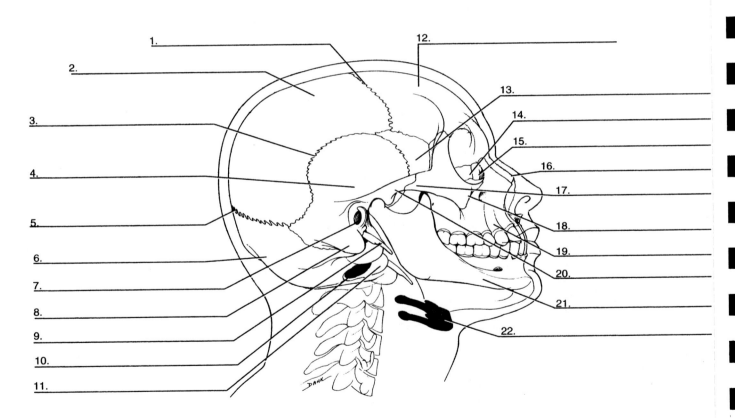

Figure 2.16 Lateral view of the skull.

LABORATORY EXERCISE 2.4

Using the listed terms, label the bones and markings on the inferior side of the skull, and then color them.

Condylar canal
External occipital
 protuberance
Foramen lacerum
Foramen magnum
Horizontal plate of
 palatine bone
Jugular foramen
Lesser palatine
 foramina

Mandibular fossa
Palatine process of
 maxilla
Zygomatic bone
Occipital bone
Temporal bone
Vomer
Stylomastoid foramen
Foramen ovale
Occipital condyle

Greater palatine
 foramen
Carotid foramen
Zygomatic arch
Styloid process

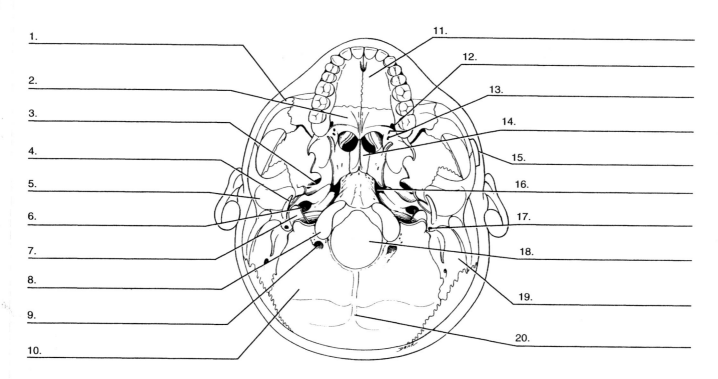

1. _____
2. _____
3. _____
4. _____
5. _____
6. _____
7. _____
8. _____
9. _____
10. _____

11. _____
12. _____
13. _____
14. _____
15. _____
16. _____
17. _____
18. _____
19. _____
20. _____

Figure 2.17 Inferior view of the skull.

LABORATORY EXERCISE 2.5

Using the listed terms, label the bones and markings on the floor, or interior side of the skull, and then color them.

Anterior cranial fossa
Foramen lacerum
Foramen magnum
Foramen ovale
Greater wing of
    sphenoid bone
Lesser wing of
    sphenoid bone

Occipital bone
Olfactory foramina
Superior orbital fissure
Temporal bone
Cribriform plate
Frontal bone
Hypoglossal canal
Posterior cranial fossa

Foramen spinosum
Sella turcica
Optic foramen
Parietal bone
Crista galli
Foramen rotundum
Middle cranial fossa

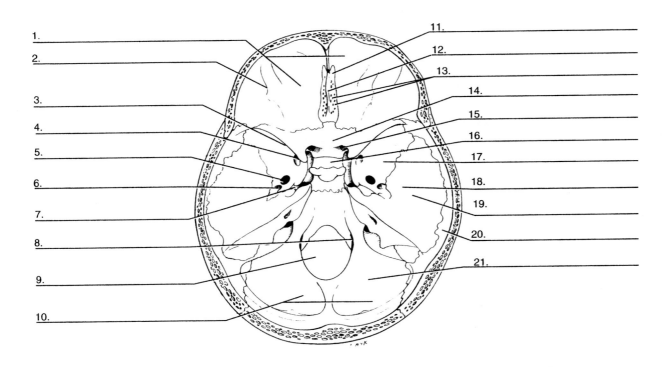

1.
2.
3.
4.
5.
6.
7.
8.
9.
10.
11.
12.
13.
14.
15.
16.
17.
18.
19.
20.
21.

Figure 2.18 Floor of the cranial cavity.

LABORATORY EXERCISE 2.6

Using the listed terms, label the bones and markings in the orbit and then color them.

Frontal bone
Inferior orbital fissure
Lacrimal bone
Maxilla
Optic foramen
Superior orbital fissure

Sphenoid bone
Supraorbital foramen
Nasal bone
Infraorbital foramen
Ethmoid bone
Zygomatic bone

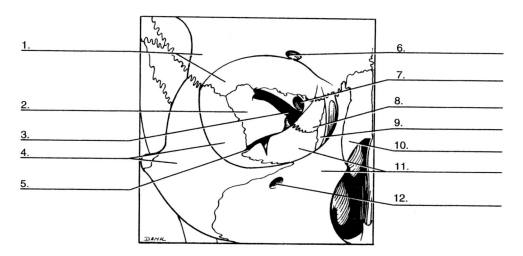

Figure 2.19 Bones in the orbit.

LABORATORY EXERCISE 2.7

Using the listed terms, label the bones and markings on the mandible and then color them.

Alveolar process
Body
Mandibular foramen
Mandibular notch
Ramus

Mental foramen
Coronoid process
Condylar process
Angle

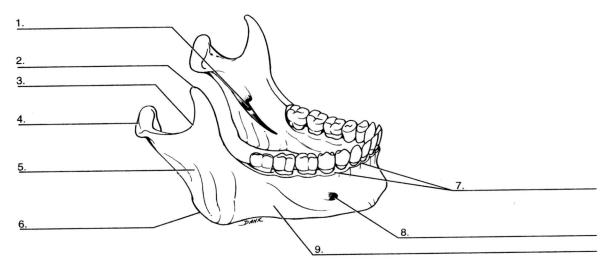

1. _____
2. _____
3. _____
4. _____
5. _____
6. _____
7. _____
8. _____
9. _____

Figure 2.20 The mandible.

Chapter 3

The Axial Skeleton

THE VERTEBRAE AND RIB CAGE

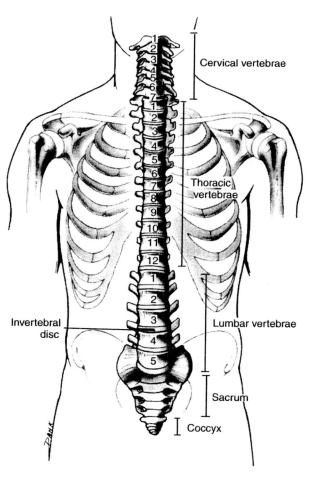

Cervical vertebrae

Thoracic vertebrae

Lumbar vertebrae

Sacrum

Coccyx

Invertebral disc

Objective

The objective of Chapter 3, "The Axial Skeleton," is for you to understand the:

1. **General characteristics of a typical vertebra**
2. **Regional characteristics of the vertebrae**
3. **Structure of the ribs and thorax**

The Vertebral Column—Description

General Characteristics [Figure 3.1]

The human vertebral column, or spine, normally consists of 26 **vertebrae**: 7 **cervical**, 12 **thoracic**, 5 **lumbar**, 1 **sacral**, and 1 **coccygeal** bone. Five sacral vertebrae are fused together into the **sacrum**, while four or five coccygeal vertebrae fuse into one or two **coccyx** bones.

The vertebrae are separated by **intervertebral discs** consisting of an outer fibrocartilage, the **annulus fibrosus**, and an inner gelatinous **nucleus pulposus**. The intervertebral discs cushion and absorb vertical shock to the vertebral column and allow the column to flex.

The vertebral column attaches the skull, thorax, and pelvis and protects the spinal cord and nerves. Between adjacent vertebrae are **intervertebral foramina**, or passageways for peripheral spinal nerves.

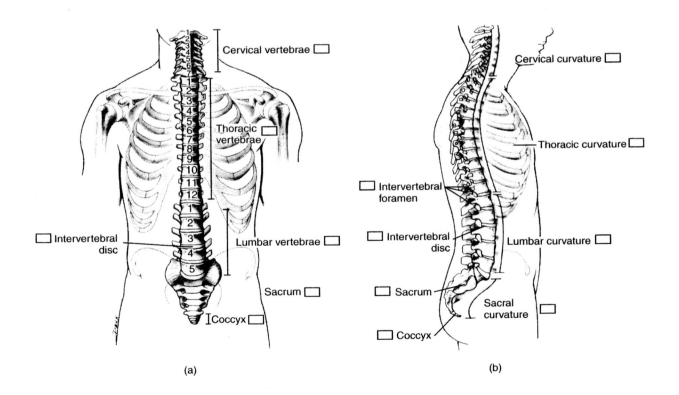

Figure 3.1 Vertebral column in (a) anterior view and (b) lateral view.

Curvatures in the Adult Vertebral Column

[Figures 3.1 and 3.2]

The adult vertebral column contains four major curvatures: **cervical**, **thoracic**, **lumbar**, and **sacral**.

The **thoracic** and **sacral** curvatures are called **primary curvatures** because they appear in the fetus and newborn. Appearing later in life, the **cervical** and **lumbar** curvatures are called **secondary curvatures**. The **cervical curvature** develops at about three months of age, when the baby begins to hold up its head. The **lumbar curvature** develops as the child starts to stand and walk.

Scoliosis is a lateral deviation of the vertebral column. **Kyphosis** (hunchback) is an exaggerated posterior curvature that usually occurs in the thoracic region. **Lordosis** is an exaggerated anterior curvature in the lumbar region.

Characteristics of a Typical Vertebra

[Figures 3.3, 3.4 and 3.5]

Although the vertebrae in the spinal column exhibit regional differences, they are similar in structure.

A typical vertebra (Figures 3.3 and 3.5) consists of a **body (centrum)**. The **vertebral arch** (neural arch) attaches posteriorly to the body. This structure consists of two pedicles that extend posteriorly from the body and join the two lamina. The vertebral arch and the vertebral body form the **vertebral foramen**. In the vertebral column the vertebral foramina of adjacent vertebrae align to form the **vertebral canal**, through which pass the spinal cord and nerves.

The vertebral arch is attached to the body by two **pedicles**. The right and left pedicles have **superior** and **inferior notches** that form passageways for the spinal nerves called **intervertebral foramina** (Figures 3.3 and 3.5). The **laminae** are flattened plates that merge superiorly to the pedicles and form the roof of the vertebral

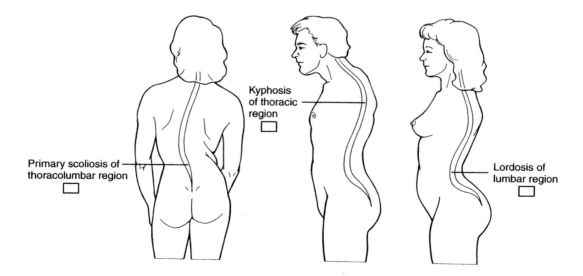

Figure 3.2 Abnormal curvatures of the vertebral column.

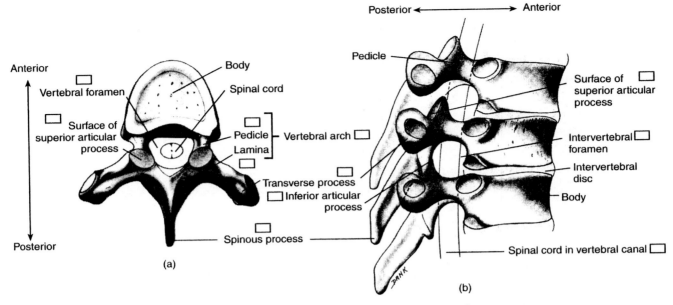

Figure 3.3 Typical vertebra. (a) Superior view. (b) Right lateral view.

arch. A **transverse process** extends laterally from each side of the vertebra at the junction of the lamina and pedicle. A single **spinous process** projects posteriorly from the junction of the two laminae.

Each vertebra has two **superior** and two **inferior articular processes**. The superior processes of one vertebra articulate with the inferior processes of the vertebra immediately superior. The articulating surfaces of these processes are called **facets**. In the cervical and thoracic vertebrae, the articular processes face superiorly and inferiorly, respectively. This arrangement facilitates lateral rotation of the vertebrae. In the lumbar vertebrae, the articular processes face medially. This limits the lateral rotation of the lumbar vertebrae and stabilizes the vertebral column.

Regional Characteristics of the Vertebrae

The Cervical Vertebrae [Figure 3.4]

There are seven cervical vertebrae (C1–C7) in the human vertebral column. Each transverse process of the cervical vertebrae contains **transverse foramina** that house the vertebral arteries. The **spinous processes** of the second through sixth cervical vertebrae are often split, or **bifid**.

The first two cervical vertebrae are atypical. The first is called the **atlas**. It lacks a body and spinous process and is a ring with an **anterior** and **posterior vertebral arch**. The atlas has concave **superior articular surfaces** that articulate with the oval-shaped **occipital condyles** of the skull. Articulation between the skull and atlas allows up and down nodding of the head.

The second cervical vertebra is called the **axis** and is identified by a prominent superior projection called the **odontoid process**, or **dens**. The dens extends superiorly into the atlas and provides a pivot around which the head and atlas rotate. The articulation between the atlas and axis enables lateral side-to-side movement of the head.

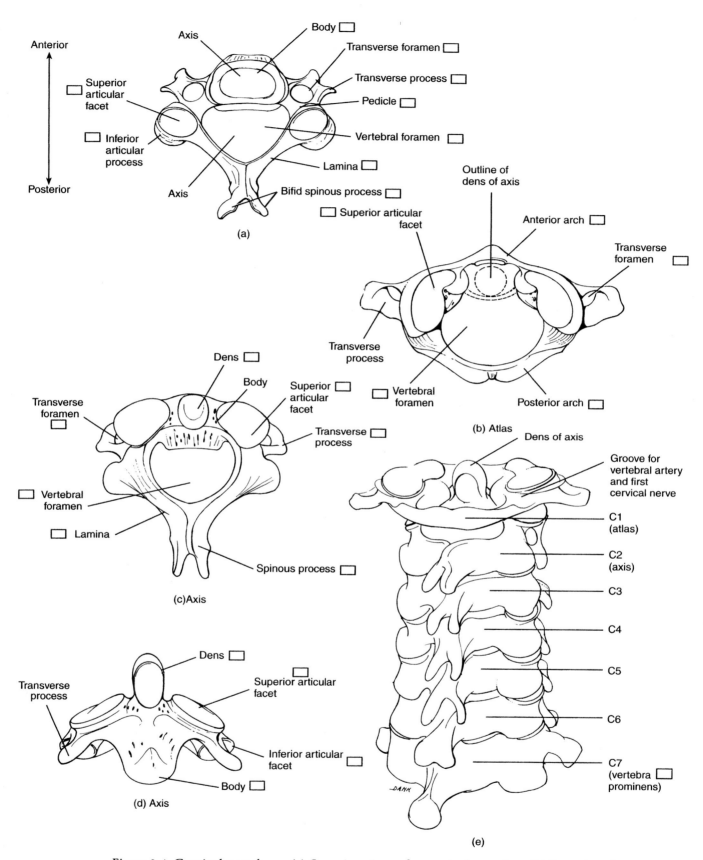

Anterior

Posterior

Axis

Body ☐

Transverse foramen ☐

Transverse process ☐

Pedicle ☐

Vertebral foramen ☐

Lamina ☐

Bifid spinous process ☐

Superior articular facet ☐

Inferior articular process ☐

Superior articular facet ☐

Axis

(a)

Outline of dens of axis

Anterior arch ☐

Transverse foramen ☐

Transverse process

Vertebral foramen ☐

Posterior arch ☐

(b) Atlas

Dens ☐

Body

Superior articular facet ☐

Transverse process ☐

Transverse foramen ☐

Vertebral foramen ☐

Lamina ☐

Spinous process ☐

(c) Axis

Dens of axis

Groove for vertebral artery and first cervical nerve

C1 (atlas)

C2 (axis)

C3

C4

C5

C6

C7 (vertebra ☐ prominens)

(e)

Dens ☐

Superior articular facet ☐

Inferior articular facet ☐

Transverse process

Body ☐

(d) Axis

Figure 3.4 Cervical vertebrae. (a) Superior view of a cervical vertebra. (b) Superior view of the atlas. (c) Superior view of the axis. (d) Anterior view of the axis. (e) The articulated cervical vertebrae in posterior view.

The seventh cervical vertebra is called the **vertebra prominens** because it has a prominent spinous process. This bony prominence is seen and felt at the base of the neck and is used as a landmark for counting the vertebra.

The Thoracic Vertebrae [Figure 3.5]

There are 12 **thoracic vertebrae** (T1–T12) in the human vertebral column. These vertebrae show more common structural features than other vertebrae.

Each thoracic vertebra has a long, laterally flattened **spinous process** directed inferiorly. The first 10 thoracic vertebrae (T1–T10) are easily distinguished by complete **costal facets** (T1) or **demifacets** (half-facets) (T2–T9) on the sides of the body. Costal facets articulate with the heads of the ribs and are found only on the thoracic vertebrae. The last two thoracic vertebrae (T11–T12) usually have only one complete facet each, located on the pedicle.

The T1–T10 vertebrae have **articular facets** on their transverse processes to articulate with the tubercles of the ribs. T11–T12 lack articular facets.

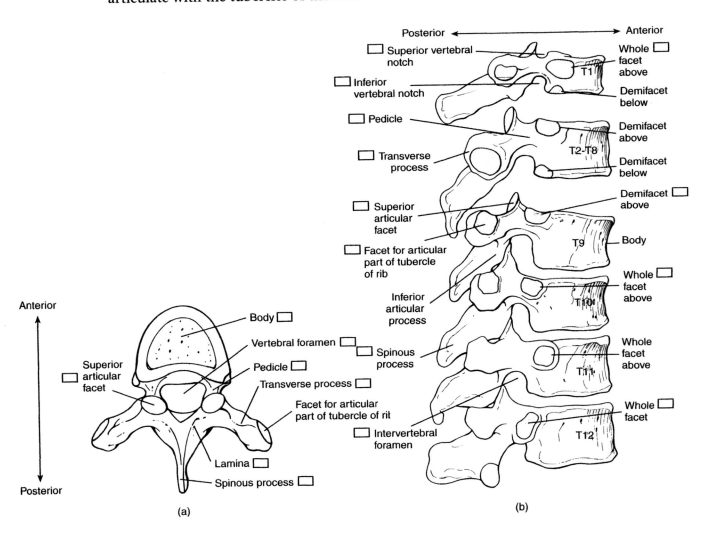

Figure 3.5 Thoracic vertebrae. (a) Superior view. (b) Articulated and right lateral view.

The Lumbar Vertebrae [Figure 3.6]

The five **lumbar vertebrae** (L1–L5) are easily identified by heavy bodies, thick spinous processes, and a lack of facets for articulation with the ribs.

The Sacrum [Figure 3.7]

The **sacrum** is a single bone formed by the fusion of five sacral vertebrae. These fusions are seen on the concave anterior surface of the sacrum as the **transverse lines**. Lateral to and at the end of these lines are four pairs of **anterior sacral (pelvic) foramina**, passageways for pelvic blood vessels and nerves.

The superior region of the sacrum is the **base**, and in its center is the body of the first sacral vertebra. The anterior edge of this body projects forward into the pelvis as the **sacral promontory**. On each side of the base are the **superior articular processes**; these articulate with the last (L5) lumbar vertebra. Laterally, the transverse processes form the two **alae** (wings) that join the sacrum to the pelvic bones. Below the alae are two **auricular surfaces**. These form an immovable sacroiliac joint with the ilium of the pelvis.

The posterior surface of the sacrum exhibits the **median sacral crest**, formed by the fusion of the spinous processes of the upper sacral vertebrae. **Posterior sacral foramina** are located on each side of the median crest and communicate with the anterior sacral foramina. The spinous processes and laminae of the fifth sacral vertebrae do not fuse. As a result, a **sacral hiatus** is found at the inferior end of the sacrum. The **sacral canal** is a tubular cavity in the sacrum that is continuous with the vertebral canal. At its inferior end, the sacrum articulates with the coccyx.

The Coccyx [Figure 3.7]

The coccyx bone is triangular and is formed from four or five fused coccygeal vertebrae.

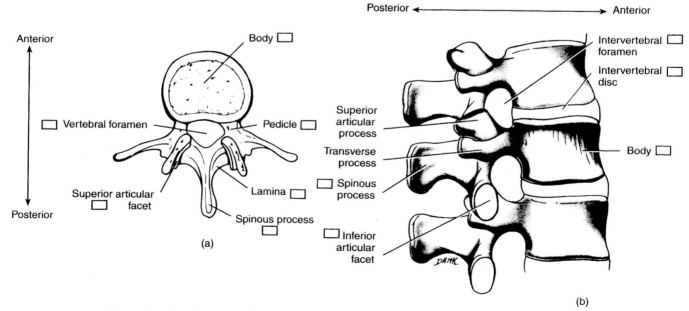

Figure 3.6 Lumbar vertebrae. (a) Superior view. (b) Articulated and right lateral view.

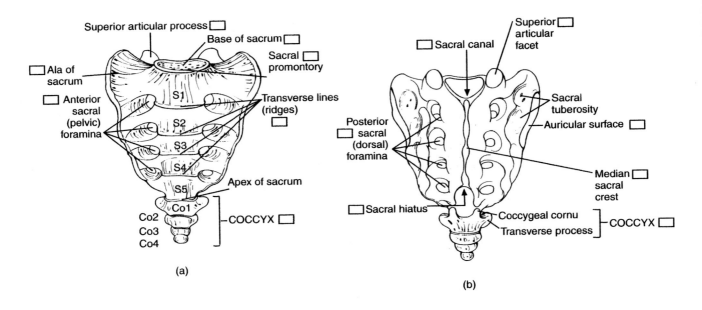

Figure 3.7 Sacrum and coccyx. (a) Anterior view. (b) Posterior view.

The Vertebral Column—Identification

[Figures 3.1 and 3.3]

Examine the adult vertebral column and its curvatures in the illustrations, in a model, and/or in a human skeleton. In a model or prepared vertebral column, identify the following and color the parts in the illustrations.

1. **cervical vertebrae**
2. **thoracic vertebrae**
3. **lumbar vertebrae**
4. **sacrum**
5. **coccyx**
6. **cervical curvature**
7. **thoracic curvature**
8. **lumbar curvature**
9. **sacral curvature**
10. **intervertebral disc**

The Typical Vertebra—Identification

[Figures 3.3 and 3.5]

Examine and color the illustrations of a typical vertebra. Then examine one of the thoracic vertebrae on a model or human skeleton and identify the following structures:

1. **body of vertebra**
2. **vertebral foramen**

3. pedicle
4. superior notch
5. inferior notch
6. intervertebral foramen
7. lamina
8. transverse process
9. superior articular process
10. inferior articular process
11. spinous process
12. vertebral arch
13. vertebral canal
14. facet or demifacet

Regional Characteristics of Vertebra—Identification

Cervical Vertebrae [Figure 3.4]

Examine and color different parts of the cervical vertebrae in the illustrations. Then examine a model or human vertebral column and identify the following structures:

1. atlas (C1)
2. vertebral arch
3. superior articular surface
4. axis (C2)
5. odontoid process (dens) (C2)
6. vertebra prominens (C7)
7. transverse foramen (C1–C7)
8. bifid spinous process (C1–C6)

Thoracic Vertebrae [Figures 3.3 and 3.5]

1. spinous process
2. facets on 10th, 11th, and 12th vertebrae
3. demifacet
4. facet on first vertebra
5. articular facets

Sacrum and Coccyx [Figure 3.7]

1. transverse lines
2. posterior sacral foramina
3. sacral promontory
4. alae
5. median sacral crest
6. sacral canal
7. anterior sacral foramina
8. base of sacrum
9. superior articular process

10. auricular surfaces
11. sacral hiatus
12. coccyx

The Bony Thorax—Description *[Figure 3.8]*

The **thorax**, or chest, is formed from the **sternum, ribs, costal cartilages**, and **thoracic vertebrae**.

The Sternum [Figure 3.8]

The **sternum**, or breastbone, is an elongated flattened bone in the anterior midline of the thorax. It is formed by the fusion of three bones: the superior **manubrium**, the middle **body**, and the inferior **xiphoid process**. The body is the largest component bone, and the xiphoid process the smallest.

The **suprasternal (jugular) notch** is the central depression in the superior portion of the manubrium. Lateral depressions in the manubrium on both sides of the suprasternal notch are the **clavicular notches**, which articulate with the medial ends of the clavicles. The costal cartilages of the first and second ribs articulate laterally with the manubrium.

The junction of the manubrium with the body is called the **sternal angle**. It is seen and felt on the surface of the sternum as a ridge and is the site of attachment of the sternum and the second rib. It is used as a starting point to count the ribs. Depressions or notches on the sides of the sternum are attachment sites of costal cartilages of the second through seventh ribs.

The small **xiphoid process** forms the most inferior portion of the sternum.

The Ribs [Figures 3.8 and 3.9]

There are usually twelve pairs of ribs classified as either **true** or **false ribs**. All ribs attach to the thoracic vertebrae. Only the first seven pairs attach to the sternum, via the hyaline **costal cartilage**; these are **true**, or **vertebrosternal**, ribs. The remaining five pairs do not attach to the sternum directly and are called **false ribs**. The cartilages of the eighth, ninth, and tenth ribs fuse and attach to the cartilage of the seventh rib. These are the **vertebrochondral ribs**. The eleventh and twelfth pair are called **floating ribs** because they do not attach to the sternum and lie in the muscles of the lateral body wall.

A typical rib is long, slender, and curved. Its superior border is broad and smooth, while its inferior border is sharp and thin. The inferior, inner surface of the typical rib has a **costal groove** that houses intercostal blood vessels and nerves.

The parts of the rib are the **head, neck, tubercle**, and **body** or **shaft**. The **head** of the typical rib contains an articular surface that fits into the facets (demifacets) on the vertebral body. Ribs that articulate between two adjacent vertebrae have a wedge-shaped head with two facets, the **superior** and **inferior facets**, separated by an **interarticular crest**. The **neck** of the rib is the constricted portion immediately lateral to the head. A **tubercle** projects from the posterior surface of the rib and consists of **nonarticular** and **articular** parts. The articular part of the tubercle

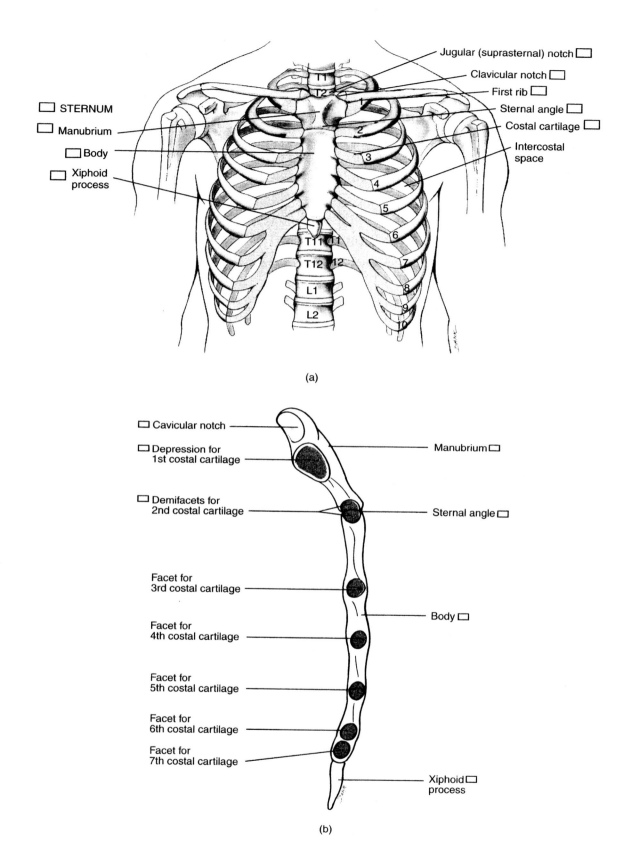

Jugular (suprasternal) notch ☐

Clavicular notch ☐

First rib ☐

Sternal angle ☐

Costal cartilage ☐

Intercostal space

☐ STERNUM

☐ Manubrium

☐ Body

☐ Xiphoid process

T1

T2

2

3

4

5

6

T11   11

T12   12

L1

L2

7

8

9

10

(a)

☐ Cavicular notch

☐ Depression for 1st costal cartilage

☐ Demifacets for 2nd costal cartilage

Facet for 3rd costal cartilage

Facet for 4th costal cartilage

Facet for 5th costal cartilage

Facet for 6th costal cartilage

Facet for 7th costal cartilage

Manubrium ☐

Sternal angle ☐

Body ☐

Xiphoid ☐ process

(b)

Figure 3.8 (a) Thoracic cage including sternum and ribs. (b) Sternum in lateral view illustrates rib and clavicle notches.

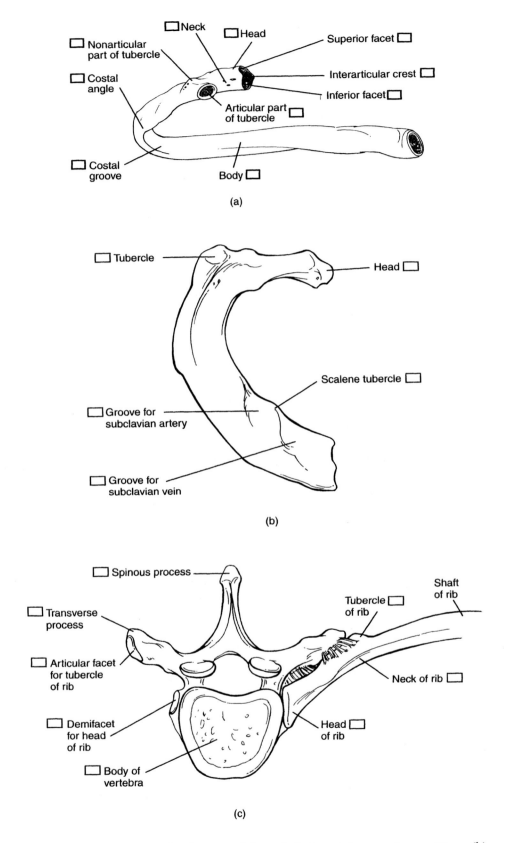

☐ Neck   ☐ Head

☐ Nonarticular part of tubercle

Superior facet ☐

☐ Costal angle

Interarticular crest ☐

Inferior facet ☐

Articular part of tubercle ☐

☐ Costal groove

Body ☐

(a)

☐ Tubercle

Head ☐

Scalene tubercle ☐

☐ Groove for subclavian artery

☐ Groove for subclavian vein

(b)

☐ Spinous process

Shaft of rib

☐ Transverse process

Tubercle ☐ of rib

☐ Articular facet for tubercle of rib

Neck of rib ☐

☐ Demifacet for head of rib

Head ☐ of rib

☐ Body of vertebra

(c)

Figure 3.9 (a) A typical rib. A left rib viewed from inferior and posterior position. (b) Atypical rib (first rib) in superior view. (c) Articulation of the head and tubercle of the rib with thoracic vertebra.

interacts with the facet on the transverse process of the same vertebra. Distal to the tubercle is a bend or **angle** from which the **body** or shaft of the rib curves anteriorly to the costal cartilage.

The first, second, tenth, eleventh, and twelfth ribs are atypical. The first rib is supero-inferiorly flattened, short, and sharply curved. The head has a single facet to articulate with the body of the first thoracic vertebra and a tubercle to articulate with the transverse process of the same vertebra. The superior surface of the first rib has an anterior groove to accommodate the **subclavian vein** and a posterior groove to accommodate the **subclavian artery**. The grooves are separated by a **scalene tubercle**, an attachment site for the anterior scalene muscle.

The second rib is similar to the first but is longer and larger. Its head, neck, and tubercle are similar to the typical rib, and it articulates with two vertebrae. The tenth through twelfth ribs are atypical in that their heads have single facets to articulate with the bodies of the corresponding vertebrae. The eleventh rib is short and has a slight angle and no neck or tubercle. The twelfth rib is short and lacks a neck, tubercle, angle, or costal groove. The last two ribs do not articulate with the transverse processes.

The Bony Thorax—Identification [Figure 3.8]

Examine and color the illustrations of the bones of the thorax. Then examine a model or a human skeleton and identify the following bones and ribs:

1. **manubrium**
2. **xiphoid process**
3. **clavicular notch**
4. **body**
5. **suprasternal notch**
6. **sternal angle**

The Ribs—Identification [Figures 3.8 and 3.9]

1. **true ribs**
2. **false ribs**
3. **floating ribs**
4. **costal cartilage**
5. **neck of rib**
6. **body of rib**
7. **head of rib**
8. **angle of rib**
9. **costal groove**
10. **articular facets**
11. **articular part of tubercle**
12. **nonarticular part of tubercle**
13. **interarticular crest**

Examine the first rib and identify the following structures:

1. **head of rib**
2. **tubercle**

3. grooves to accommodate subclavian vein and artery
4. articular facet
5. scalene tubercle

Examine the thorax on a model or human skeleton. Identify and examine the eleventh and twelfth ribs.

CHAPTER SUMMARY AND CHECKLIST

I. THE VERTEBRAL COLUMN

A. General Characteristics
1. Seven cervical vertebrae
2. Twelve thoracic vertebrae
3. Five lumbar vertebrae
4. One sacral and one coccygeal bone
5. Intervertebral discs between vertebrae
6. Intervertebral foramina between vertebrae

B. Curvature of Adult Vertebral Column
1. Four major curvatures
2. Primary thoracic and sacral curvatures
3. Secondary cervical and lumbar curvatures
4. Cervical curvature develops at about three months
5. Lumbar curvature develops when child stands
6. Scoliosis: lateral deviation of vertebral column
7. Kyphosis: hunchback; posterior curvature
8. Lordosis: exaggerated lumbar curvature

II. GENERAL CHARACTERISTICS OF A TYPICAL VERTEBRA

A. The Typical Vertebra
1. Body is an oval anterior region that bears weight
2. Vertebral arch and body form vertebral foramen
3. Pedicles form lateral boundaries of vertebral arch
4. Laminae form roof over vertebral arch
5. Transverse process extends laterally from vertebra
6. Spinous process is a posterior projection
7. Contains superior and inferior articular processes

III. REGIONAL CHARACTERISTICS OF THE VERTEBRAE

A. Cervical Vertebrae
1. Smallest vertebrae because they hold only the head
2. Transverse foramina are a characteristic feature
3. First vertebra lacks body and is called the atlas
4. Superior articular surfaces articulate with skull
5. Second vertebra has dens and is called the axis
6. Seventh vertebra is vertebra prominens
7. Spinous process is bifid

B. **Thoracic Vertebrae**
   1. Spinous process is long and flattened
   2. First 10 vertebrae have:
      a. facets or demifacets on bodies
      b. articular facets on transverse processes
   3. Last two vertebrae have a single facet on pedicles

C. **Lumbar Vertebrae**
   1. Large and massive vertebrae
   2. Do not articulate with ribs

D. **Sacrum**
   1. Five sacral vertebrae fuse to form sacrum
   2. Fusion seen at transverse lines
   3. Anterior edge of body projects as sacral promontory
   4. Transverse processes fused to form two alae
   5. Auricular surfaces form sacroiliac joint
   6. Median sacral crest formed by spinous processes
   7. Sacral hiatus formed at end of sacrum

E. **Coccyx**
   1. Vestigial tailbone
   2. Formed by fusion of coccygeal vertebrae

IV. **THE BONY THORAX**

A. **The Sternum**
   1. The elongated breastplate
   2. Formed from manubrium, body, and xiphoid process
   3. Suprasternal and clavicular notches on manubrium
   4. Junction of body and manubrium forms sternal angle
   5. Xiphoid process is most inferior projection

B. **The Ribs**
   1. First seven pairs of ribs are true ribs
   2. Remaining five pairs are false ribs
   3. Last two pairs are floating ribs
   4. Typical rib exhibits costal groove
   5. Head of typical rib articulates with two vertebrae
   6. Rib has a constricted neck, an angle, and a body
   7. Atypical ribs 1, 2, 10, 11, and 12

Laboratory Exercises 3

NAME _____

LAB SECTION _____ DATE _____

LABORATORY EXERCISE 3.1

Part I

Vertebral Column

1. Which part of the vertebral column contains:

   a. seven vertebrae _____

   b. fused vertebra _____

   c. twelve vertebrae _____

   d. five vertebrae _____

2. What is found between individual vertebrae? _____

3. What constitutes the primary curvature? _____

4. What constitutes the secondary curvature? _____

5. When does the cervical curvature develop? _____

6. When does the lumbar curvature develop? _____

7. Abnormal lateral deviation of the vertebral column is

   called _____

8. Abnormal posterior curvature of the vertebral column is

   called _____

9. Exaggerated anterior lumbar curvature is called _____

10. Which vertebra is the vertebra prominens? _____

Part II

Various Vertebrae

Examine the vertebra, label it, and color its different parts.

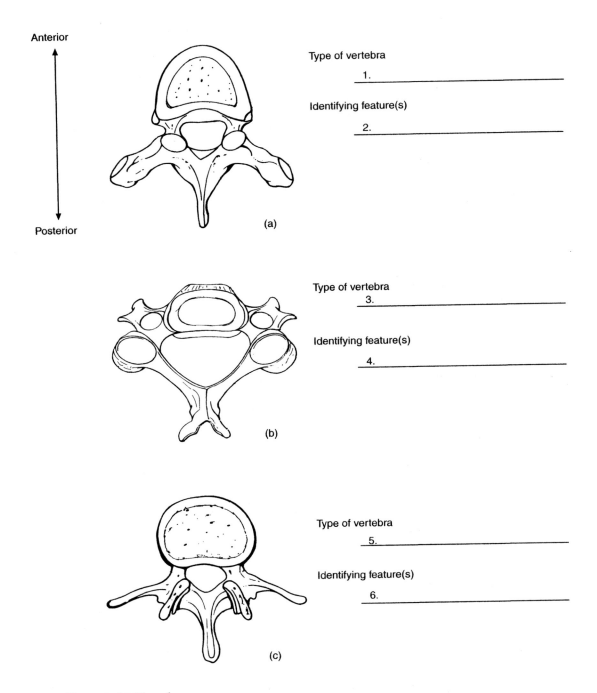

Anterior

Posterior

Type of vertebra

1. _____

Identifying feature(s)

2. _____

(a)

Type of vertebra

3. _____

Identifying feature(s)

4. _____

(b)

Type of vertebra

5. _____

Identifying feature(s)

6. _____

(c)

Figure 3.10 Vertebrae.

Part III

The Vertebrae, Thoracic Cage, and Ribs

Using the listed terms, label the vertebrae and then color them.

Body of vertebra   Lamina
Demifacet   Intervertebral foramen
Inferior articular facet Whole facet
Pedicle   Vertebral foramen
Superior articular facet Spinous process
Superior vertebral notch Inferior vertebral notch
Transverse process   Facet for tubercle of rib

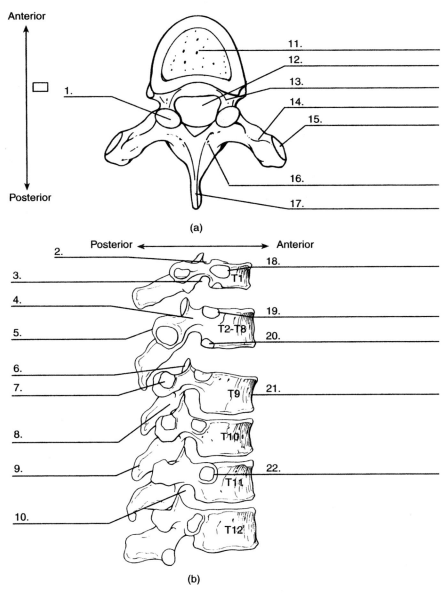

Figure 3.11 Vertebrae.

LABORATORY EXERCISE 3.2

Using the listed terms, label the structures in the thoracic cage and then color them.

Body
Costal cartilage
Jugular (suprasternal) notch
Manubrium
Clavicular notch

Intercostal space
First rib
Sternal angle
Xiphoid process

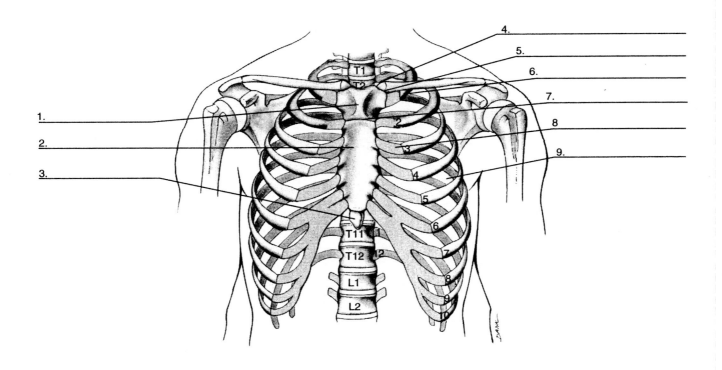

Figure 3.12 Thoracic cage.

Using the listed terms, label the structures on the typical rib and first rib and then color them.

Costal groove　　　　　　　　　　Body of rib
Groove for subclavian artery　　　Groove for subclavian vein
Head of rib　　　　　　　　　　　Superior facet
Neck of rib　　　　　　　　　　　Inferior facet
Nonarticulating part of tubercle　Costal angle
Tubercle　　　　　　　　　　　　Articular part of tubercle
Interarticular crest　　　　　　　Scalene tubercle

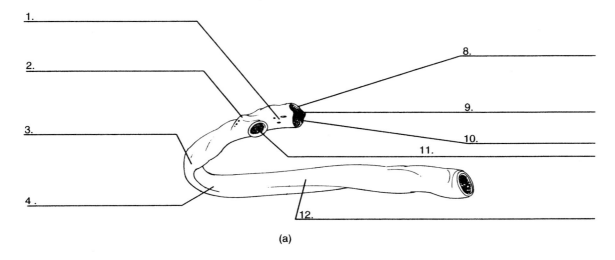

(a)

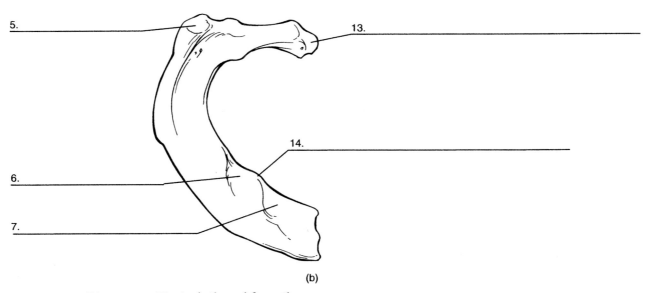

(b)

Figure 3.13 Typical rib and first rib.

Chapter 4

The Appendicular Skeleton

THE PECTORAL GIRDLE AND UPPER EXTREMITY

CLAVICLE

Acromial extremity

Acromioclavicular joint

Sternal extremity

Sternoclavicular joint

STERNUM

Scapula

Humerus

ULNA

RADIUS

CARPALS

METACARPALS

PHALANGES

Objective

The objective of Chapter 4, "The Appendicular Skeleton," is for you to become acquainted with the bones of the:

1. **Pectoral girdle**
2. **Upper extremity**
3. **Hand**

The Appendicular Skeleton— Description [Figure 4.1]

The appendicular skeleton consists of the bones of the **upper extremities** and **pectoral girdle** and the bones of the **lower extremities** and **pelvic girdle**. The pectoral and pelvic girdles attach the bones of the upper and lower extremities to the axial skeleton.

The **pectoral (shoulder) girdle** consists of the paired **scapulae** (singular, **scapula**) and **clavicles**. Connections at the clavicles of the pectoral girdle attach to the axial skeleton at the sternum.

Each **upper extremity** consists of the **humerus (upper arm bone)**, **ulna** and **radius (forearm)** bones, **carpal** bones of the wrist, and **metacarpal** and **phalangeal** bones of the hand.

The Pectoral (Shoulder) Girdle—Description

The Clavicle [Figures 4.1 and 4.2]

The clavicle (collarbone) is S-shaped and binds the shoulder to the axial skeleton. Its location is superior to the first rib. The medial end of the clavicle, the **sternal extremity**, is enlarged and blunt and has a **costal tuberosity** on its inferior surface. The medial end articulates with the sternum at the **sternoclavicular joint**. The lateral end, called the **acromial extremity**, is broad and flat and has a **conoid tubercle** on its inferior surface. The lateral end articulates with the acromion process of the scapula at the **acromioclavicular joint**.

The Scapula [Figure 4.3]

The scapula (shoulder blade) is a large, triangular flat bone on the posterior thorax over ribs two to seven. The posterior surface of the scapula bears a sharp ridge, called the **spine**, that extends diagonally across the bone and terminates laterally as an expanded anterior process called the **acromion process**. The acromion process articulates with the clavicle at the acromioclavicular joint. Above the scapular spine is the **supraspinous fossa**, and below the spine is the **infraspinous fossa**. On the anterior (costal) surface of the scapula is a concave area called the **subscapular fossa**. These fossae serve as surfaces for attachment of shoulder muscles. Inferior to the acromion is the **glenoid cavity** or **fossa** into which fits the head of the humerus. Superior and anterior to the glenoid cavity is the **coracoid process**; tendons of the arm muscles attach to this thumblike process.

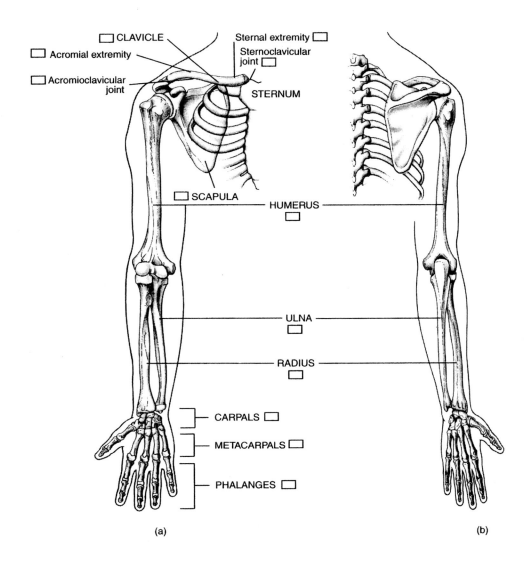

Figure 4.1 Right pectoral (shoulder) girdle and upper extremity. (a) Anterior view. (b) Posterior view.

Each scapula has three borders: the **medial (vertebral) border**, the **axillary** or **lateral border**, and the **superior border**. The **superior angle** of the scapula is formed at the junction of the superior and vertebral borders. The **scapular notch** is located on the superior border of the scapula near the coracoid process. The **inferior angle** is formed at the junction of the medial and axillary borders.

The Pectoral (Shoulder) Girdle—Identification

[Figures 4.1, 4.2, and 4.3]

Examine and color the illustrations of the clavicle and scapula. Then examine the clavicle and scapula of a model or human skeleton, and identify the following structures:

LATERAL   MEDIAL

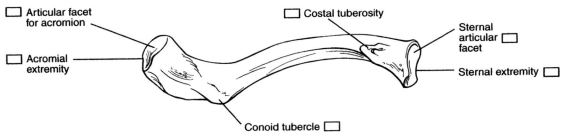

☐ Articular facet
   for acromion

☐ Costal tuberosity

Sternal
articular ☐
facet

☐ Acromial
   extremity

Sternal extremity ☐

Conoid tubercle ☐

Figure 4.2 The right clavicle, inferior view with articulating surfaces.

The Clavicle

1. sternal extremity
2. conoid tubercle
3. acromial extremity
4. costal tuberosity

The Scapula

1. spine
2. coracoid process
3. acromion process

4. glenoid cavity or fossa
5. scapular notch
6. medial border
7. superior border
8. axillary border
9. superior angle
10. inferior angle
11. supraspinous fossa
12. infraspinous fossa
13. subscapular fossa

The Upper Extremity

The Humerus (Arm Bone)—Description

[Figures 4.4 and 4.5]

The humerus (arm bone) is the longest bone of the upper extremity and a typical long bone. The proximal end of the humerus is the smooth **head,** which fits into the glenoid cavity of the scapula. A slight narrowing below the head is the **anatomical neck.** The rough projection immediately lateral to the neck is the **greater tubercle;** the **lesser tubercle** lies anterior to the greater tubercle. The **intertubercular (bicipital) groove** or **sulcus** separates the two tubercles. Distal to the tubercles is the **surgical neck,** a frequent site of fractures in the humerus.

The central portion of the **body (shaft)** of the humerus is round but it gradually flattens distally. On the lateral side, at the middle of the bone, is a roughened area called the **deltoid tuberosity,** an attachment site for the tendon of the deltoid muscle. A **radial groove,** formed by the radial nerve, runs obliquely down the posterior side of the bone.

The distal end of the humerus has **medial** and **lateral epicondyles.** Between the two epicondyles are two articular surfaces. The lateral **capitulum** articulates with the head of the radius and the medial **trochlea** articulates with the proximal end of the ulna.

Superior to the trochlea on the anteroinferior surface of the humerus is a depression called the **coronoid fossa** (Fig. 4.4). This fossa receives the coronoid process of the ulna when the elbow is flexed. A small **radial fossa,** lateral to the coronoid fossa, receives the head of the radius when the elbow is flexed. On the posteroinferior surface of the humerus is another, larger depression called the

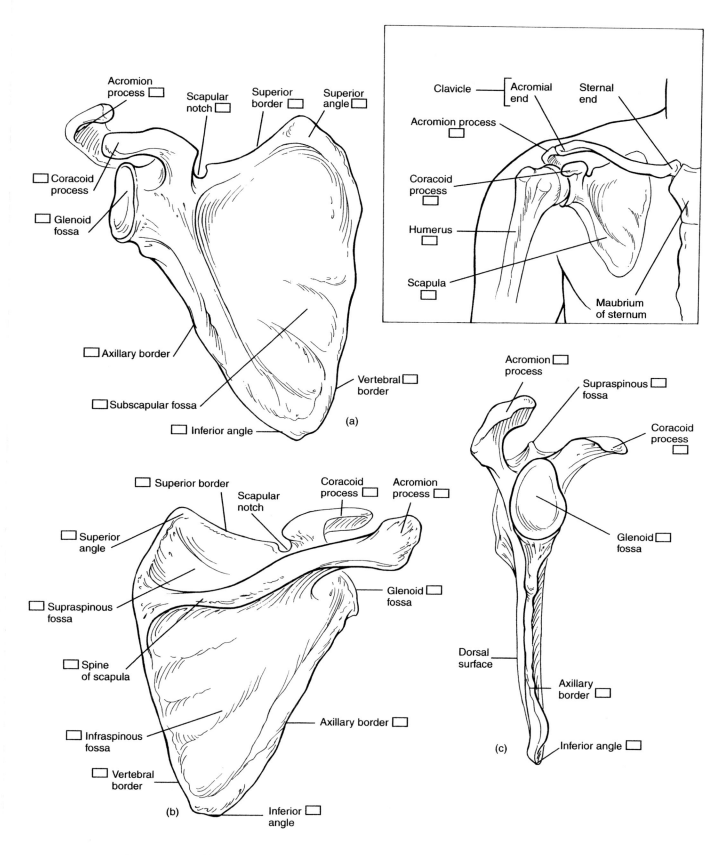

Figure 4.3 Different views of the scapula. (a) Anterior view. (b) Posterior view. (c) Lateral view.

olecranon fossa. This fossa receives the olecranon process of the ulna when the elbow is extended.

The Humerus—Identification [Figures 4.4 and 4.5]

Examine and color the illustrations of the humerus. Then examine the humerus of a model or human skeleton and identify the following structures:

1. head
2. greater tubercle
3. lesser tubercle
4. intertubercular groove
5. anatomical neck
6. surgical neck
7. body of humerus
8. radial groove
9. deltoid tuberosity
10. lateral epicondyle
11. medial epicondyle
12. capitulum
13. trochlea
14. radial fossa
15. coronoid fossa
16. olecranon fossa

The Forearm: Ulna and Radius—Description

[Figures 4.4 and 4.5]

Two parallel bones, the **ulna** and **radius**, form the forearm skeleton. The ulna is longer and larger than the radius. In the anatomical position the ulna lies medially (little finger side) and the radius laterally (on the thumb side). The proximal ends of these two bones articulate with the humerus, and their distal ends articulate with the carpal or wrist bones.

ULNA [Figures 4.4 and 4.5]

The proximal end of the ulna fits over the trochlea of the humerus. A larger, posterior extension of the ulna is the **olecranon process**. The smaller, anterior projection of the bone is the **coronoid process**. The depression between the two projections, where the ulna articulates with the humerus, is the **trochlear (semilunar) notch**. On the lateral side of the coronoid process is a small depression called the **radial notch**, which articulates with the head of the radius. The distal end of the ulna bears a round projection called the **styloid process**.

RADIUS [Figures 4.4 and 4.5]

The head of the radius articulates with the capitulum of the humerus and the radial notch of the ulna. Medial and inferior to the head is the **radial tuberosity**. The distal end of the radius articulates medially with the ulna at the **ulnar notch** and inferiorly with two wrist bones. A lateral **styloid process** also lies at the distal end of the radius.

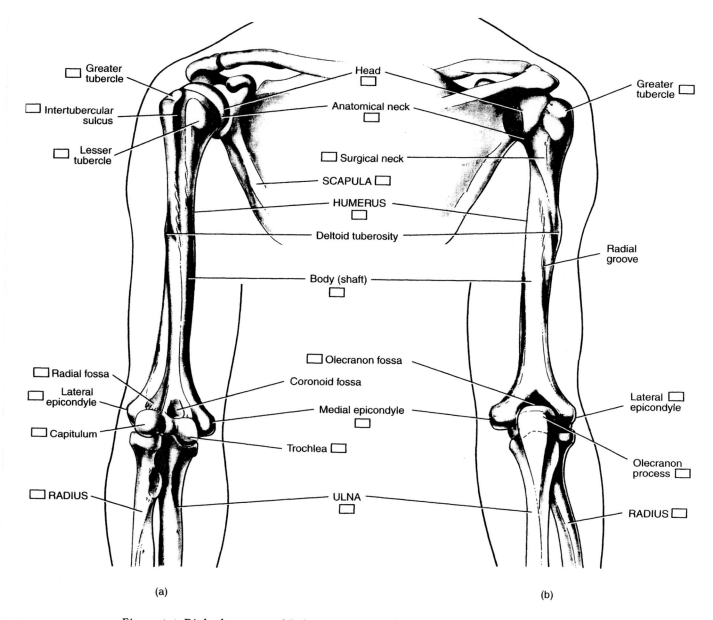

Greater tubercle ☐

Intertubercular sulcus ☐

Lesser tubercle ☐

Head ☐

Anatomical neck ☐

Surgical neck ☐

SCAPULA ☐

HUMERUS ☐

Deltoid tuberosity

Body (shaft) ☐

Greater tubercle ☐

Radial groove

Radial fossa ☐

Lateral epicondyle ☐

Capitulum ☐

☐ RADIUS

Olecranon fossa ☐

Coronoid fossa

Medial epicondyle ☐

Trochlea ☐

ULNA ☐

Lateral epicondyle ☐

Olecranon process ☐

RADIUS ☐

(a)

(b)

Figure 4.4 Right humerus. (a) Anterior view. (b) Posterior view.

The Ulna and Radius—Identification

[Figures 4.4 and 4.5]

Examine and color the illustrations of the ulna and radius. Then examine the ulna and radius on a model or human skeleton and identify the following structures:

Ulna

1. olecranon process
2. coronoid process
3. trochlear notch
4. radial notch
5. head of ulna
6. styloid process

Radius

1. head of radius
2. radial tuberosity
3. ulnar notch
4. styloid process

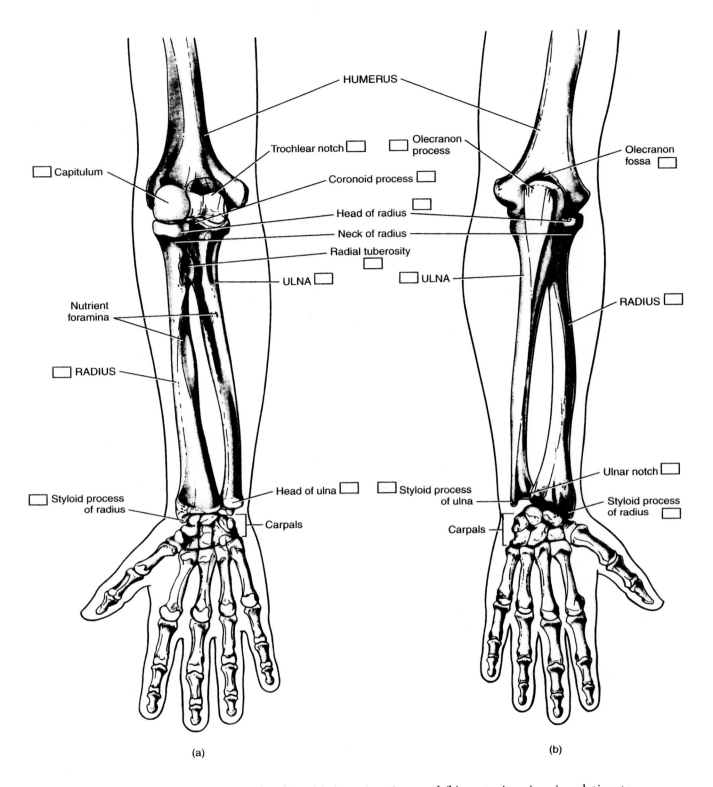

HUMERUS

Capitulum ☐

Trochlear notch ☐

Olecranon process ☐

Olecranon fossa ☐

Coronoid process ☐

Head of radius ☐

Neck of radius

Radial tuberosity ☐

Nutrient foramina

ULNA ☐

ULNA ☐

RADIUS ☐

RADIUS ☐

Ulnar notch ☐

Styloid process of radius ☐

Head of ulna ☐

Styloid process of ulna ☐

Styloid process of radius ☐

Carpals

Carpals

(a)

(b)

Figure 4.5 Right ulna and radius. (a) Anterior view and (b) posterior view in relation to humerus and hand.

The Hand: Carpals, Metacarpals, and Phalanges—Description *[Figures 4.1 and 4.6]*

The skeleton of the hand consists of the **carpal bones** (wrist), **metacarpal bones** (palm), and **phalanges** (bones of the digits).

The **carpus** (wrist) consists of eight small bones, the **carpals**, arranged in two transverse rows of four each. With the hand in the anatomical position, the proximal row of carpals, from lateral (thumb) to medial position, are the **scaphoid**, **lunate**, **triquetral**, and **pisiform**. The carpals in the distal row (lateral to medial) are **trapezium**, **trapezoid**, **capitate**, and **hamate**.

The **metacarpus** (palm) consists of five **metacarpal bones**. Each metacarpal bone consists of a proximal **base**, a middle **shaft**, and a distal **head**. These bones are numbered I to V, starting with the lateral bone (thumb). The bases articulate with the distal row of carpal bones, whereas the heads articulate with the phalanges of the fingers.

Each hand contains 14 **phalanges** (singular, **phalanx**), or finger bones. Each finger has three phalanges: **proximal**, **middle**, and **distal**. An exception is the thumb, which has two phalanges, a proximal and distal phalanx. Each phalanx consists of a **base**, **shaft**, and **head**.

The Carpals, Metacarpals, and Phalanges—Identification *[Figures 4.1 and 4.6]*

The best way to study the bones of the hand and their structural relationships is on a model, where the hand bones are connected by a wire or string. Examine the

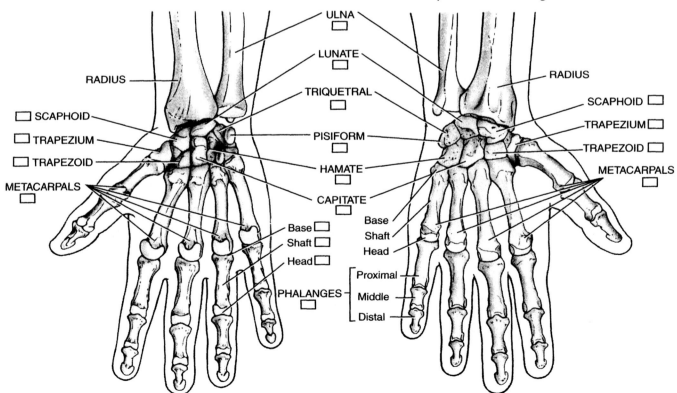

Figure 4.6 Right wrist and hand. (a) Anterior view. (b) Posterior view.

illustrations of the hand bones and color them. Then examine the bones of the model or human skeleton and identify the following:

Carpals

1. scaphoid
2. lunate
3. trapezium
4. trapezoid
5. capitate
6. hamate
7. triquetral
8. pisiform

Metacarpals

1. metacarpal I to V
2. base
3. shaft
4. head

Phalanges

1. proximal
2. middle
3. distal

CHAPTER SUMMARY AND CHECKLIST

I. THE PECTORAL (SHOULDER) GIRDLE

A. The Clavicle
1. Lies superior to the first rib
2. Medial end is sternal extremity
3. Lateral end is acromial extremity
4. Conoid tubercle on inferior lateral surface
5. Costal tuberosity on inferior medial surface

B. The Scapula
1. Posterior surface exhibits sharp ridge called a spine
2. Anterior expanded process is acromion
3. Above spine is the supraspinous fossa
4. Below spine is the infraspinous fossa
5. Anterior surface contains the subscapular fossa
6. Glenoid cavity lies inferior to acromion
7. Superior to glenoid cavity is the coracoid process
8. Medial border is nearest vertebral column
9. Border near arm is the axillary border
10. Superior edge is the superior border

II. THE UPPER EXTREMITY

A. The Humerus (Arm Bone)
1. Proximal end (head) fits into glenoid cavity of the scapula
2. Construction below head is anatomical neck
3. Rough projections are greater and lesser tubercles
4. Deltoid tuberosity in midregion of the shaft
5. Distal end contains medial and lateral epicondyles
6. Capitulum is lateral rounded projection
7. Trochlea is medial projection
8. Olecranon, radial, and coronoid fossae lie inferiorly

B. The Ulna and Radius (Forearm)
1. Ulna and radius form the forearm of the skeleton
2. Ulna is larger and longer than radius

3. Ulna is medial to radius
4. Proximal ends articulate with humerus
5. Distal ends articulate with wrist bones

C. **The Ulna**
   1. Proximally, posterior projection is the olecranon process
   2. Anterior projection is the coronoid process
   3. Trochlear notch is for ulna articulation with humerus
   4. Radial notch is where radius articulates with ulna
   5. Head and styloid process are distal

D. **The Radius**
   1. Proximal head articulates with capitulum of humerus
   2. Articulates with ulna at ulnar notch

E. **The Carpals, Metacarpals, and Phalanges (Hand)**
   1. There are eight carpal bones of the wrist
   2. There are five metacarpal bones of the hand
   3. Metacarpal bones are numbered I to V
   4. Metacarpal bones have base, shaft, and head
   5. There are 14 phalanges of the fingers
   6. Phalanges are proximal, middle, and distal

Laboratory Exercises 4

LABORATORY EXERCISE 4.1
The Appendicular Skeleton

Using the listed terms, label the scapula and then color the component bones.

Glenoid fossa
Acromion process
Superior border
Axillary border
Inferior angle
Supraspinous fossa
Infraspinous fossa

Coracoid process
Scapular notch
Superior angle
Subscapular fossa
Vertebral border
Spine of scapula

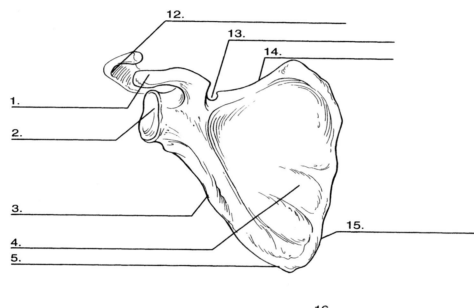

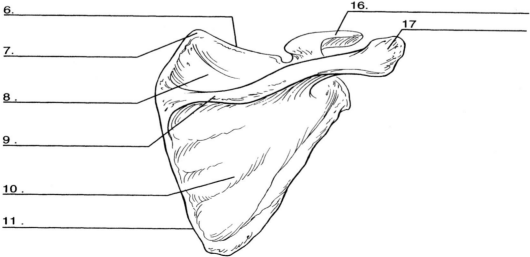

Figure 4.7 (Top) Anterior and (bottom) posterior views of the scapula.

LABORATORY EXERCISE 4.2

Using the listed terms, label the humerus and then color the component bones.

Greater tubercle
Intertubercular sulcus
Anatomical neck of humerus
Deltoid tuberosity
Radial fossa
Lateral epicondyle
Capitulum
Radial groove

Lesser tubercle
Head of humerus
Surgical neck
Body of humerus
Coronoid fossa
Medial epicondyle
Trochlea
Olecranon fossa

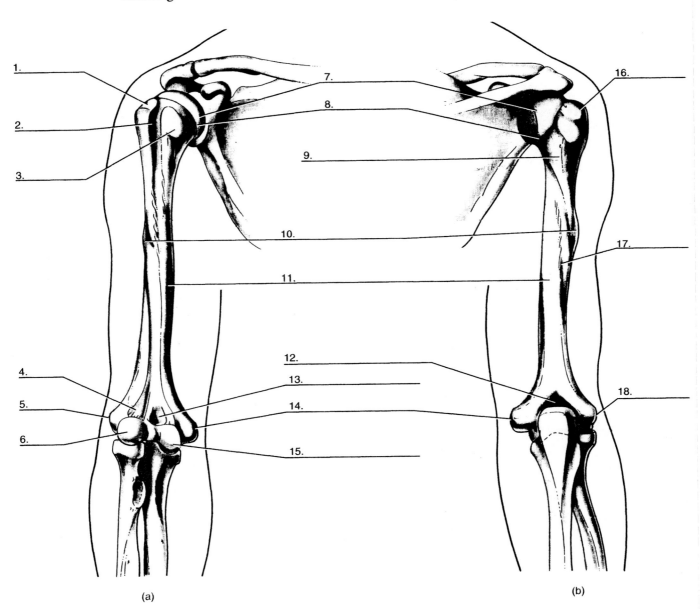

(a)   (b)

Figure 4.8 (a) Anterior and (b) posterior views of the right humerus.

LABORATORY EXERCISE 4.3

Using the listed terms, label the ulna and radius and then color the component bones.

Coronoid process
Head of radius
Radial tuberosity
Radius
Head of ulna
Ulnar notch

Olecranon process
Neck of radius
Ulna
Styloid process of radius
Styloid process of ulna

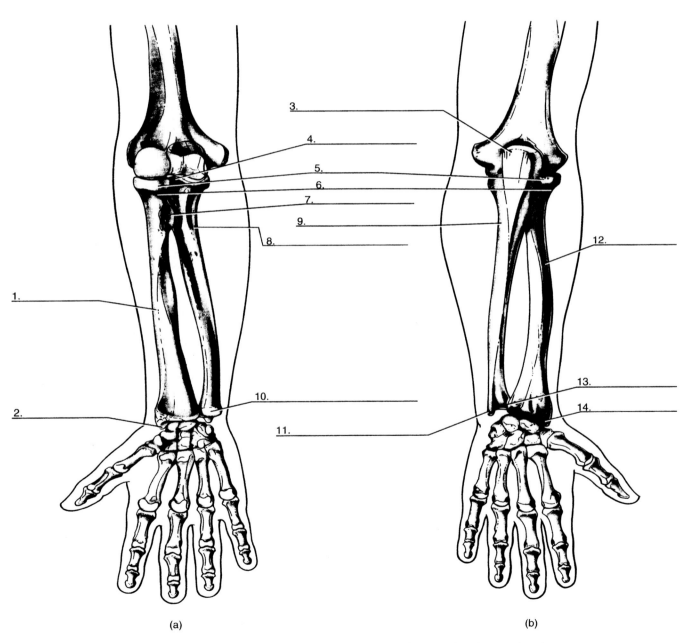

(a)   (b)

Figure 4.9 (a) Anterior and (b) posterior views of the right ulna and radius.

LABORATORY EXERCISE 4.4

Using the listed terms, label the wrist and hand and then color the component bones.

Scaphoid
Lunate
Trapezoid
Hamate
Metacarpals

Trapezium
Triquetral
Pisiform
Capitate
Phalanges

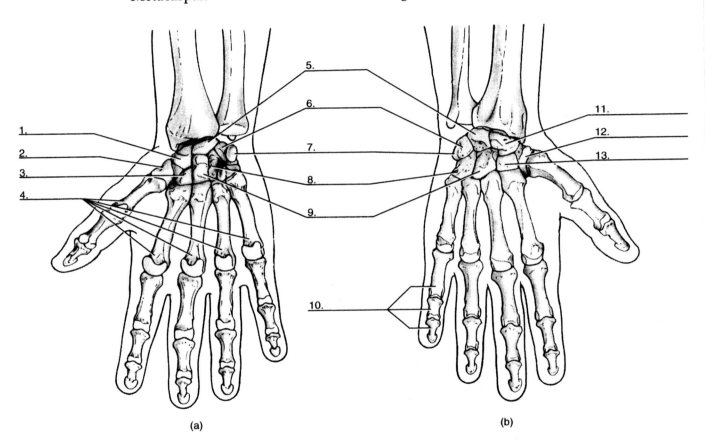

Figure 4.10 (a) Anterior and (b) posterior views of the right wrist and hand.

Chapter 5

The Appendicular Skeleton

THE PELVIC GIRDLE AND LOWER EXTREMITY

Objective
The Pelvic (Hip) Girdle—Description
    The Ilium
    The Ischium
    The Pubis
The Pelvic Cavity—Description
Gender Differences in Pelvic Bones—
    Description
The Pelvic Girdle—Identification
    The Ilium
    The Ischium
    The Pubis
The Pelvic Cavity—Identification
Gender Differences in the Pelvic Bones—
    Identification
The Lower Extremity—Description
    The Femur
    The Patella
    The Femur and Patella—Identification
    The Tibia
    The Fibula
    The Tibia and Fibula—Identification
The Ankle and Foot Bones—Description
    The Tarsal Bones
    The Metatarsal Bones
    The Phalanges
The Ankle and Foot Bones—Identification
Chapter Summary and Checklist
Laboratory Exercises

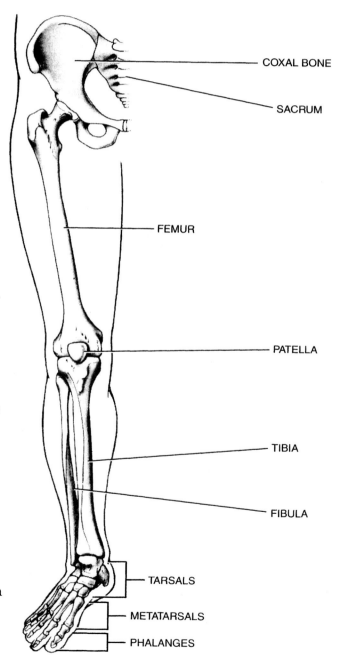

COXAL BONE

SACRUM

FEMUR

PATELLA

TIBIA

FIBULA

TARSALS

METATARSALS

PHALANGES

Objective

The objective of Chapter 5, "The Appendicular Skeleton," is to familiarize you with the bones of the:

1. Pelvic girdle
2. Male and female pelvis
3. Thigh, leg, ankle, and foot

The Pelvic (Hip) Girdle—Description

[Figures 5.1, 5.2, and 5.3]

The **pelvic girdle**, or **bony pelvis**, is formed by a pair of **coxal**, or **hip bones**. The coxal bones are joined anteriorly at the **symphysis pubis** and posteriorly to the **sacrum**.

In early life each coxal bone consists of the **ilium**, **ischium**, and **pubis**. In adults, the these three bones fuse into one coxal bone. The fusion occurs in the lateral socket, called the **acetabulum**. The head of the femur fits into the acetabulum for articulation with the coxal bone.

The Ilium [Figure 5.1]

The ilium is the largest and most superior of the three coxal bones. Its superior border, the **iliac crest**, forms the prominence of the hip. Its anterior end is called the **anterior superior iliac spine** and its posterior end is the **posterior superior iliac spine**. Just below these two spines are the **anterior inferior iliac spine** and **posterior inferior iliac spine**.

Under the posterior inferior iliac spine is the **greater sciatic notch**, through which passes the sciatic nerve. The smooth, concave internal surface of the ilium is the **iliac fossa**, to which iliac muscle attaches. Posterior to the iliac fossa is the roughened **auricular surface**, which attaches to the sacrum and forms the sacroiliac joint. Descending diagonally and anteriorly from the auricular surface is a ridge called the **iliopectineal**, or **arcuate line**. The lateral surface of the ilium has three ridges, to which attach the gluteal muscles of the buttocks. These ridges are the **inferior**, **anterior**, and **posterior gluteal lines** (Figure 5.1).

The Ischium [Figure 5.1]

The ischium forms the posterior inferior region of the coxal bone. The **ischial spine**, located inferior to the greater sciatic notch, is a prominent projection on the posterior surface of the ischium. Inferior to this spine is the **lesser sciatic notch**, which serves as a passageway for blood vessels and nerves. The **ischial tuberosity** is an inferior bony projection. The **ramus** is an anterior projection from the ischial tuberosity that joins with the pubis to surround the large **obturator foramen**.

The Pubis [Figures 5.1 and 5.2]

The pubis is the smallest of the three coxal bones. It forms the anterior and inferior portion of the coxal bone. The pubis consists of a **superior ramus** and an **inferior ramus** that support the body of the pubis. The **pubic crest** is located on

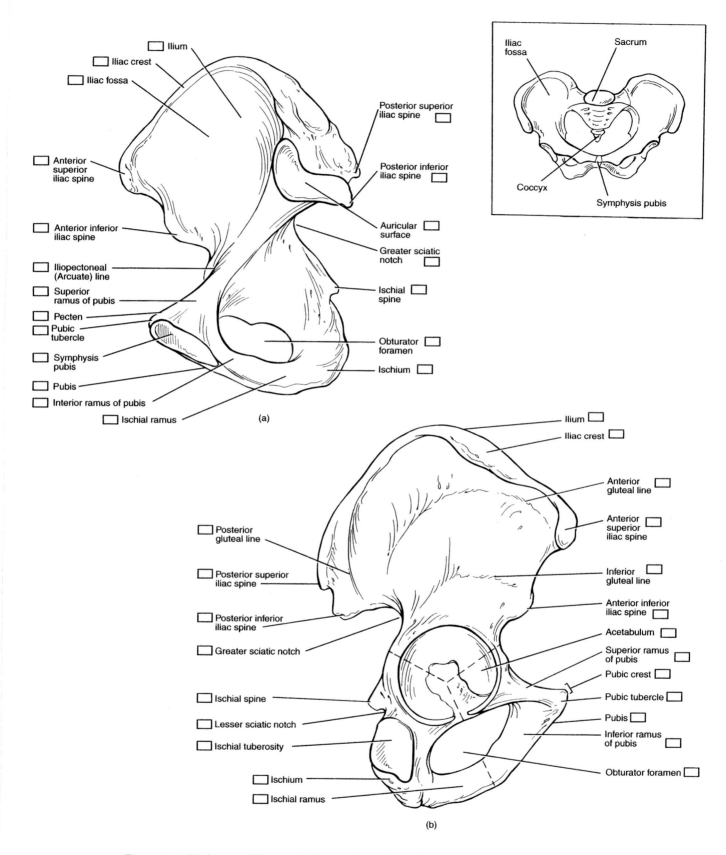

Ilium
Iliac crest
Iliac fossa
Posterior superior iliac spine
Posterior inferior iliac spine
Anterior superior iliac spine
Auricular surface
Anterior inferior iliac spine
Greater sciatic notch
Iliopectoneal (Arcuate) line
Ischial spine
Superior ramus of pubis
Pecten
Pubic tubercle
Obturator foramen
Symphysis pubis
Ischium
Pubis
Interior ramus of pubis
Ischial ramus
(a)

Iliac fossa
Sacrum
Coccyx
Symphysis pubis

Ilium
Iliac crest
Anterior gluteal line
Posterior gluteal line
Anterior superior iliac spine
Posterior superior iliac spine
Inferior gluteal line
Posterior inferior iliac spine
Anterior inferior iliac spine
Greater sciatic notch
Acetabulum
Superior ramus of pubis
Pubic crest
Ischial spine
Pubic tubercle
Lesser sciatic notch
Pubis
Ischial tuberosity
Inferior ramus of pubis
Ischium
Obturator foramen
Ischial ramus
(b)

Figure 5.1 Right coxal bone. (a) Internal surface. (b) External surface. Dotted lines indicate junctions of the three bones, ilium, ischium, and pubis.

the anterior border of the superior ramus. Lateral to the pubic crest is the **pubic tubercle**. The **symphysis pubis** is a fibrocartilagenous joint between the two coxal bones. Inferior to this joint, the inferior pubic rami plus the ischial rami form the **pubic arch**. Width of the angle in this arch distinguishes the male and female pelvis.

The Pelvic Cavity—Description *[Figures 5.2 and 5.3]*

The human pelvis is divided into a **false (greater) pelvis** and a **true (lesser) pelvis** by a horizontal plane that passes from the sacral promontory, along the **iliopectineal (arcuate) lines** on the inner surface of the ilium, to the upper margin of the symphysis pubis. The bony boundary formed by this plane is the **pelvic brim**. The expanded cavity superior to the pelvic brim is the false pelvis. The cavity of the false pelvis is also part of the abdominal cavity and contains abdominal organs. The pelvic brim surrounds the **pelvic inlet** of the true (lesser) pelvis, which is located inferior to the pelvic brim. The true pelvis contains the pelvic organs. The inferior opening of the true pelvis is the **pelvic outlet**.

Gender Differences in Pelvic Bones— Description *[Figures 5.2 and 5.3]*

Male Pelvis

Larger and more massive
Attachment sites of tendons and
    muscles more pronounced
Heart-shaped pelvic inlet
Smaller pelvic outlet
Pubic arch angle less than 90
    degrees
Obturator foramen large

Female Pelvis

Lighter, wider, shallower, and less
    funnel shaped
Ischial tuberosities situated farther
    apart
Larger, rounder pelvic inlet
Larger, rounder pelvic outlet
Pubic arch angle exceeds 90 degrees
Obturator foramen small and oval

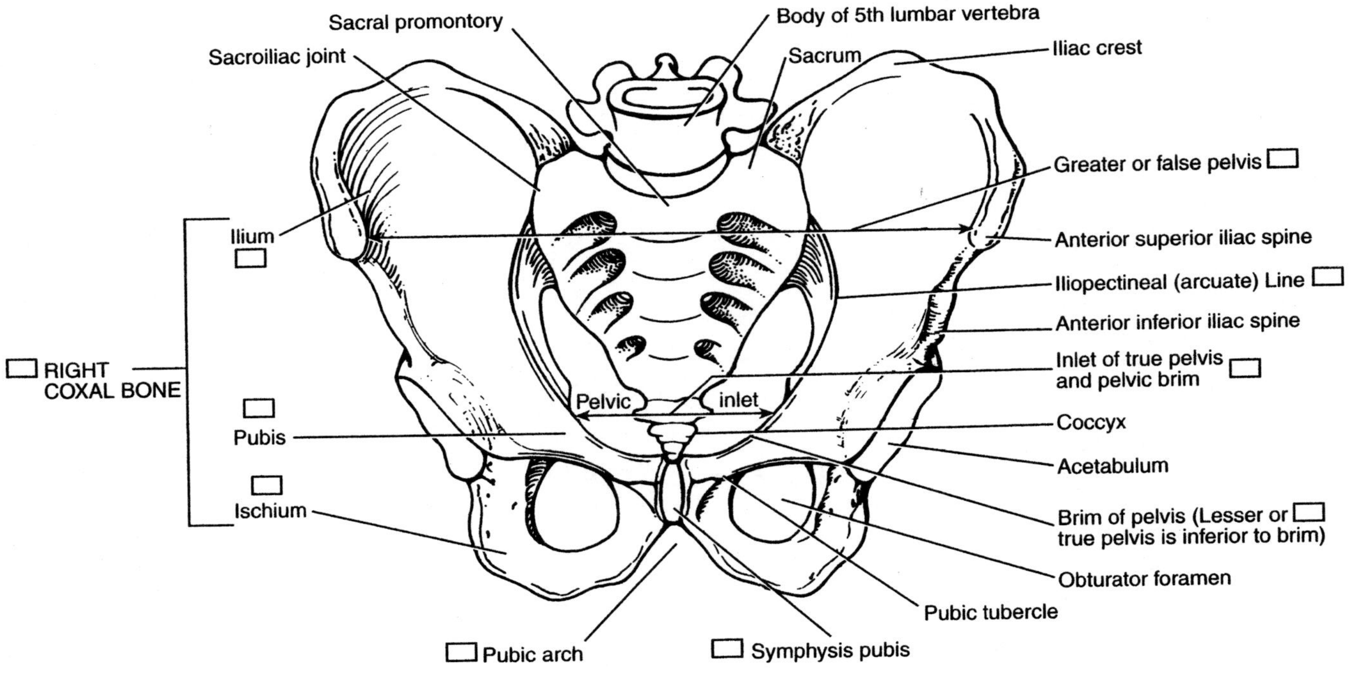

Figure 5.2 Pelvic (hip) girdle of a male in anterior view.

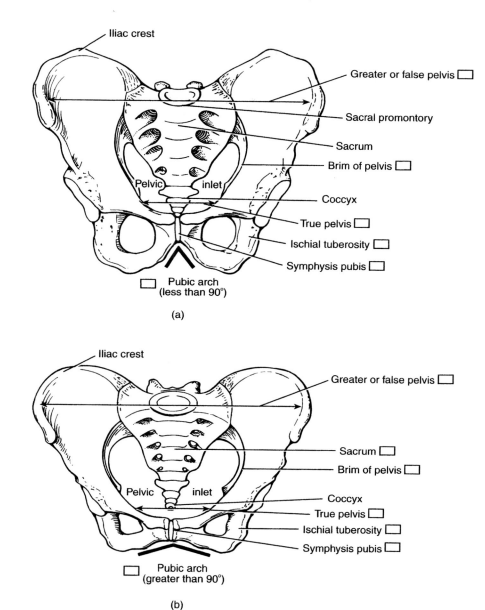

Figure 5.3 (a) Male and (b) female pelvis in anterior view.

The Pelvic Girdle—Identification

THE ILIUM *[Figures 5.1, 5.2, and 5.3]*

Examine and color the illustrations of bones in the pelvic girdle. Then examine the pelvis on a model or human skeleton and identify the following structures:

1. **acetabulum**
2. **iliac fossa**
3. **iliac crest**
4. **anterior and posterior superior iliac spine**
5. **anterior and posterior inferior iliac spine**
6. **auricular surface**
7. **greater sciatic notch**
8. **inferior, anterior, and posterior gluteal lines**
9. **iliopectineal (arcuate) line**

THE ISCHIUM [Figures 5.1, 5.2, and 5.3]

1. ischial spine
2. lesser sciatic notch
3. ischial tuberosity
4. obturator foramen

THE PUBIS [Figures 5.1, 5.2, and 5.3]

1. pubic crest
2. superior ramus
3. inferior ramus
4. pubic arch
5. pubic tubercle
6. symphysis pubis

The Pelvic Cavity—Identification *[Figures 5.2 and 5.3]*

Examine the illustrations of the pelvic cavity. Then examine a model or human pelvis and identify the following structures:

1. false (greater) pelvis
2. true (lesser) pelvis
3. iliopectineal (arcuate) lines
4. pelvic brim
5. pelvic inlet
6. pelvic outlet

Gender Differences in the Pelvic Bones— Identification *[Figures 5.2 and 5.3]*

Using the listed characteristics, compare the female and male pelvis.

Male Pelvis	*Female Pelvis*
large and heavy	light and wide
deeper cavity	shallow and less funnel shaped
heart-shaped pelvic inlet	rounder pelvic inlet
pubic arch < 90 degrees	pubic arch > 90 degrees
obturator foramen large	obturator foramen oval

The Lower Extremity—Description

Each lower extremity is attached to the pelvis and consists of the following bones: the **femur** in the thigh, the **patella** (kneecap), the **fibula** and **tibia** in the leg, the **tarsals** (ankle), the **metatarsals** (foot), and the **phalanges** (toes).

The Femur [Figures 5.4 and 5.5]

The femur, or thighbone, is the longest, heaviest, and strongest bone in the human body. The proximal rounded **head** of the femur fits into and articulates with the acetabulum of the coxal bone. The distal end of the femur articulates with the tibia of the leg. The femur head bears a shallow central depression called the **fovea capitis**. A short ligament at the head of the femur attaches the femur to the acetabulum.

The **neck** of the femur is the constricted region distal to the head. At the junction of the neck and shaft are two prominent projections: the larger and lateral **greater trochanter**, and the smaller and medial **lesser trochanter**. Anteriorly, a narrow **intertrochanteric line** runs obliquely between the greater trochanter and lesser trochanter. Posteriorly, the **intertrochanteric crest** is a prominent ridge between the trochanters.

The **body**, or **shaft**, of the femur exhibits a prominent vertical ridge on its posterior surface called the **linea aspera**. Superiorly, at the base of the greater trochanter, the linea aspera expands into the **gluteal tuberosity**.

The distal region of the femur expands to form the **medial** and **lateral condyles**, which articulate with the tibia. An anterior depression between the condyles is the **patellar surface**, which articulates with the patella, or kneecap. A deeper posterior depression between the condyles is the **intercondylar fossa**.

Superior to the condyles on the lateral and medial sides are the **lateral** and **medial epicondyles**.

The Patella [Figures 5.5 and 5.6]

The patella, or kneecap, is a sesamoid bone that develops within the tendons of the quadriceps femoris muscle. It is triangular in shape and situated in front of the knee joint. The pointed inferior end of the patella is the **apex** and the wide superior end is the **base**. The posterior surface of the patella is the smooth **articular area**, which is separated into two **facets** by a vertical ridge. These facets articulate with the medial and lateral condyles of the femur.

The Femur and Patella—Identification
[Figures 5.4 and 5.5]

Examine and color the illustrations of the femur and patella. Then examine them on a model of a human skeleton and identify the following structures:

1. head
2. fovea capitis
3. neck
4. greater trochanter
5. lesser trochanter
6. intertrochanteric crest
7. intertrochanteric line
8. body, or shaft
9. gluteal tuberosity
10. linea aspera
11. medial epicondyles
12. lateral epicondyles
13. medial condyle
14. lateral condyle
15. intercondylar fossa
16. patellar surface

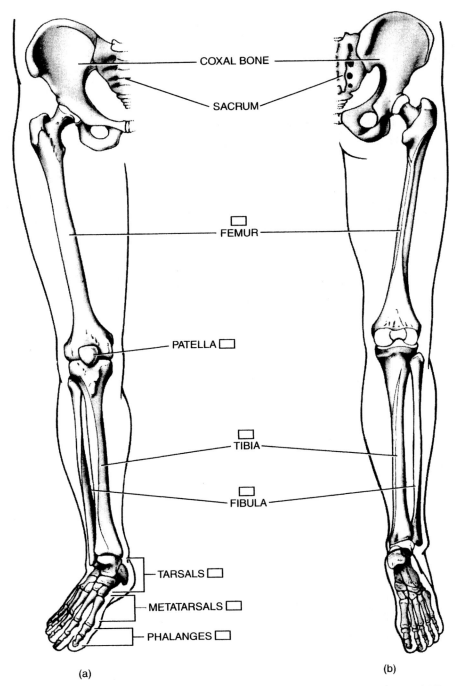

COXAL BONE

SACRUM

FEMUR ☐

PATELLA ☐

TIBIA ☐

FIBULA ☐

TARSALS ☐

METATARSALS ☐

PHALANGES ☐

(a)   (b)

Figure 5.4 Right pelvic (hip) girdle and lower extremity. (a) Anterior view. (b) Posterior view.

The Tibia [Figures 5.4, 5.5, and 5.7]

The leg contains two bones, the medial **tibia** and the lateral **fibula**. The tibia is the larger of the two. The proximal end of the tibia has a **medial condyle** and a **lateral condyle**, which articulate with the corresponding condyles of the femur. Between the concave surfaces of condyles is ridge called the **intercondylar eminence**. A prominent **tibial tuberosity** is located immediately inferior to the condyles on the anterior surface of the tibia. A sharp ridge, the **anterior crest**, extends along the anterior surface of the tibia.

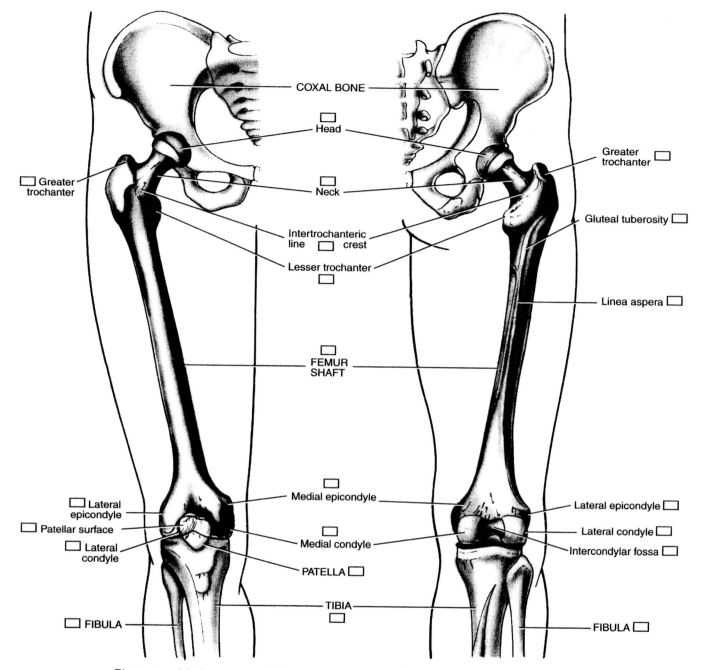

COXAL BONE

Head □

Greater trochanter □

Greater trochanter □

Neck □

Gluteal tuberosity □

Intertrochanteric line □ crest

Lesser trochanter □

Linea aspera □

FEMUR SHAFT □

Medial epicondyle □

Lateral epicondyle □

Lateral epicondyle □

Patellar surface □

Medial condyle □

Lateral condyle □

Lateral condyle □

Intercondylar fossa □

PATELLA □

TIBIA

FIBULA □

FIBULA □

Figure 5.5 (a) Anterior and (b) posterior views of the right femur in relation to the coxal bone, patella, tibia, and fibula.

The distal end of the tibia expands into the **medial malleolus**. The distended end articulates with the talus bone of the ankle and forms the ankle joint. A **fibular notch** on the lateral surface of the tibia articulates with the fibula.

The Fibula [Figures 5.4, 5.5, and 5.7]

The fibula lies laterally to the tibia. It does not articulate with either the femur or the patella. The proximal end of the fibula expands into a **head** that articulates with the lateral condyle of the tibia. The distal end of the fibula terminates in the

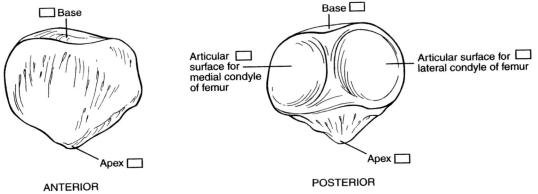

Base ☐

Articular ☐
surface for
medial condyle
of femur

Base ☐

Articular surface for ☐
lateral condyle of femur

Apex ☐

Apex ☐

ANTERIOR

POSTERIOR

Figure 5.6 Anterior and posterior views of the patella.

lateral malleolus. It forms a prominent bulge on the lateral side of the ankle and articulates with the talus bone. The inferior portion of the fibula also articulates with the tibia at the fibular notch.

The Tibia and Fibula—Identification

[Figures 5.4, 5.5, and 5.7]

Examine and color the illustrations of the tibia and fibula. Then examine them on a model or human skeleton and identify the following structures:

Tibia

1. medial condyle
2. lateral condyle
3. intercondylar eminence
4. fibular notch
5. tibial tuberosity
6. anterior crest
7. medial malleolus

Fibula

1. head
2. lateral malleolus

The Ankle and Foot Bones—Description

The ankle and foot contain a total of 26 bones, which are the **tarsals, metatarsals,** and **phalanges.**

The Tarsal Bones [Figures 5.4 and 5.8]

There are seven tarsal bones. The **talus** is the most superior tarsal bone. It articulates with the tibia and fibula to form the ankle joint and fits between the malleoli (medial and lateral malleolus) of the tibia and fibula. Inferior to the talus is the **calcaneus**, the largest and strongest tarsal bone which forms the heel of the foot. The calcaneus protrudes posteriorly and is the site of Achilles tendon, which attaches the calf muscles.

Anterior to the talus are the medial **navicular** and the lateral **cuboid** bones. Anterior to the navicular bone are the three **cuneiform** bones that articulate with the metatarsals. The cuneiform bones are called the **medial (first)**, **intermediate (second)**, and **lateral (third)**.

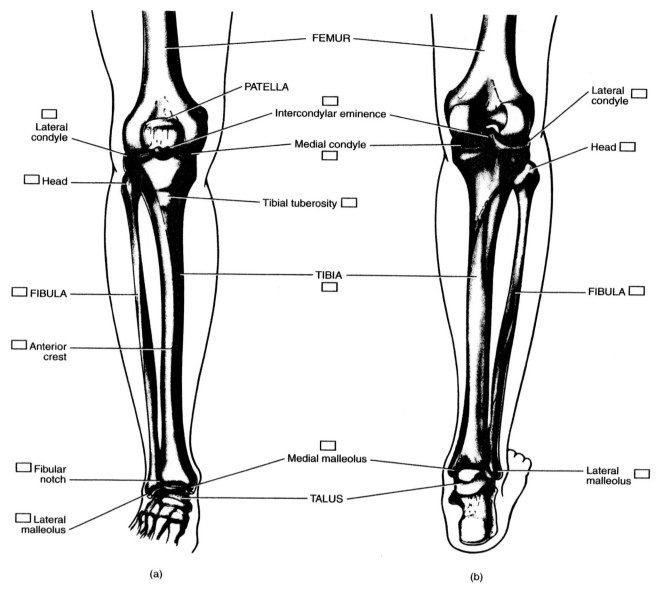

FEMUR

PATELLA

Intercondylar eminence

Lateral condyle ☐

Medial condyle ☐

Lateral condyle ☐

Lateral condyle ☐

Head ☐

☐ Head

Tibial tuberosity ☐

☐ FIBULA

FIBULA ☐

TIBIA ☐

☐ Anterior crest

☐ Fibular notch

Medial malleolus ☐

☐ Lateral malleolus

TALUS

Lateral malleolus ☐

(a)

(b)

Figure 5.7 (a) Anterior and (b) posterior views of the right tibia and fibula.

The Metatarsal Bones [Figures 5.4 and 5.8]

The metatarsal bones are numbered from 1 to 5, starting on the medial side. The first metatarsal bone articulates with the proximal phalanges of the big toe. The metatarsal bones articulate with the cuneiform and cuboid bones proximally and with the proximal phalanges distally.

The Phalanges (Toes) [Figures 5.4 and 5.8]

There are 14 phalanges of the toes. The big toe has only two phalanges, the proximal and distal. The remaining toes have three each, the **proximal, middle**, and **distal** phalanges. The metatarsal bone and phalanges in the foot have a similar arrangement to the metacarpal bones and phalanges in the hand. The phalanges have a **base, shaft**, and **head**.

The Ankle and Foot Bones—Identification

[Figures 5.4 and 5.8]

Examine and color the illustrations of the ankle and foot bones. Then examine them in the ankle and foot of a model or human skeleton and identify the following:

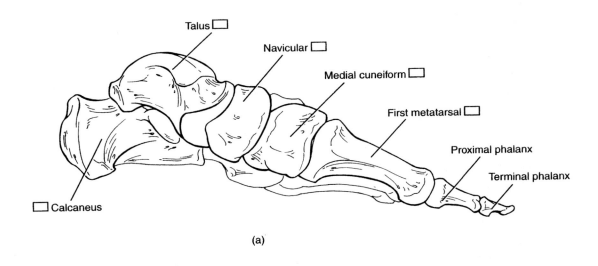

(a)

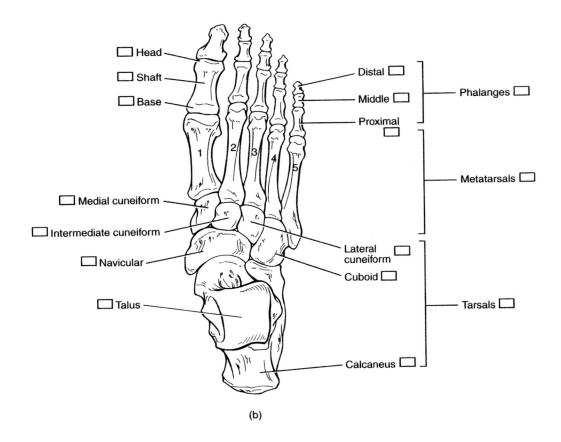

(b)

Figure 5.8 (a) Lateral and (b) superior views of the foot bones.

The Tarsal Bones
1. talus
2. calcaneus
3. navicular
4. cuneiform bones (medial, intermediate, lateral)
5. cuboid

The Metatarsal Bones
metatarsal bones 1, 2, 3, 4, and 5

The Phalanges
1. proximal phalanges
2. middle phalanges
3. distal phalanges
4. base
5. shaft
6. head

CHAPTER SUMMARY AND CHECKLIST

I. THE PELVIC GIRDLE

   A. **Composition of the Pelvic Girdle**
     1. Formed by a pair of coxal bones
     2. Develops from ilium, ischium, and pubis bones
     3. Bones joined anteriorly at symphysis pubis
     4. Posteriorly, pelvis joins to sacrum
     5. Acetabulum formed by fusion of the three coxal bones

   B. **The Ilium**
     1. Largest and most superior of pelvic bones
     2. Iliac spines are sites of muscle attachment
     3. Greater sciatic notch is for sciatic nerve passage
     4. Iliac fossa is site for iliac muscle origin
     5. Gluteal ridges are for gluteal muscle attachments

   C. **The Ischium**
     1. Forms posterior inferior region of pelvis
     2. The ramus is anterior projection
     3. Ischium and pubis surround obturator foramen

   D. **The Pubis**
     1. Smallest bone of the girdle
     2. Consists of two rami, superior and inferior
     3. Pubic crest and tubercle lie on superior ramus
     4. Contributes to formation of symphysis pubis
     5. Inferior rami contribute to pubic arch

II. THE PELVIC CAVITY

   A. **Anatomical Regions**
     1. True and false pelvis
     2. Superior iliac bones and sacrum form false pelvis
     3. Pelvic brim surrounds pelvic inlet of true pelvis
     4. Inferior opening is pelvic outlet

   B. **Differences in Pelvic Bones**
     1. Male pelvis
       a. larger and more massive
       b. muscle attachments more pronounced
       c. heart-shaped pelvic inlet
       d. pubic arch angle less than 90 degrees

2. Female pelvis
   a. lighter, wider, and shallower
   b. rounder pelvic inlet and outlet
   c. pubic arch angle greater than 90 degrees

III. **THE LOWER EXTREMITY**

   A. **The Femur**
      1. Longest, heaviest, and strongest bone in the body
      2. Round head fits into acetabulum of pelvis
      3. Fovea capitis on head of femur for ligament
      4. Neck is constricted region distal to head
      5. Greater and lesser trochanters on shaft near neck
      6. Body, or shaft, has a vertical ridge called linea aspera
      7. Distally expanded into medial and lateral condyles
      8. Patellar surface between condyles
      9. Lateral and medial epicondyles above condyles

   B. **The Patella**
      1. Develops within the tendon of a muscle
      2. Pointed apex, wide base, and articular area

   C. **The Tibia**
      1. Larger leg bone, supports most of the body's weight
      2. Medial and lateral condyles articulate with femur
      3. Medial malleolus is prominent on side of ankle
      4. Fibular notch on lateral surface of bone

   D. **The Fibula**
      1. Slender bone located lateral to the tibia
      2. Lateral malleolus is prominent on side of ankle

IV. **THE ANKLE AND FOOT BONES**

   A. **The Tarsal Bones**
      1. Seven tarsal bones
      2. Talus bone is the most superior tarsal bone
      3. Calcaneus is the largest and strongest tarsal bone
      4. Calcaneus is the heel and protrudes posteriorly

   B. **The Metatarsal Bones**
      1. There are five metatarsal bones in the foot
      2. Metatarsal bones are numbered 1 to 5
      3. First metatarsal bone articulates with big toe
      4. Articulate proximally with cuneiform and cuboid bones

   C. **The Phalanges**
      1. There are 14 phalanges in the toes
      2. Phalanges are proximal, middle, and distal
      3. Phalanges have base, shaft, and head
      4. Articulate proximally with metatarsal bones

Laboratory Exercises 5

NAME _____

LAB SECTION _____ DATE _____

LABORATORY EXERCISE 5.1

Part I

Bones of Lower Extremity and Pelvis

1. List the names of component bones of the coxal bone.

    _____   _____   _____

2. List five characteristics of the male pelvis.

    _____

    _____

    _____

    _____

    _____

3. List five characteristics of the female pelvis.

    _____

    _____

    _____

    _____

    _____

4. What bone develops within the tendons in the leg? _____

5. What three sets of bones comprise the foot bones?

    _____

    _____

    _____

6. What bone forms the heel of the foot? _____

Part II

Pelvic Bones

Using the listed terms, label the hip bone and then color the component structures.

Ilium
Pubic tubercle
Pubis bone
Ischial spine
Greater sciatic notch
Posterior inferior iliac
     spine
Anterior superior iliac
     spine

Superior ramus of
     pubis
Iliac crest
Iliac fossa
Auricular surface
Ischial ramus
Ischial bone
Symphysis pubis
Obturator foramen

Iliopectineal (arcuate)
     line
Posterior superior iliac
     spine
Anterior inferior iliac
     spine
Inferior ramus of
     pubis

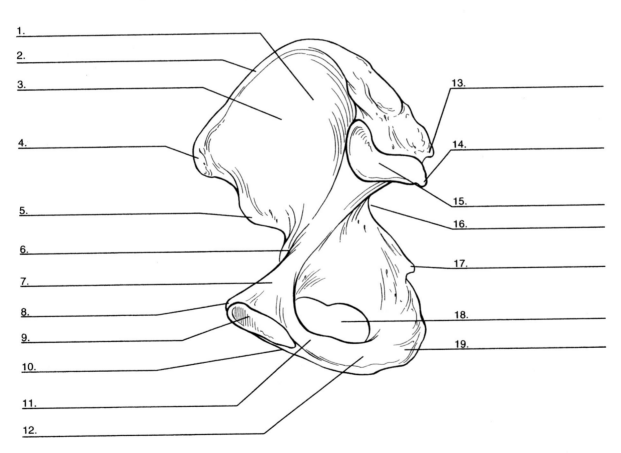

1. _____
2. _____
3. _____
4. _____
5. _____
6. _____
7. _____
8. _____
9. _____
10. _____
11. _____
12. _____

13. _____
14. _____
15. _____
16. _____
17. _____
18. _____
19. _____

Figure 5.9 Internal surface of the coxal (hip) bone.

LABORATORY EXERCISE 5.2

Using the listed terms, label the parts of the pelvic bone and then color them.

Iliac bone (Ilium)
Pubic bone (Pubis)
Ischial bone (Ischium)
Symphysis pubis
Anterior superior iliac spine
Anterior inferior iliac spine
Acetabulum
Obturator foramen

Iliac crest
Greater or false pelvis
Pubic arch
Pubic tubercle
Iliopectineal (arcuate) line
Pelvic inlet (true pelvis)
Pelvic brim

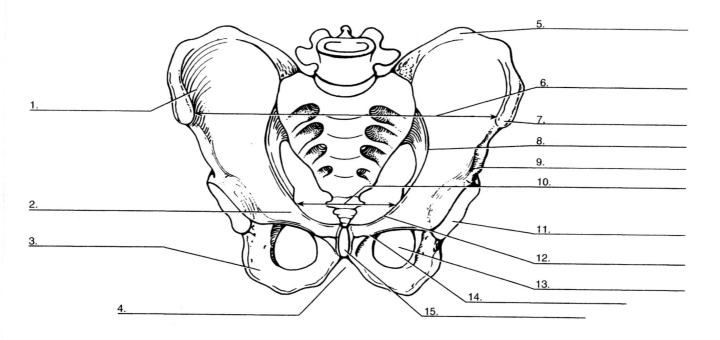

Figure 5.10 Anterior view of a male pelvic (hip) bone.

LABORATORY EXERCISE 5.3

Using the listed terms, label the parts of the femur and then color them.

Greater trochanter
Lateral epicondyle
Lesser trochanter
Gluteal tuberosity
Patella
Medial condyle
Head of femur

Neck of femur
Intertrochanteric crest
Femur shaft
Linea aspera
Lateral condyle
Intercondylar fossa
Medial epicondyle

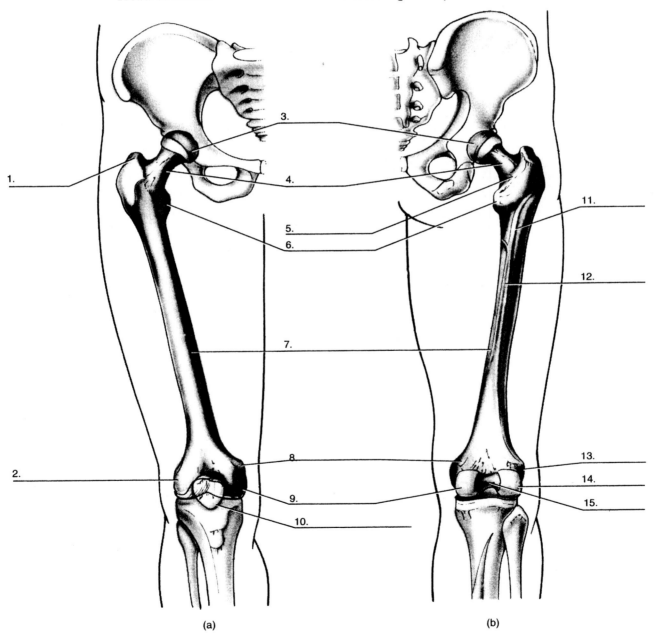

(a)   (b)

Figure 5.11 (a) Anterior and (b) posterior views of the femur.

LABORATORY EXERCISE 5.4

Using the listed terms, label the parts of the tibia and fibula and then color them.

Lateral condyle
Intercondylar eminence
Anterior crest of tibia
Medial malleolus
Tibia
Head of fibula

Tibial tuberosity
Fibular notch
Lateral malleolus
Fibula
Medial condyle

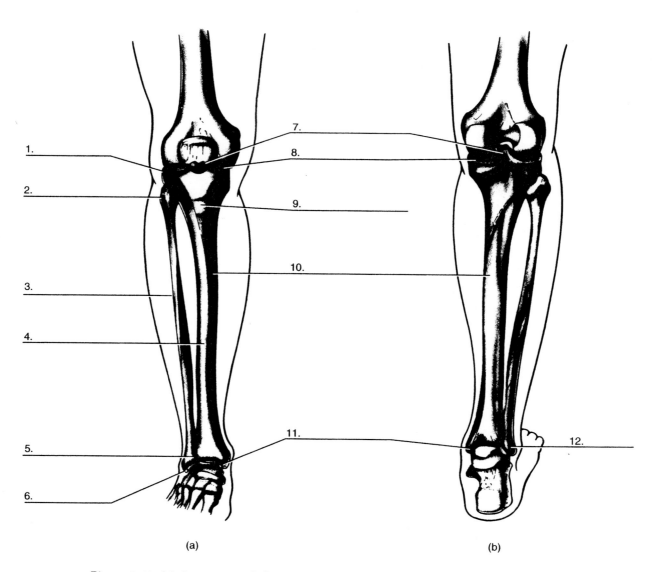

(a)   (b)

Figure 5.12 (a) Anterior and (b) posterior views of right tibia and fibula.

LABORATORY EXERCISE 5.5

Using the listed terms, label the component foot bones and then color them.

Talus   Calcaneus
Navicular   Medial cuneiform
Intermediate cuneiform   Lateral cuneiform
Cuboid   Metatarsals
Tarsals   Phalanges
Distal, middle, and proximal phalanges
Head, shaft, and base of phalanges

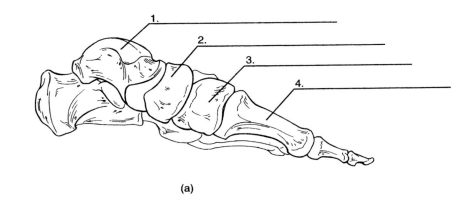

(a)

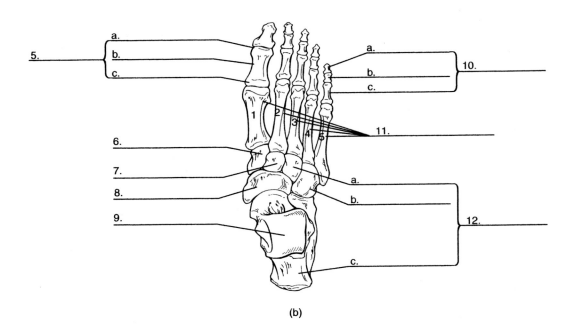

(b)

Figure 5.13 (a) Lateral and (b) superior views of the bones of the foot.

Part Three

The Muscular System

6

Introduction to the Joints and Muscles

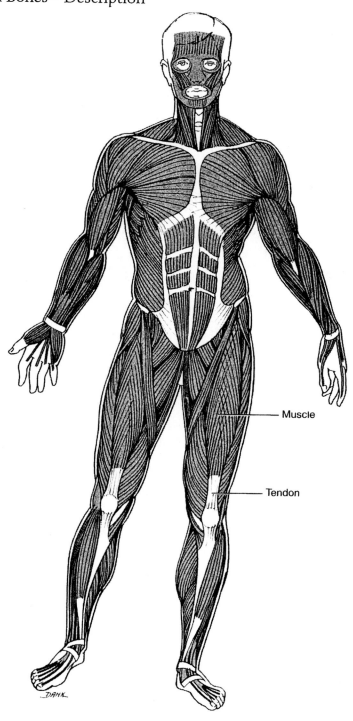

Muscle

Tendon

Objective

The objective of Chapter 6, "Introduction to the Joints and Muscles," is to acquaint you with:

1. **The classification of fibrous, cartilaginous, and synovial joints**
2. **Skeletal muscles**
3. **The types of muscular movement**
4. **The criteria for naming muscles**

Joints, or Articulations, between Bones— Description *[Figures 6.1 and 6.2]*

There are three types of joints, or articulations, in the human body: fibrous, cartilaginous, and synovial. A joint is an area or point of contact between bones, between bones and cartilage, or between bones and teeth.

Fibrous Joints

Fibrous joints have no cavity between bones. Instead, heavy fibrous connective tissue binds the bones. The three types of fibrous joints are **suture**, **syndesmosis**, and **gomphosis**.

Type of Joint **Fibrous Joint**	*Description*	*Example*
Suture	Immovable, found between skull bones only	Coronal, sagittal, lambdoidal, and squamous sutures
Syndesmosis	Slightly movable, connect bones with ligament or terosseous membrane	Between tibia and fibula or ulna and radius
Gomphosis	Immovable, between teeth and sockets	In alveolar processes of mandible or maxillary bones

Cartilaginous Joints

Like the fibrous joint, the **cartilaginous joint** lacks a cavity. A layer of hyaline cartilage or fibrocartilage connects the bones. The two kinds of cartilaginous joints are **synchondrosis** and symphysis.

Type of Joint **Cartilaginous Joint**	*Description*	*Example*
Synchondrosis	Connects bones with hyaline cartilage; joint is immovable	Epiphyseal plates; joint between manubrium and first rib
Symphysis	Connects bones with fibrocartilage disk	Between vertebral bodies, pubic symphysis, and anubrium and body of sternum

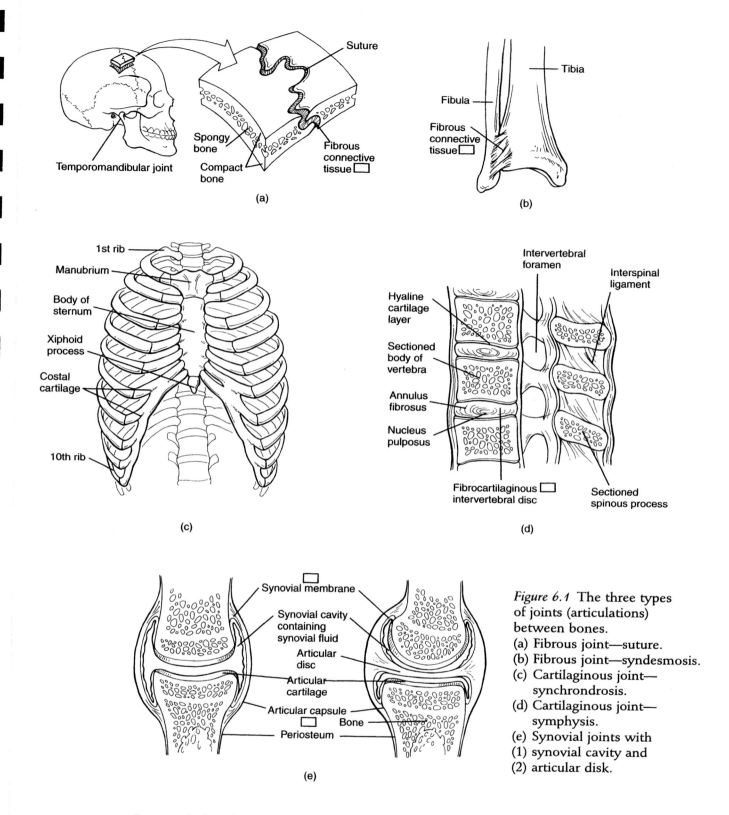

Suture

Temporomandibular joint

Spongy bone

Compact bone

Fibrous connective tissue □

(a)

Tibia

Fibula

Fibrous connective tissue □

(b)

1st rib

Manubrium

Body of sternum

Xiphoid process

Costal cartilage

10th rib

(c)

Intervertebral foramen

Interspinal ligament

Hyaline cartilage layer

Sectioned body of vertebra

Annulus fibrosus

Nucleus pulposus

Fibrocartilaginous □ intervertebral disc

Sectioned spinous process

(d)

Synovial membrane □

Synovial cavity containing synovial fluid

Articular disc

Articular cartilage

Articular capsule □

Bone

Periosteum

(e)

Figure 6.1 The three types of joints (articulations) between bones.
(a) Fibrous joint—suture.
(b) Fibrous joint—syndesmosis.
(c) Cartilaginous joint— synchrondrosis.
(d) Cartilaginous joint— symphysis.
(e) Synovial joints with
(1) synovial cavity and
(2) articular disk.

Synovial Joints

The **synovial** joint is the most common type of joint in the body; it has a synovial cavity and synovial fluid. It is freely movable. The bones have articular cartilage at their ends, and the connective tissue capsule and ligaments tightly bind the interacting bones. The joint cavity may contain articular disks of fibrocartilage

Major Synovial Joints and Their Location

[Figure 6.1 and 6.2]

The Skull

- **Temporomandibular joint:** Formed between the condylar process of the mandible and the mandibular fossa of the temporal bone of the skull.

The Vertebral Column

- **Atlanto-occipital joint:** Formed between the occipital condyle of the skull and the superior articular surface of the atlas.
- **Atlantoaxial joint:** Formed between the articular surfaces of the atlas and axis of the cervical vertebrae.
- **Intervertebral joint:** The joint formed between the vertebral arches is synovial, and that formed between the vertebral bodies is cartilaginous.

The Thorax

- **Costovertebral joint:** Formed between the rib and the vertebra. A typical rib articulates with the vertebra at two sites: the head of the rib with the body of the vertebra, and the tubercle of the rib with the transverse process of the vertebra.
- **Sternocostal joint:** Formed between the costal cartilage of ribs 2 to 7 and the sternum.

The Pectoral Girdle

- **Sternoclavicular joint:** Formed between the medial end of the clavicle and the clavicular notch of the sternum.
- **Acromioclavicular joint:** Formed between the lateral end of the clavicle and the acromion process of the scapula.
- **Glenohumeral (shoulder) joint:** Formed between the head of the humerus and the glenoid cavity of the scapula.

The Upper Extremity

- **Elbow joint:** Formed between the inferior portion of the humerus and the proximal end of the ulna and radius.
- **Radioulnar joint:** Formed between the radius and ulnar bones at their proximal and distal extremities. Along the shaft of the bones, a fibrous interosseous membrane unites the two bones in syndesmosis.
- **Radiocarpal (wrist) joint:** Formed between the distal end of the radius and the carpal bones of the wrist.
- **Intercarpal joints:** Formed between the adjacent carpal bones.
- **Carpometacarpal joints:** Formed between the distal carpal bones and the bases of the five metacarpal bones.
- **Metacarpophalangeal joints:** Formed between the metacarpal bones and the proximal ends of the phalanges.

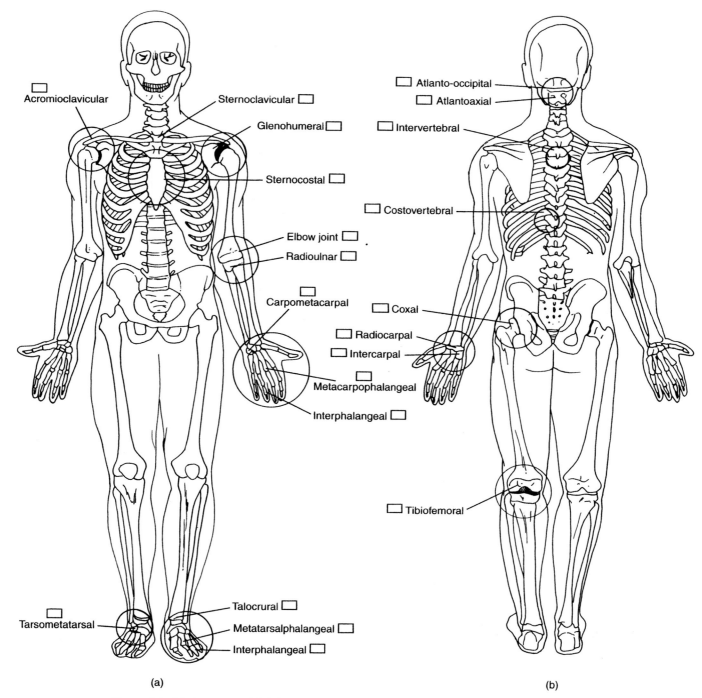

Figure 6.2 Major synovial joints.

- **Interphalangeal joints:** Formed between the adjacent phalangeal bones of the fingers.

The Lower Extremity

- **Hip (coxal) joint:** Formed between the acetabulum of the coxal bone and the head of the femur.
- **Knee (tibiofemoral) joint:** Formed by two condyloid joints between each condyle of the femur and the condyles of the tibia, and the patella and femur.

- **Tibiofibular joint**: Formed proximally and distally between the fibula and an articular surface of the tibia. A fibrous interosseous membrane joins the shafts of the two bones in syndesmosis.
- **Talocrural (ankle) joint**: Formed by the distal ends of the tibia and fibula, which form a socket, and the upper region of the talus.
- **Tarsometatarsal joints**: Formed between the tarsal bones and the bases of the five metatarsal bones.
- **Metatarsophalangeal joints**: Formed between the metatarsal and proximal phalangeal bones.
- **Interphalangeal joints**: Formed between the adjacent phalangeal bones in the foot.

Joints, or Articulation, Between Bones— Identification *[Figures 6.1 and 6.2]*

Examine the illustrations of various types of joints and their location. Then examine the human skeleton and identify the following:

FIBROUS JOINTS

1. **suture**
2. **syndesmosis**
3. **gomphosis**

CARTILAGINOUS JOINTS

1. **synchondrosis**
2. **symphysis**

SYNOVIAL JOINTS

1. **temporomandibular**
2. **atlanto-occipital**
3. **atlantoaxial**
4. **intervertebral**
5. **costovertebral**
6. **sternocostal**
7. **sternoclavicular**
8. **glenohumeral (shoulder)**
9. **acromioclavicular**
10. **elbow joint**
11. **radioulnar**
12. **radiocarpal**
13. **intercarpal**
14. **carpometacarpal**
15. **metacarpophalangeal**
16. **interphalangeal**
17. **hip (coxal)**
18. **knee (tibiofemoral)**

19. tibiofibular
20. talocrural
21. tarsometatarsal
22. metatarsophalangeal
23. interphalangeal

The Skeletal Muscles—Introduction

The skeletal muscles are specialized to enable **contraction**. During contractions, one bone moves relative to another at the joint. Most skeletal muscles are attached to the **periosteum** (connective tissue) of the bone at both ends by a dense, tough, fibrous connective tissue called the **tendon**. Some tendons are flat and sheetlike; these are called **aponeuroses**. **Fascia** is the term used to describe all other fibrous connective tissue, that is not specifically named.

The attachments of muscle with bone are called the **origin** and the **insertion**. The *origin* of the muscle is the fixed and proximal attachment site. The *insertion* is the movable, distal attachment. Most muscles of the face originate from bone and insert into the skin. Surface markings observed on different bones are important sites of origin and insertion of the tendons. Most movements in the body are paired, one motion usually complementing another.

Types of Muscular Movements at Joints—
Description *[Figures 6.3, 6.4, and 6.5]*

Reviewing the description of the anatomical position in Chapter 1 will help you to understand muscular movements. Body movement is always related to the anatomical position and its terminology.

Angular Movement

Angular movement increases or decreases the angle between two bones. The four types of angular movement are **flexion**, **extension**, **abduction**, and **adduction**.

Flexion decreases the angle between bones and brings them closer together. Examples include bending the elbow or the knee. In the ankle, flexion raises the foot and is called **dorsiflexion** (Figures 6.3 and 6.5).

Extension, the opposite of flexion, increases the angle between two bones. Examples include straightening the arm or leg. In the ankle, extension moves the foot inferiorly and is called **plantar flexion** (Figures 6.3 and 6.5).

Abduction moves the body part away from the midline in a lateral direction. Examples include moving the arm or leg in a lateral direction (Figure 6.3). Abduction in the fingers and toes means moving them laterally from the midline of the hand or foot (Figure 6.4).

Adduction is the opposite of abduction. Movement is toward the midline of the body or, in case of the digits, toward the midline of the hand or foot. Examples include assuming the anatomical position of the limbs or digits (Figures 6.3 and 6.4).

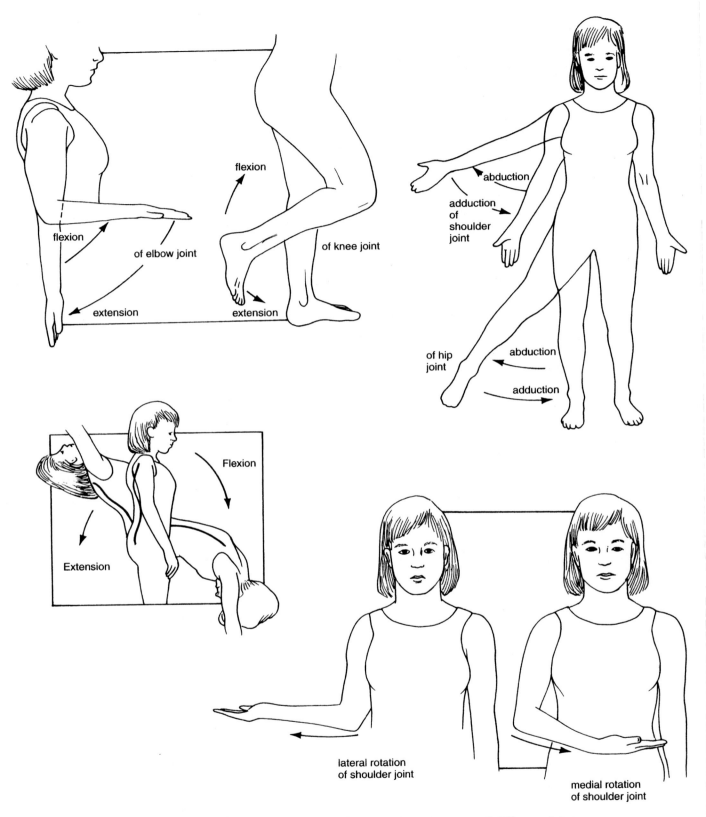

Figure 6.3 **Anatomical terminology in reference to movement of different joints.**

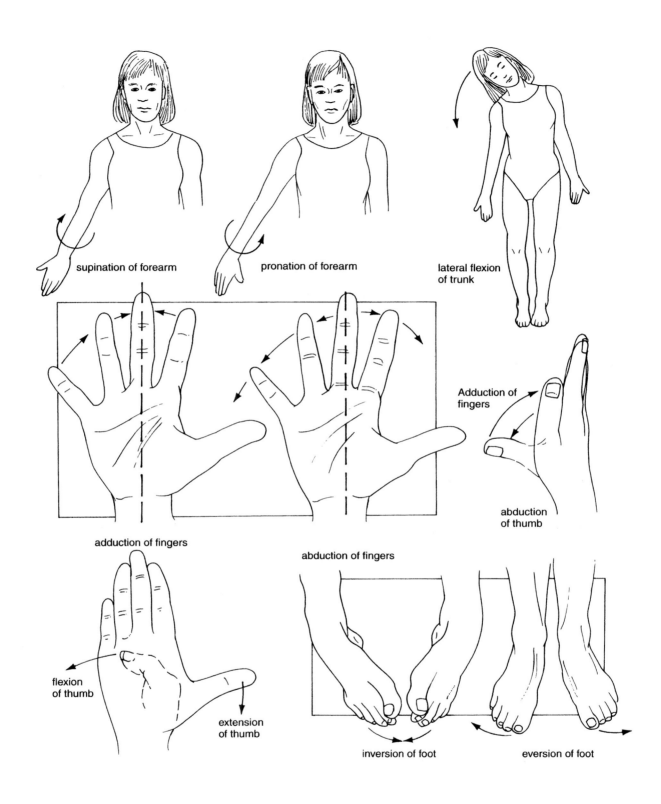

Labels within the figure:
- supination of forearm
- pronation of forearm
- lateral flexion of trunk
- Adduction of fingers
- abduction of thumb
- adduction of fingers
- abduction of fingers
- flexion of thumb
- extension of thumb
- inversion of foot
- eversion of foot

Figure 6.4 Anatomical terminology in reference to movement at different joints.

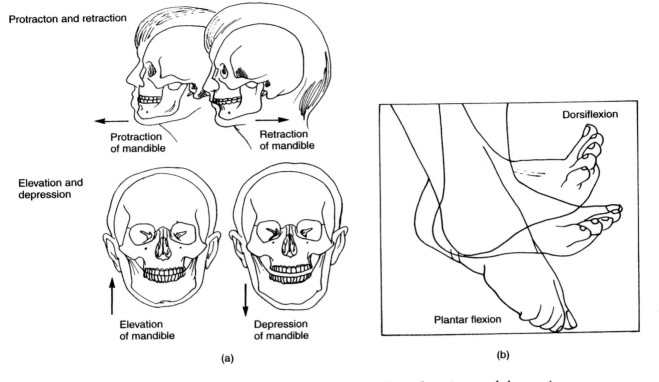

Protraction and retraction

Protraction
of mandible

Retraction
of mandible

Elevation and
depression

Elevation
of mandible

Depression
of mandible

(a)

Dorsiflexion

Plantar flexion

(b)

Figure 6.5 (a) The mandible in protraction, retraction, elevation, and depression.
(b) The ankle in dorsiflexion and plantar flexion.

Circular Movement

Rotation involves turning a bone around its own longitudinal axis or around
another bone. Turning the head from side to side in a "no" motion is rotation.
Rotation of the anterior surface of the bone toward the midline of the body is
medial rotation, and rotation of the same bone away from the midline is **lateral
rotation** (Figure 6.3).

Special Movements

Inversion is moving the sole of the foot inward, or medially (Figure 6.4).

Eversion, the opposite of inversion, is moving the sole of the foot to face outward, or laterally (Figure 6.4).

Protraction is moving a body part forward and parallel to the ground.
Examples include thrusting the lower jaw forward (Figure 6.5).

Retraction returns the protracted part of the body, as in the lower jaw thrust
forward, to its original position (Figure 6.5).

Elevation moves a body part in a superior direction. Examples include raising
the mandible to close the mouth and raising the scapulae to shrug the shoulders
(Figure 6.5).

Depression is the opposite of elevation. Examples include opening the mouth
or lowering the scapulae to their original position (Figure 6.5).

Supination is a specialized movement of the forearm in which the palm of the
hand is rotated anteriorly (forward). When this movement is completed, the hand
is in the anatomical position and the ulna and radius are parallel (Figure 6.4).

Pronation is the opposite of supination. The forearm is rotated so that the palm faces posteriorly (Figure 6.4).

Criteria for Naming the Skeletal Muscles

1. **Shape** of the muscle. Numerous muscles have a geometrical shape, for example, the **trapezius**, **deltoid**, and **rhomboid**.
2. **Location** of the muscle. For example, **iliacus**, **tibialis**, **occipitalis**, and **temporalis**.
3. **Direction** of muscle fibers. For example, **rectus abdominis** (straight or parallel to the midline), **transversus abdominis** (across or perpendicular to the midline), and **external oblique** (oblique to the midline).
4. **Size** of the muscle. For example, **maximus** (largest), **minimus** (smallest), **longus** (long), and **brevis** (short) designate muscle size. For example, **gluteus maximus**, **gluteus minimus**, **palmaris longus**, and **palmaris brevis**.
5. **Number of origins**. The terms **biceps**, **triceps**, and **quadriceps** (two, three, and four heads) denote the number of origins of these muscles. For example, **biceps brachii**, **triceps brachii**, and **quadriceps femoris**.
6. **Origin** or **insertion**. The best example is the **sternocleidomastoid** muscle, which originates on the sternum and clavicle and inserts on the mastoid process of the temporal bone.
7. **Action** of the muscle. Their names begin with **extensor, flexor, adductor, abductor, supinator**, and **pronator**. For example, **extensor pollicis longus** and **flexor digitorum superficialis**.

Several of the criteria are used in combination to name a particular muscle. For example, **extensor pollicis longus** describes the action of this muscle (*extensor*), the area of insertion (*pollicis* = thumb), and the size of the muscle (*longus*). A shorter extensor muscle of the thumb is the **extensor pollicis brevis**.

The major muscles in the next chapters will be presented in groups according to the following anatomical locations: Chapters 7 and 8, **muscles of the appendicular skeleton**; and Chapters 9 and 10, **muscles of the axial skeleton**. The muscles of the appendicular skeleton are attached to the pectoral and pelvic girdles and move the upper and lower extremities. The muscles of the axial skeleton are attached to the bones of the axial skeleton and include the muscles of the head, face, neck, and trunk.

CHAPTER SUMMARY AND CHECKLIST

I. **ARTICULATIONS BETWEEN BONES**

A. **Structural Classification of Bony Joints**
1. Fibrous joint: connective tissue binds bones
2. Cartilaginous joint: cartilage binds bones
3. Synovial joint: presence of a synovial cavity

B. **Fibrous Joints**
1. Sutures: immovable joints between skull bones
2. Syndesmoses: interosseous membranes between bones

3. Gomphoses: joints between teeth and sockets

C. Cartilaginous Joints
1. Synchondroses: epiphyseal plates between bones
2. Symphyses: fibrocartilage between bones

D. Synovial Joints
1. Most common joints in the body
2. Freely movable joints with presence of a synovial cavity
3. Bones covered by articular cartilage
4. Joints surrounded by capsule and ligaments

II. THE SKELETAL MUSCLES

A. Introduction
1. Skeletal muscles enable voluntary movement
2. Most movement occurs across synovial joints
3. A tendon attaches most muscles to periosteum of bone
4. Sheetlike tendons are aponeuroses
5. Origin of muscle is the more fixed, proximal end
6. Insertion is the more movable, distal end
7. Muscle attachment on bones produces surface marks

III. TYPE OF MUSCULAR MOVEMENTS AT JOINTS

A. Angular Movement
1. Flexion: decreases angle between bones
2. Extension: increases angle between bones
3. Abduction: moves body part away from midline
4. Adduction: moves body part toward midline

B. Circular Movement
1. Rotation: turning bone around on its own axis

C. Special Movements
1. Inversion: moves sole of foot inward or medially
2. Eversion: moves sole of foot outward or laterally
3. Protraction: moves body part forward
4. Retraction: returns protracted body part to normal position
5. Elevation: moves body part in superior direction
6. Depression: moves body part in inferior direction
7. Supination: palm of hand rotated anteriorly
8. Pronation: palm of hand rotated posteriorly

IV. CRITERIA FOR NAMING THE SKELETAL MUSCLES

A. Specific Characteristics
1. Shape
2. Location
3. Direction of muscle fibers
4. Size
5. Number of origin sites per muscle
6. Origin and/or insertion of the muscle
7. Action in movement

Laboratory Exercises 6

LABORATORY EXERCISE 6.1

Part I

Joints

Match the description of joints on the left with the type of joint on the right.

Fibrocartilage joint between bones _____	A. Suture
Cartilage joint between growing bones _____	B. Symphysis
Joint between skull and mandible _____	C. Atlanto-occipital
Joint between skull and atlas _____	D. Temporomandibular
Immovable joint between skull bones _____	E. Synchondrosis
Joint between ulna and radius _____	F. Synovial
Head of femur and acetabulum	G. Radioulnar
form this joint _____	H. Hip joint
Most common joint in the body _____	I. Glenohumeral
Ribs and sternum form this joint _____	J. Costosternal
Joint between humerus and scapula _____	

Part II

Movements

Supply the type of movement described.

1. Moving body part away from midline laterally _____

2. Turning bone along its long axis _____

3. Moving body part in superior direction _____

4. Decreasing the angle between two bones _____

5. Increasing the angle between two bones _____

6. Moving body part toward midline _____

7. Rotation of palm anteriorly (forward) _____

8. Movement of sole inward _____

9. Movement of sole outward _____

10. Moving jaw thrust forward to original position _____

11. Moves the jaw forward, parallel to ground _____

12. Rotation of palm posteriorly _____

Part III

Naming of Muscles

Supply a muscle named for the following criteria:

1. Shape _____

2. Action _____

3. Size _____

4. Location _____

5. Direction of fibers _____

6. Number of origins _____

7. Origin or insertion _____

7

The Muscles of the Appendicular Skeleton

MUSCLES OF THE PECTORAL GIRDLE AND UPPER EXTREMITY

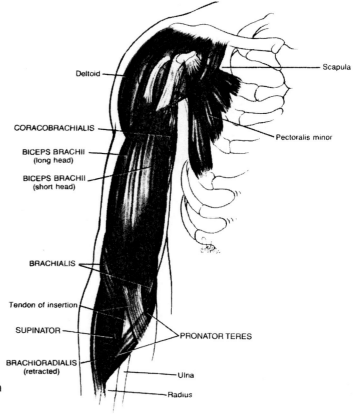

121

Objective

The objective of Chapter 7, "The Muscles of the Appendicular Skeleton," is for you to understand the origin, insertion, action, and innervation of the muscles of the:

1. **Pectoral girdle**
2. **Upper extremity**
3. **Forearm**
4. **Wrist, hand, and fingers**

Muscles of the Pectoral Girdle—Description

[Figures 7.1 and 7.2]

The muscles that move the pectoral girdle are the **subclavius, serratus anterior, pectoralis minor, trapezius, levator scapulae, rhomboid (rhomboideus) major**, and **rhomboid minor**. All of these muscles originate at the thorax (rib cage) and insert into the scapula. They are subdivided into anterior and posterior groups.

The **pectoralis major** (anterior group) and **latissimus dorsi** (posterior group) originate at the thorax but insert into the upper extremity (humerus).

Anterior Muscles That Attach to the Pectoral Girdle

The anterior muscles include the **subclavius, serratus anterior**, and **pectoralis minor**.

SERRATUS ANTERIOR [Figures 7.1 and 7.2]

Description: Wraps around and extends along the lateral wall of the thorax, between the ribs and the ventral surface of the scapula. Its origins have a serrated or "sawtooth" appearance.
Origin: Outer surfaces of the first eight or nine ribs.
Insertion: Ventral surface of vertebral (medial) border of the scapula.
Action: Protracts and rotates the scapula superiorly, holds it against the chest wall.
Innervation: Long thoracic nerve.

PECTORALIS MINOR [Figures 7.1 and 7.2]

Description: A thin muscle on the upper anterior thoracic wall, located deep to the larger pectoralis major muscle.
Origin: Outer surfaces of the third, fourth, and fifth ribs, near the cartilages.
Insertion: Medial border of the corocoid process of the scapula.
Action: Draws scapula forward and downward.
Innervation: Medial pectoral nerve from the brachial plexus, a complex network of nerves in the axilla.

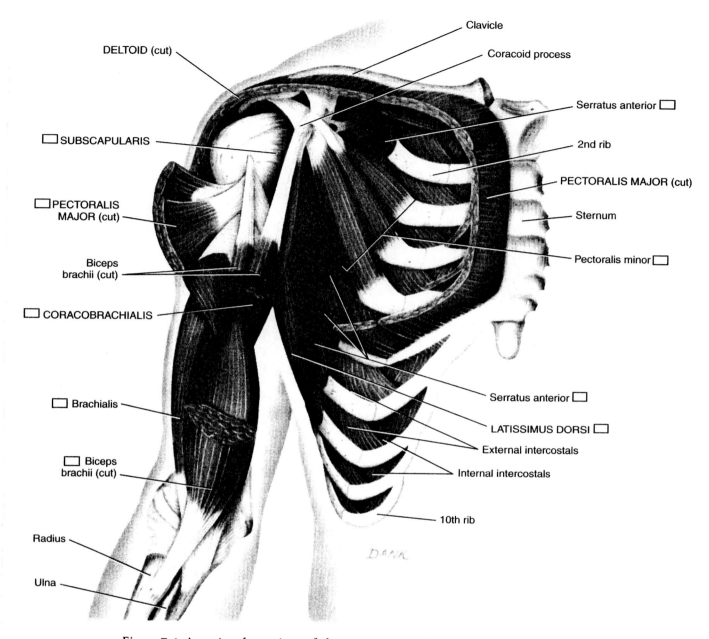

DELTOID (cut)

Clavicle

Coracoid process

☐SUBSCAPULARIS

Serratus anterior ☐

2nd rib

PECTORALIS MAJOR (cut)

☐PECTORALIS
MAJOR (cut)

Sternum

Biceps
brachii (cut)

Pectoralis minor ☐

☐ CORACOBRACHIALIS

☐ Brachialis

Serratus anterior ☐

LATISSIMUS DORSI ☐

☐ Biceps
brachii (cut)

External intercostals

Internal intercostals

Radius

10th rib

DANK

Ulna

Figure 7.1 Anterior deep view of the anterior muscles that move the arm (humerus) and pectoral girdle.

Posterior Muscles That Attach to the Pectoral Girdle

The posterior muscles include the **trapezius, levator scapulae, rhomboid (rhomboideus) major,** and **rhomboid minor.**

TRAPEZIUS [Figures 7.3 and 7.4]

Description: A large muscle resembling a trapezoid that overlays the posterior regions of the neck, shoulders, and thorax on both sides of the midline. This muscle attaches the pectoral girdle to the skull and vertebral column.

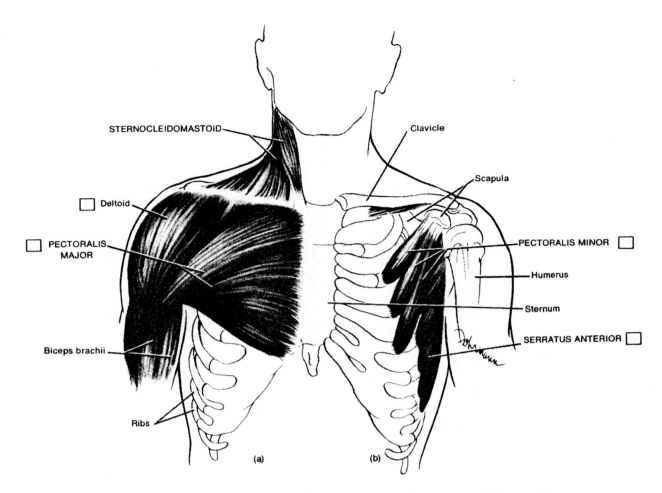

Figure 7.2 (a) Anterior superficial view and (b) anterior deep view of the anterior muscles that move the pectoral (shoulder) girdle.

Origin: Occipital protuberance of the occipital bone, nuchal ligament, and spinous processes of seventh cervical and all thoracic vertebrae.
Insertion: Clavicle, acromion, and spine of the scapula.
Action: Superior rotation and elevation of the scapula, elevation of the clavicle, and extension of the head.
Innervation: Spinal accessory nerve (cranial nerve XI).

LEVATOR SCAPULAE [Figures 7.3 and 7.4]

Description: Located deep to the trapezius muscle in the dorsal and lateral regions of the neck.
Origin: Atlas, axis, and third and fourth cervical vertebrae.
Insertion: Superior vertebral border of the scapula.
Action: Elevation and inferior rotation of the scapula and pulling the scapula medially.
Innervation: Third and fourth cervical nerves.

RHOMBOID MAJOR [Figures 7.3 and 7.4]

Description: Also lies deep to the trapezius muscle and inferior to the levator scapulae. The muscle has a rhomboid shape and its fibers pass obliquely from the vertebrae to the scapula.

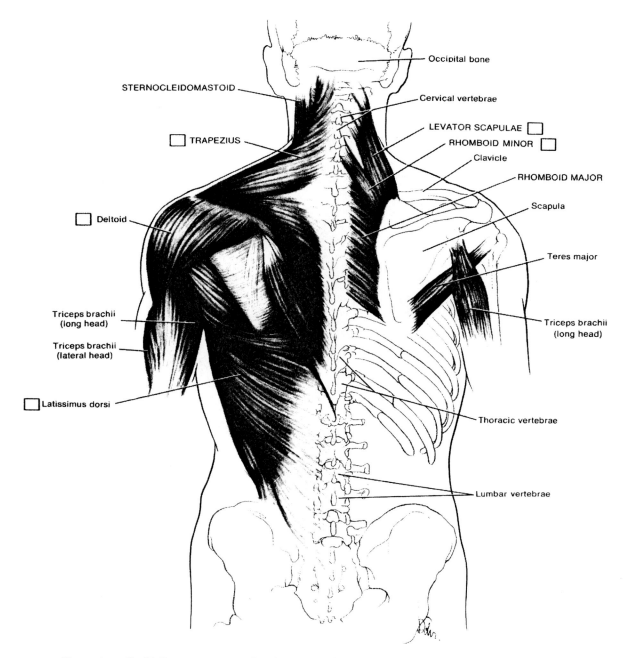

Figure 7.3 (Left) Posterior superficial view and (right) posterior deep view of posterior muscles that move the pectoral (shoulder) girdle.

Origin: Spinous processes of the second, third, fourth, and fifth thoracic vertebrae.
Insertion: Vertebral border of the scapula below the spine of the scapula.
Action: Elevation, inferior rotation, and adduction of the scapula.
Innervation: Dorsal scapular nerve from the brachial plexus.

Muscles of the Upper Extremity—Description

Nine muscles cross the shoulder joint and insert into the periosteum of the humerus (arm bone). Seven of the nine originate on the scapula and are considered

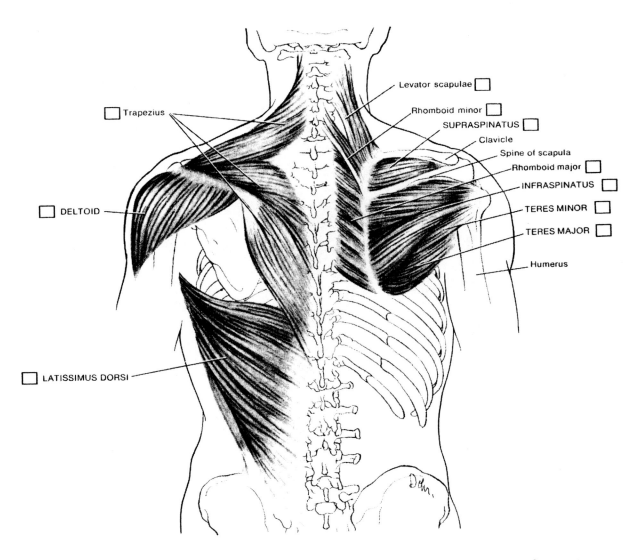

Levator scapulae ☐
Rhomboid minor ☐
SUPRASPINATUS ☐
Clavicle
Spine of scapula
Rhomboid major ☐
INFRASPINATUS ☐
TERES MINOR ☐
TERES MAJOR ☐
Humerus

Trapezius ☐

DELTOID ☐

LATISSIMUS DORSI ☐

Figure 7.4 (Left) Posterior superficial view and (right) posterior deep view of posterior muscles that move the arm (humerus) and pectoral girdle.

scapular muscles. These are the **deltoid, supraspinatus, infraspinatus, teres major, teres minor, subscapularis,** and **coracobrachialis.** The other two muscles originate primarily on the axial skeleton. The **axial muscles** are the **pectoralis major** and **latissimus dorsi**; they insert onto the humerus.

Axial Muscles That Move the Humerus (Arm)

The muscles that move the humerus are the **pectoralis major** and **latissimus dorsi.**

PECTORALIS MAJOR [Figures 7.1 and 7.2]

Description: A large, superficial fan-shaped muscle located on the anterior region of the thorax. It overlies the smaller pectoralis minor.
Origin: Clavicle, sternum, cartilages of the second to sixth ribs, and the aponeurosis of the external oblique muscle of the abdomen.

Insertion: Greater tubercle of the humerus.
Action: Flexes, adducts, and rotates the humerus medially.
Innervation: Pectoral nerves from the brachial plexus.

LATISSIUMS DORSI [Figures 7.3 and 7.4]

Description: A large, flat, wide muscle that extends superiorly from the lower back toward the axilla. Its muscle fibers extend over the lumbar region and lower half of the posterior thorax.
Origin: Spines of lower six thoracic vertebrae, lumbar vertebrae, sacrum, and iliac crest.
Insertion: Floor of the intertubercular sulcus of the humerus.
Action: Draws the shoulder inferiorly and posteriorly. Also involved in powerful extension, adduction, and medial rotation of the humerus.
Innervation: Thoracodorsal nerve from the brachial plexus.

Scapular Muscles That Move the Humerus (Arm)

The scapular muscles that move the humerus include the **deltoid, supraspinatus, infraspinatus, teres major, teres minor, subscapularis,** and **coracobrachialis.** The supraspinatus, infraspinatus, teres minor, and subscapularis are collectively called the **rotator cuff** because their main function is to hold the head of the humerus in the glenoid cavity of the scapula.

DELTOID [Figures 7.2, 7.3, and 7.4]

Description: A large triangular muscle that covers the shoulder joint.
Origin: Clavicle, acromion, and spine of the scapula.
Insertion: Deltoid tuberosity of the humerus.
Action: Contracts all fibers producing abduction of the humerus. Partial contraction of the muscle produces flexion, extension, and lateral or medial rotation of the humerus.
Innervation: Axillary nerve from the brachial plexus.

SUBSCAPULARIS [Figure 7.1]

Description: A deep muscle of the shoulder located in the subscapular fossa.
Origin: Subscapular fossa.
Insertion: Lesser tubercle of the humerus.
Action: Rotates the humerus medially and stabilizes the head of the humerus in the glenoid cavity.
Innervation: Subscapular nerve from the brachial plexus.

SUPRASPINATUS [Figure 7.4]

Description: Occupies the entire supraspinous fossa, superior to the spine of the scapula.
Origin: Supraspinous fossa of the scapula.
Insertion: Greater tubercle of the humerus.
Action: Abducts the humerus.
Innervation: Suprascapular nerve from the brachial plexus.

INFRASPINATUS [Figure 7.4]

Description: A thick muscle that occupies most of the infraspinous fossa, inferior to the spine of the scapula.
Origin: Infraspinous fossa of the scapula.
Insertion: Greater tubercle of the humerus.
Action: Rotates the humerus laterally.
Innervation: Suprascapular nerve from the brachial plexus.

TERES MAJOR [Figures 7.3 and 7.4]

Description: A flattened but thick muscle. Its fibers run superiorly and laterally from the inferior region of the scapula toward the humerus.
Origin: Posterior surface of the inferior angle of the scapula.
Insertion: Intertubercular sulcus of the humerus.
Action: Adducts, extend, and rotates the humerus medially.
Innervation: Lower subscapular nerve from the brachial plexus.

TERES MINOR [Figure 7.4]

Description: A small muscle situated inferior to the infraspinatus muscle and superior to the teres major. Its fibers course obliquely, superiorly and laterally toward the humerus.
Origin: Lateral (axillary) border of the scapula.
Insertion: Greater tubercle of the humerus.
Action: Rotates the humerus laterally.
Innervation: Axillary nerve.

CORACOBRACHIALIS [Figures 7.1 and 7.5]

Description: A small muscle that extends from the scapula to the middle of the humerus along its upper medial surface.
Origin: Apex of the corocoid process of the scapula.
Insertion: Middle of the medial surface of the shaft of the humerus.
Action: Flexes and adducts the humerus.
Innervation: Musculocutaneous nerve.

Muscles of the Pectoral Girdle and Upper Extremity—Identification [Figures 7.1–7.5]

Examine and color the illustrations of the axial and scapular muscles. Then examine the upper extremity of a model or prepared human cadaver and identify the following muscles:

1. latissimus dorsi
2. trapezius
3. deltoid
4. rhomboid major
5. rhomboid minor
6. levator scapulae
7. supraspinatus

8. subscapularis
9. infraspinatus
10. teres minor
11. teres major
12. serratus anterior
13. pectoralis minor
14. pectoralis major
15. coracobrachialis

Muscles That Move the Forearm (Radius and Ulna)—Description

Muscles that move the forearm cross the elbow joint either from the scapula or from the humerus to the radius or ulna. In the elbow joint, contraction of these muscles will produce only **flexion** or **extension** of the forearm. Thus, muscles are divided into **flexors** and **extensors**. Muscles on the anterior surface of the humerus flex the elbow, and muscles on the posterior surface extend the elbow.

The Anterior Flexor Muscles of the Forearm

The flexor muscles of the forearm are the **biceps brachii**, **brachialis**, and **brachioradialis**.

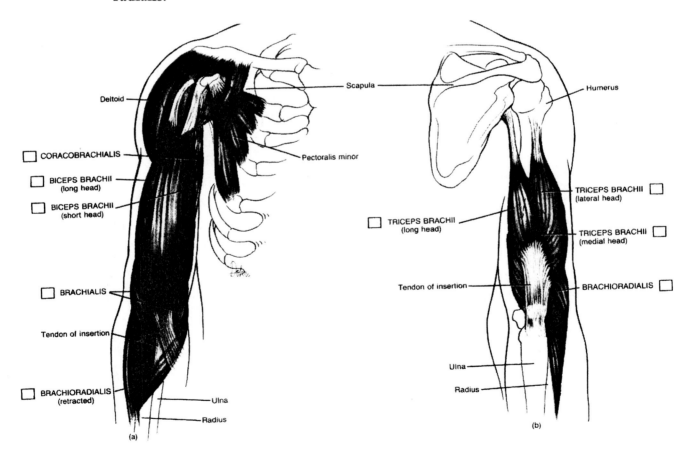

Figure 7.5 (a) Anterior view and (b) posterior view of muscles that move the forearm (radius and ulna).

BICEPS BRACHII [Figures 7.1, 7.2, 7.5, and 7.6]

Description: This long muscle is located on the anterior surface of the humerus. The biceps brachii has two heads: the long head and short head.
Origin: The **short head** originates from the coracoid process of the scapula; the **long head** originates from a small tuberosity superior to the glenoid fossa of the scapula.
Insertion: By a common tendon into the radial tuberosity on the radius bone.
Action: Flexes the humerus, flexes the forearm, and supinates the hand.
Innervation: Musculocutaneous nerve.

BRACHIALIS [Figures 7.1, 7.5, and 7.6]

Description: Located deep to the biceps brachii muscle.
Origin: Distal half of the anterior surface of the humerus.
Insertion: Coronoid process and tuberosity of the ulna.
Action: Flexes the forearm.
Innervation: Musculocutaneous nerve.

BRACHIORADIALIS [Figures 7.5, 7.6, and 7.7]

Description: The most superficial muscle on the lateral or radial surface of the forearm. It forms the lateral side of the cubital fossa. The fibers of the brachioradialis muscle extend from the distal end of the humerus to the distal end of the radius.
Origin: Lateral supracondylar ridge at distal end of the humerus.
Insertion: Lateral side of base of the styloid process of the radius.
Action: Flexes the forearm.
Innervation: Radial nerve.

The Posterior Extensor Muscles of the Forearm

The extensor muscles of the forearm are the **triceps brachii** and **anconeus**.

TRICEPS BRACHII [Figures 7.3, 7.5, and 7.7]

Description: The only large muscle situated on the posterior surface of the humerus; it extends the entire length of the bone. This muscle arises by three heads: the long, medial, and lateral heads. The triceps brachii opposes the actions of the biceps brachii muscle.
Origin: The **long head** originates from the infraglenoid tuberosity of the scapula, the **medial head** arises from the posterior surface of the humerus and distal to the radial groove, and the **lateral head** from the posterior and lateral surface of the humerus.
Insertion: By a common tendon into the olecranon process of the ulna.
Action: All three heads the extend the forearm.
Innervation: Radial nerve.

ANCONEUS [Figure 7.7]

The anconeus is a small muscle that originates from the lateral epicondyle of the humerus and inserts into the olecranon process of the ulna. Its function is to assist in the extension of the forearm. It is innervated by the radial nerve.

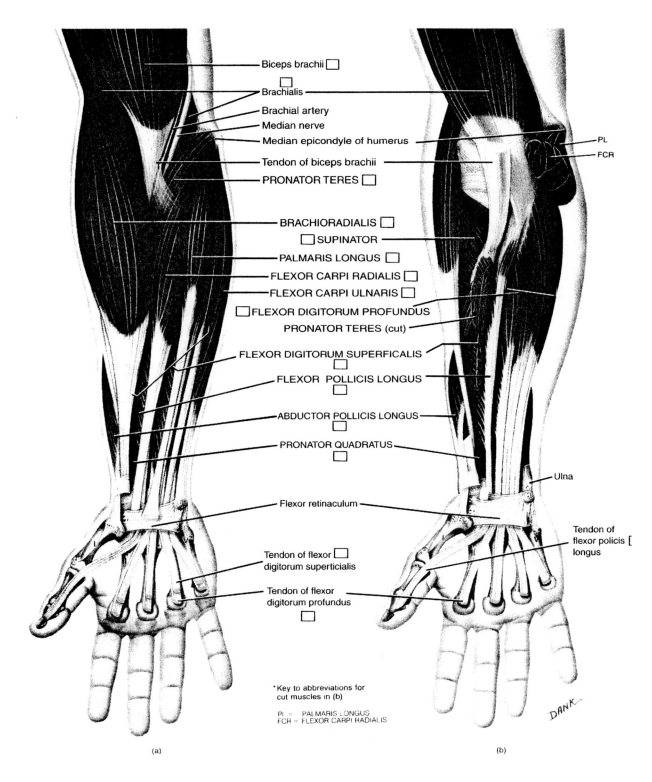

Biceps brachii ☐
Brachialis ☐
Brachial artery
Median nerve
Median epicondyle of humerus
Tendon of biceps brachii
PRONATOR TERES ☐

BRACHIORADIALIS ☐
☐ SUPINATOR
PALMARIS LONGUS ☐
FLEXOR CARPI RADIALIS ☐
FLEXOR CARPI ULNARIS ☐
☐ FLEXOR DIGITORUM PROFUNDUS
PRONATOR TERES (cut)
FLEXOR DIGITORUM SUPERFICALIS ☐
FLEXOR POLLICIS LONGUS ☐
ABDUCTOR POLLICIS LONGUS ☐
PRONATOR QUADRATUS ☐

Flexor retinaculum

Tendon of flexor ☐
digitorum superticialis

Tendon of flexor
digitorum profundus ☐

PL
FCR

Ulna

Tendon of
flexor policis [
longus

DANK

*Key to abbreviations for
cut muscles in (b)

PI = PALMARIS LONGUS
FCR = FLEXOR CARPI RADIALIS

(a)   (b)

Figure 7.6 (a) Superficial anterior view and (b) deep anterior view of anterior forearm muscles that move the wrist, hand, and fingers.

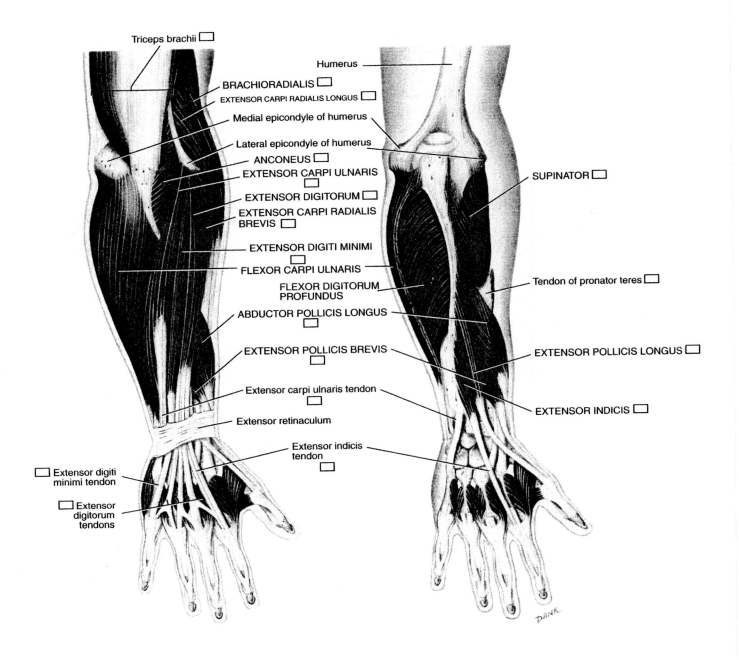

Triceps brachii ☐
Humerus
BRACHIORADIALIS ☐
EXTENSOR CARPI RADIALIS LONGUS ☐
Medial epicondyle of humerus
Lateral epicondyle of humerus
ANCONEUS ☐
SUPINATOR ☐
EXTENSOR CARPI ULNARIS ☐
EXTENSOR DIGITORUM ☐
EXTENSOR CARPI RADIALIS BREVIS ☐
EXTENSOR DIGITI MINIMI ☐
FLEXOR CARPI ULNARIS
FLEXOR DIGITORUM PROFUNDUS
Tendon of pronator teres ☐
ABDUCTOR POLLICIS LONGUS ☐
EXTENSOR POLLICIS BREVIS ☐
EXTENSOR POLLICIS LONGUS ☐
Extensor carpi ulnaris tendon ☐
EXTENSOR INDICIS ☐
Extensor retinaculum
Extensor indicis tendon ☐
☐ Extensor digiti minimi tendon
☐ Extensor digitorum tendons

Figure 7.7 (Left) Superficial posterior and (right) deep posterior views of the muscles that move the wrist, hand, and fingers.

Muscles That Move the Forearm (Radius and Ulna)—Identification *[Figures 7.5–7.7]*

Examine the illustrations and color the anterior and posterior muscles on and around the humerus. Then examine the arm on a model or prepared human cadaver and identify the following anterior and posterior muscles:

1. **biceps brachii**
2. **brachialis**

3. **brachioradialis**
4. **triceps brachii**
5. **anconeus**

Forearm Muscles That Move the Wrist, Hand, and Fingers—Description

The forearm extends from the elbow to the wrist. The forearm muscles are situated on the anterior and the posterior surfaces and move the wrist, hand, and fingers.

The muscles on the **anterior** surface of the forearm are **flexors** of the wrist and the fingers, and some of these muscles are **pronators** of the forearm. Most of the anterior muscles originate on the humerus and insert into the carpals, metacarpals, and phalanges. The muscles on the **posterior** surface of the forearm function as the **extensors** of the wrist and fingers and as **supinators**. Most posterior forearm muscles originate on the humerus and insert onto the metacarpals and phalanges. The anterior and posterior muscles are further subdivided into superficial and deep groups.

The bellies (fleshy portion) of the forearm muscles are located proximally. Distally, the muscles taper into long narrow tendons that insert onto the wrist or hand bones. Near the distal ends of the radius and ulna or the wrist, the tendons of the flexor and extensor muscles are securely bound by thick, strong collagenous bands called the **retinacula** (singular, **retinaculum**). The **flexor retinaculum** is on the anterior, or palmar, surface of the carpal bones, and the **extensor retinaculum** is on the posterior surface of the carpal bones.

Anterior Muscles of the Forearm (Flexors and Pronators)—Description

The superficial forearm muscles are the **pronator teres** (which rotates the radius and turns the palm downward) and three flexors: the **flexor carpi ulnaris,** the **flexor carpi radialis,** and the **palmaris longus.** The **flexor digitorum superficialis** is located in the middle of the anterior muscle group. The deep muscles of the forearm include the **flexor digitorum profundus,** the **flexor pollicis longus,** and the forearm pronator muscle, the **pronator quadratus.**

Superficial Pronator Muscle

PRONATOR TERES [Figure 7.6]

> *Description:* Located on the upper medial side of the forearm, the pronator teres passes inferolaterally across the forearm. The lateral border of the muscle forms the medial boundary of the cubital fossa. The pronator teres lies deep to the brachioradialis muscle.
> *Origin:* Medial epicondyle of the humerus and coronoid process of the ulna.
> *Insertion:* Upper lateral shaft of the radius.
> *Action:* Pronates the forearm.
> *Innervation:* Median nerve.

Superficial Flexor Muscles of the Forearm

FLEXOR CARPI RADIALIS [Figure 7.6]

Description: A slender muscle the belly of which is located on the medial or ulnar side of the pronator teres. The flexor carpi radialis passes diagonally across the forearm to the wrist.
Origin: Medial epicondyle of the humerus.
Insertion: Second and third metacarpal bone.
Action: Flexes the hand at the wrist and assists in its abduction.
Innervation: Median nerve.

PALMARIS LONGUS [Figure 7.6]

Description: A long, slender muscle situated medial to the flexor carpi radialis. The palmaris longus is not present in all people and is highly variable in appearance.
Origin: Medial epicondyle of the humerus.
Insertion: Palmar connective tissue (aponeurosis).
Action: Flexes the hand at the wrist.
Innervation: Median nerve.

FLEXOR CARPI ULNARIS [Figures 7.6 and 7.7]

Description: The flexor carpi ulnaris is the most medial of the superficial flexor muscles of the forearm. The muscle extends interiorly on the medial or ulnar side of the forearm toward the wrist.
Origin: Medial epicondyle of the humerus and olecranon process of the ulna.
Insertion: Pisiform bone of the wrist and base of the fifth metacarpal bone.
Action: Flexes the wrist and adducts the hand.
Innervation: Ulnar nerve.

FLEXOR DIGITORUM SUPERFICIALIS [Figure 7.6]

Description: Lies deep to the flexor carpi radialis, palmaris longus, and flexor carpi ulnaris. The flexor digitorum superficialis is the largest superficial flexor muscle and arises by two heads. The flexor digitorum superficialis is in the intermediate muscle layer and is most visible in the distal half of the forearm.
Origin: Medial epicondyle of the humerus, coronoid process of the ulna, and anterior surface of the radius.
Insertion: By four tendons into the palmar surface of the middle phalanges of the medial four digits (fingers 2 to 5).
Action: Flexes the middle phalanges of the medial four digits (fingers 2 to 5).
Innervation: Median nerve.

Deep Flexor Muscles of the Forearm

FLEXOR DIGITORUM PROFUNDUS [Figure 7.6]

Description: Situated deep to the superficial flexor muscles, especially the flexor digitorum superficialis. The flexor digitorum profundus muscle terminates in four distinct tendons, which continue into the hand deep to the tendons of the flexor digitorum superficialis. The tendons pass under the **flexor retinaculum**.

Origin: Anteromedial surface of the ulna and the interosseous membrane (the thick collagenous membrane between the ulna and radius).
Insertion: By four tendons into the palmar phalanges of the medial four digits (fingers 2 to 5).
Action: Flexes the distal phalanges of each finger and other phalanges.
Innervation: Ulnar nerve innervates the medial half of the muscle, while median nerve innervates the lateral half.

FLEXOR POLLICIS LONGUS [Figure 7.6]

Description: A deep, lateral muscle situated on the radial side of the forearm, the flexor pollicis longus runs parallel to the flexor digitorum profundus.
Origin: Anterior surface of the radius and lateral half of the interosseous membrane.
Insertion: Base of the distal phalanx of the thumb.
Action: Flexes the thumb.
Innervation: Median nerve.

Deep Pronator Muscle

PRONATOR QUADRATUS [Figure 7.6]

Description: A small, deep, quadrangular muscle of the anterior surface of the forearm. The pronator quadratus extends across the anterior surfaces of the distal parts of the ulna and radius immediately proximal to the wrist.
Origin: Distal fourth of the anterior surface of the ulna.
Insertion: Anterior surface of the distal fourth of the radius.
Action: Pronates the forearm.
Innervation: Median nerve.

Anterior Muscles of the Forearm (Flexors and Pronators)—Identification [Figure 7.6]

Examine the illustrations of the anterior superficial and deep muscles of the forearm. Then examine the prepared anterior forearm of a model or human cadaver and identify the following muscles:

1. palmaris longus
2. flexor carpi radialis
3. flexor carpi ulnaris
4. flexor pollicis longus
5. flexor digitorum superficialis
6. flexor digitorum profundus
7. pronator teres
8. pronator quadratus

Posterior Muscles of the Forearm—Extensors and Supinators

The posterior muscles of the forearm extend the hand at the wrist and the digits. The extensor muscles are divided into those in the **superficial layer** and those in the **deep layer**.

Superficial Extensor Muscles of the Posterior Forearm—Description

The superficial layer of muscles consists of three wrist extensors: the **extensor carpi radialis longus**, the **extensor carpi radialis brevis**, and the **extensor carpi ulnaris**. The two finger extensors are the **extensor digitorum** and **extensor digiti minimi**.

EXTENSOR CARPI RADIALIS LONGUS [Figure 7.7]

Description: This long, tapered muscle runs along the lateral side of the forearm and connects the humerus with the hand. The extensor carpi radialis longus runs parallel to and is often overlapped by or combined with the brachioradialis.
Origin: Lateral supracondylar ridge of the humerus.
Insertion: Base of the second metacarpal bone.
Action: Extend and abducts the wrist.
Innervation: Radial nerve.

EXTENSOR CARPI RADIALIS BREVIS [Figure 7.7]

Description: A shorter muscle than extensor carpi radialis longus, situated medial to the extensor carpi radialis longus, which partially covers it.
Origin: Lateral epicondyle of the humerus.
Insertion: Base of the third metacarpal bone.
Action: Extend and abducts the wrist.
Innervation: Radial nerve.

EXTENSOR CARPI ULNARIS [Figure 7.7]

Description: A long, thin muscle located on the ulnar side of the forearm. The extensor carpi ulnaris is the most medial extensor muscle on the posterior surface of the forearm.
Origin: Lateral epicondyle of the humerus.
Insertion: Base of the fifth metacarpal bone.
Action: Extend and adducts the wrist.
Innervation: Radial nerve.

EXTENSOR DIGITORUM [Figure 7.7]

Description: The principal extensor muscle of the fingers. Its location is in the center of the forearm on the posterior surface. The extensor digitorum divides into four tendons proximal to the wrist, which pass deep to the extensor retinaculum. These tendons then continue to each finger.
Origin: Lateral epicondyle of the humerus.

Insertion: By four tendons into distal phalanges of fingers 2 to 5.
Action: Extend the fingers and wrist.
Innervation: Radial nerve.

EXTENSOR DIGITI MINIMI [Figures 7.6 and 7.7]

Description: A long, slender muscle situated on the medial side of the extensor digitorum.
Origin: Lateral epicondyle of the humerus.
Insertion: Tendon of the extensor digitorum on the fifth phalanx (little finger).
Action: Extend the little finger.
Innervation: Deep radial nerve.

Deep Muscles of the Posterior Forearm

The deep muscle layer consists of the **supinator, abductor pollicis longus, extensor pollicis brevis, extensor pollicis longus,** and **extensor indicis.**

SUPINATOR [Figures 7.6 and 7.7]

Description: A broad muscle whose fibers extend inferiorly and laterally to wrap around the upper radius. The supinator is a deep muscle of the posterior region of the elbow.
Origin: Lateral epicondyle of the humerus and crest of the ulna.
Insertion: Lateral surface of the proximal third of the radius.
Action: Supinates the forearm.
Innervation: Deep radial nerve.

Deep Extensor Muscles of the Posterior Forearm

ABDUCTOR POLLICIS LONGUS [Figures 7.6 and 7.7]

Description: Lies distal to the supinator and parallel to the extensor pollicis longus muscles.
Origin: Posterior surfaces of the ulna and radius and interosseous membrane.
Insertion: Base of the first metacarpal bone.
Action: Abducts and extends the thumb.
Innervation: Deep radial nerve.

EXTENSOR POLLICIS BREVIS [Figure 7.7]

Description: Lies distal and medial to the abductor pollicis longus.
Origin: Posterior surface of the radius and interosseous membrane.
Insertion: Base of the proximal phalanx of the thumb.
Action: Extend the thumb and abducts the wrist.
Innervation: Deep radial nerve.

EXTENSOR POLLICIS LONGUS [Figure 7.7]

Description: Longer than the extensor pollicis brevis.

Origin: Posterior surface of the ulna and the interosseous membrane.
Insertion: Base of the distal phalanx of the thumb.
Action: Extend the distal phalanx of the thumb.
Innervation: Deep radial nerve.

EXTENSOR INDICIS [Figure 7.7]

Description: A narrow, elongated muscle that lies medial to and alongside the extensor pollicis longus.
Origin: Posterior surface of the ulna and interosseous membrane.
Insertion: Tendon of the extensor digitorum of index finger (second digit).
Action: Extends the index finger.
Innervation: Deep radial nerve.

Posterior Muscles of the Forearm (Extensors and Supinators)—Identification [Figure 7.7]

Examine the illustrations of the posterior superficial and deep muscles of the forearm. Then examine the prepared posterior forearm of a model or human cadaver and identify the following muscles:

1. extensor carpi radialis longus
2. extensor carpi radialis brevis
3. extensor carpi ulnaris
4. extensor pollicis longus
5. extensor pollicis brevis
6. adductor pollicis longus
7. extensor digiti minimi
8. extensor digitorum
9. extensor indicis
10. supinator

Muscles of the Hand—Description

Flexing and extending the hand and digits are performed by the anterior and posterior muscles of the forearm. Precise movement of the fingers is produced by tiny intrinsic muscles in the hand. Their small size make it difficult to prepare them for demonstration.

CHAPTER SUMMARY AND CHECKLIST

   I. **Muscles of the Pectoral Girdle**

   A. **Description**
      1. Those that arise from axial skeleton
      2. Those that attach to clavicle or scapula
      3. Those that move the pectoral girdle
      4. Divided into anterior and posterior groups

B. **Anterior Muscles of the Pectoral Girdle**
1. Serratus anterior
2. Pectoralis minor

C. **Posterior Muscles of the Pectoral Girdle**
1. Trapezius
2. Levator scapulae
3. Rhomboid major
4. Rhomboid minor

II. **Muscles of the Upper Extremity**

A. **Description**
1. Nine muscles cross shoulder to insert into humerus
2. Seven muscles originate on the scapula
3. Two muscles originate on the axial skeleton

B. **Axial Muscles That Move the Humerus**
1. Pectoralis major
2. Latissimus dorsi

C. **Scapular Muscles That Move the Humerus**
1. Deltoid
2. Subscapularis
3. Supraspinatus
4. Infraspinatus
5. Teres major
6. Teres minor
7. Coracobrachialis

III. **Muscles That Move the Ulna and Radius (Forearm)**

A. **Description**
1. Originate on either the humerus or scapula
2. Cross elbow and attach to radius or ulna
3. Flexors: along anterior humerus
4. Extensors: along posterior humerus

B. **Anterior Flexor Muscles of the Forearm**
1. Biceps brachii
2. Brachialis
3. Brachioradialis

C. **Posterior Extensor Muscles of the Forearm**
1. Triceps brachii
2. Anconeus

IV. **Anterior Muscles of the Forearm That Move Wrist, Hand, and Fingers**

A. **Description**
1. Muscles situated on anterior and posterior surfaces
2. Most anterior muscles flex wrist and fingers
3. Some anterior muscles pronate the forearm

4. Most anterior muscles originate on the humerus
5. Most posterior muscles extend wrist and fingers
6. Posterior muscles contain a supinator
7. Most posterior muscles originate on the humerus
8. Divided into superficial and deep groups
9. Tendons surrounded by retinacula

B. Superficial Anterior Muscles of the Forearm
1. Pronator teres
2. Flexor carpi radialis
3. Palmaris longus
4. Flexor carpi ulnaris
5. Flexor digitorum superficialis

C. Deep Anterior Muscles of the Forearm
1. Flexor digitorum profundus
2. Flexor pollicis longus
3. Pronator quadratus

V. Posterior Muscles of the Forearm That Move Wrist, Hand, and Fingers

A. Superficial Extensor Muscles
1. Extensor carpi radialis longus
2. Extensor carpi radialis brevis
3. Extensor carpi ulnaris
4. Extensor digitorum
5. Extensor digiti minimi

B. Deep Muscles
1. Supinator

C. Deep Extensor Muscles
1. Abductor pollicis longus
2. Extensor pollicis brevis
3. Extensor pollicis longus
4. Extensor indicis

Laboratory Exercises 7

NAME _____

LAB SECTION _____ DATE _____

LABORATORY EXERCISE 7.1

Part I

Muscles of the Pectoral Girdle and Upper Extremity

Supply the muscles that fit the description:

1. Protract and rotate the scapula superiorly _____

2. Superior rotation and elevation of scapula _____

3. Elevation, inferior rotation, and adduction of scapula _____

4. Flex, adduct, and rotate the humerus medially _____

5. Extend, adduct, and rotate the humerus medially _____

6. Abduct the humerus _____

7. Rotate the humerus laterally _____

8. Exhibits a sawtooth appearance _____

9. Overlays posterior regions of neck, shoulders, and thorax on both sides of the midline _____

10. Insert on the spine of the scapula

    a. _____

    b. _____

11. Originate from scapular fossae

    a. _____

    b. _____

    c. _____

Part II

Muscles that Move the Forearm, Wrist, Hand, and Fingers

1. Which three muscles flex the humerus and forearm and supinate the hand?

   a. _____

   b. _____

   c. _____

2. Which two muscles extend the forearm?

   a. _____

   b. _____

3. Which two muscles pronate the forearm?

   a. _____

   b. _____

4. Which two muscles flex the hand at the wrist?

   a. _____

   b. _____

5. Which two muscles flex the middle and distal phalanges?

   a. _____

   b. _____

6. Which four muscles move the thumb?

   a. _____

   b. _____

   c. _____

   d. _____

7. Which two muscles extend and abduct the wrist?

    a. _____

    b. _____

8. Which muscles partly or wholly originate on the humerus or its parts?

    a. _____

    b. _____

    c. _____

Part III

Muscles of the Thorax, Upper Arm, and Lower Arm

Using the listed terms, label the anterior muscles of the thorax and then color them.

Deltoid   Pectoralis major
Pectoralis minor   Serratus anterior

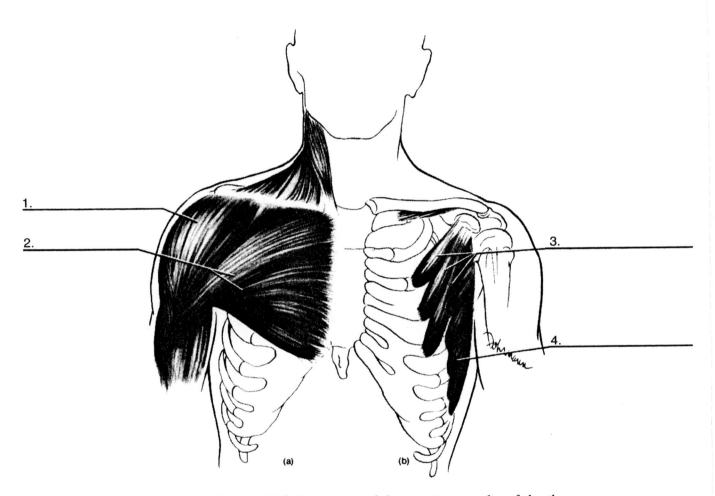

1.
2.
3.
4.

(a)   (b)

Figure 7.8 (a) Superficial and (b) deep views of the anterior muscles of the thorax.

LABORATORY EXERCISE 7.2

Using the listed terms, label the muscles of the upper arm and then color them.

Coracobrachialis
Biceps brachii (long and short heads)
Triceps brachii (long, lateral, and medial heads)
Brachialis

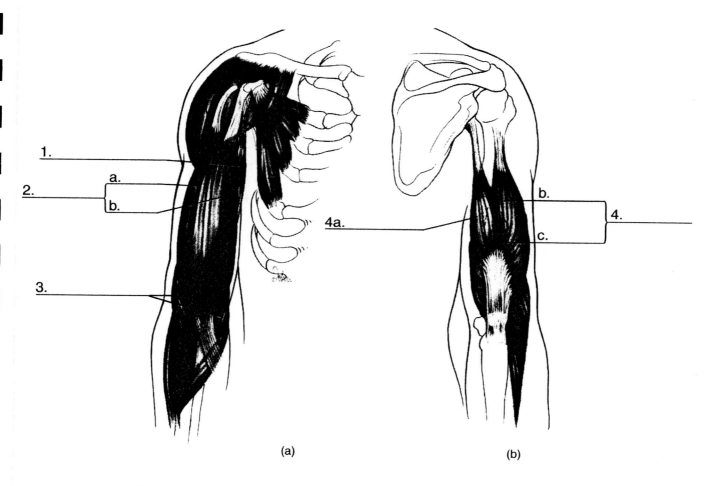

(a)   (b)

Figure 7.9 (a) Anterior and (b) posterior views of the muscles of the upper arm (humerus).

LABORATORY EXERCISE 7.3

Using the listed terms, label the posterior muscles of the thorax and then color them.

Trapezius
Levator scapulae
Supraspinatus
Teres minor
Deltoid

Rhomboid minor
Infraspinatus
Latissimus dorsi
Rhomboid major
Teres major

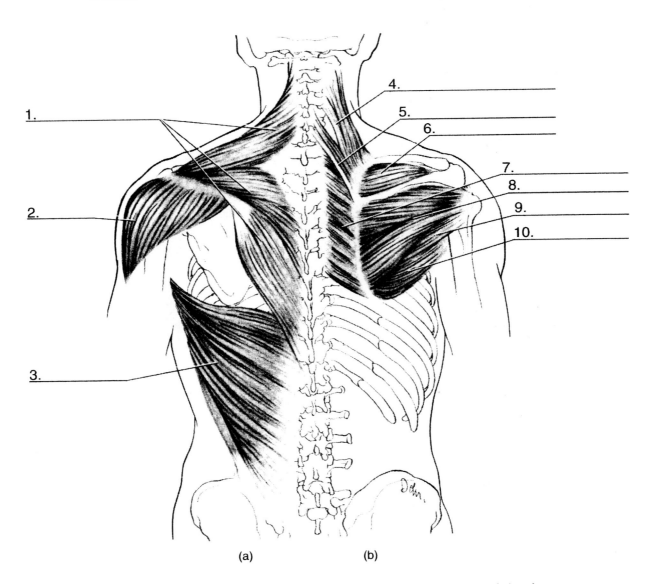

(a)   (b)

Figure 7.10 (a) Superficial and (b) deep views of the posterior muscles of the thorax.

LABORATORY EXERCISE 7.4

Using the listed terms, label the anterior forearm muscles and then color them.

Pronator teres
Palmaris longus
Flexor carpi ulnaris
Flexor digitorum superficialis
Adductor pollicis longus
Supinator

Brachioradialis
Flexor carpi radialis
Flexor digitorum profundus
Flexor pollicis longus
Pronator quadratus

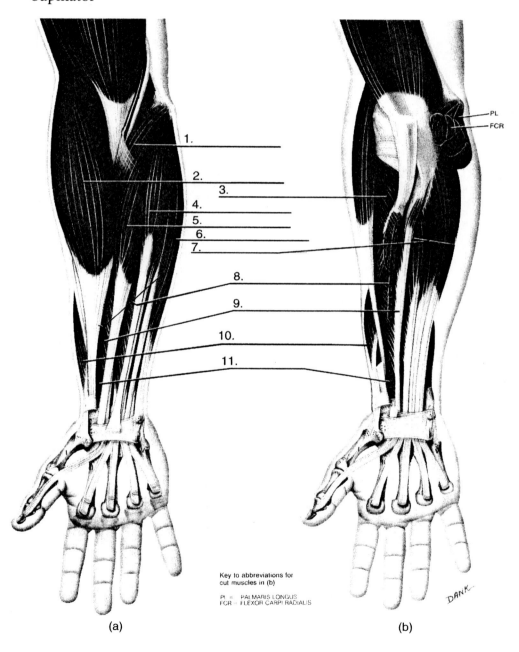

Key to abbreviations for
cut muscles in (b)

PL = PALMARIS LONGUS
FCR = FLEXOR CARPI RADIALIS

(a)

(b)

Figure 7.11 (a) Superficial and (b) deep views of the anterior muscles of the forearm.

LABORATORY EXERCISE 7.5

Using the listed terms, label the anterior forearm muscles and then color them.

Extensor carpi radialis longus
Extensor carpi radialis brevis
Extensor digit minimi
Flexor digitorum profundus
Extensor pollicis brevis
Extensor indicis

Extensor carpi ulnaris
Extensor digitorum
Flexor carpi ulnaris
Abductor pollicis longus
Extensor pollicis longus
Supinator

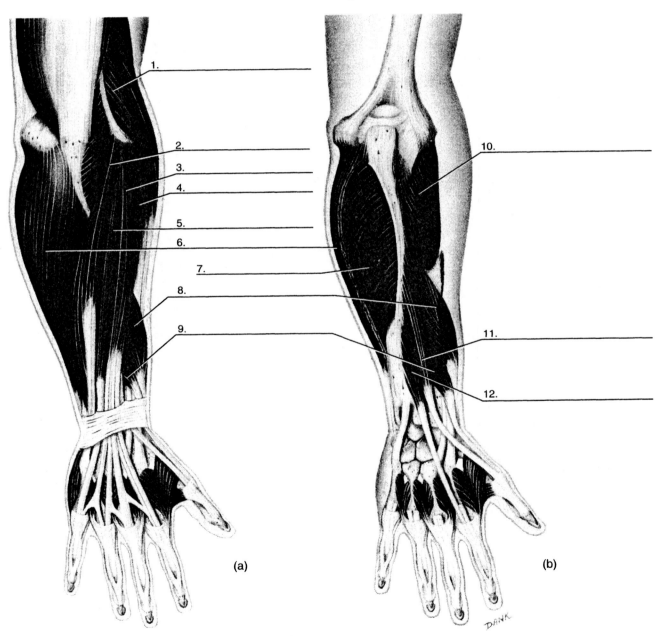

Figure 7.12 (a) Superficial and (b) deep views of the posterior muscles of the forearm.

8

The Muscles of the Appendicular Skeleton

MUSCLES OF THE PELVIC GIRDLE AND LOWER EXTREMITY

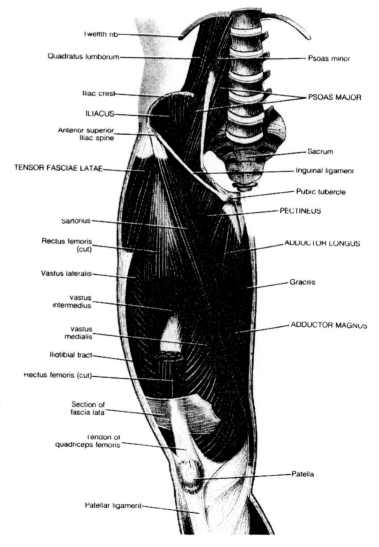

Twelfth rib
Quadratus lumborum
Iliac crest
ILIACUS
Anterior superior Iliac spine
TENSOR FASCIAE LATAE
Sartorius
Rectus femoris (cut)
Vastus lateralis
vastus intermedius
vastus medialis
Iliotibial tract
Rectus femoris (cut)
Section of fascia lata
Tendon of quadriceps femoris
Patellar ligament

Psoas minor
PSOAS MAJOR
Sacrum
Inguinal ligament
Pubic tubercle
PECTINEUS
ADDUCTOR LONGUS
Gracilis
ADDUCTOR MAGNUS
Patella

Objective

The objective of Chapter 8, "The Muscles of the Appendicular Skeleton," is to present the origin, insertion, action, and innervation of the muscles of the:

1. **Pelvic girdle and lower extremity**
2. **Thigh**
3. **Lower leg**

Muscles That Move the Thigh—Description

Most muscles that move the thigh are attached to the pelvic bones and thigh bone (femur). These muscles are divided into three groups: **anterior, posterior,** and **medial.**

Anterior Thigh Muscles of the Iliac Region— Description

The muscles that flex the thigh at the hip joint are the **iliacus** and the **psoas major,** often described jointly as **iliopsoas** because the two merge into one distally.

PSOAS MAJOR [Figure 8.1]

Description: Descends from the posterior abdominal region into the thigh by passing deep to the inguinal ligament. The psoas major muscle lies lateral to the lumbar vertebrae and brim of the pelvis.
Origin: Transverse processes and bodies of lumbar vertebrae.
Insertion: Lesser trochanter of the femur.
Action: Flexes the thigh at the hip.
Innervation: Lumbar nerves.

ILIACUS [Figure 8.1]

Description: A large, flat, fan-shaped muscle located laterally to the psoas major muscle in the pelvis. The iliacus muscle fills the iliac fossa of the pelvic girdle.
Origin: Iliac fossa and iliac crest.
Insertion: Lesser trochanter of the femur via the tendon of the psoas major.
Action: Flexes the thigh at the hip.
Innervation: Femoral nerve.

Anterior Muscles of the Thigh—Description

The other anterior muscles that move the thigh are the **tensor fasciae latae, sartorius,** and **quadriceps femoris.**

TENSOR FASCIAE LATAE [Figure 8.1]

Description: Located superficially on the lateral surface of the hip. The tensor fasciae latae muscle arises from the ilium and attaches to the connective tissue fascia, called the **iliotibial tract** of the **fascia lata.** The fascia lata surrounds the entire leg.

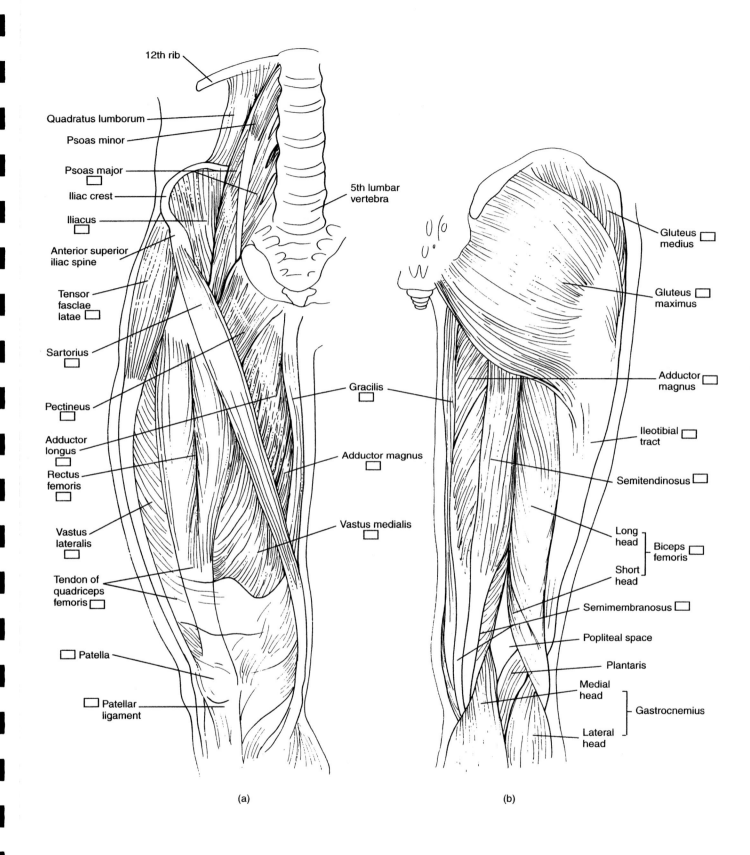

Figure 8.1 (a) Anterior view of the superficial muscles that move the thigh. (b) Posterior view of superficial muscles that move the thigh.

Origin: Outer lip of the iliac crest.
Insertion: Between the two layers of the iliotibial tract of fascia lata, which extends to the lateral condyle of the tibia.
Action: Flexes and abducts the thigh.
Innervation: Superior gluteal nerve.

SARTORIUS [Figure 8.1]

Description: Narrow and straplike, the sartorius is the longest muscle in the body and the most superficial in the anterior region. It obliquely crosses the anterior thigh from the lateral to the medial side.
Origin: Anterior superior iliac spine.
Insertion: Medial surface of the body of the tibia.
Action: Flexes the thigh and leg and rotates the thigh laterally. It is named after the latin word *sartor*, or "tailor," because it is used for sitting cross-legged (the tailor's position).
Innervation: Femoral nerve.

QUADRICEPS FEMORIS—DESCRIPTION [Figure 8.1]

Consists of four distinct muscles: **rectus femoris, vastus lateralis, vastus medialis**, and **vastus intermedius**. All have separate origins, but the tendons of all four merge to insert via a common tendon on the **tibia**. The quadriceps femoris is the great extensor muscle of the leg and covers the anterior and lateral surfaces of the femur.

QUADRICEPS FEMORIS

Origins: Originates from the anterior inferior iliac spine and the superior region of the acetabulum. It occupies a superficial position and is located in the middle of the anterior thigh. **Vastus medialis** originates from the medial surface of the linea aspera of the femur. This muscle forms the characteristic bulge in the lower medial region of the thigh. **Vastus intermedius** originates from the anterior and lateral surfaces of the upper two-thirds of the shaft of the femur. It lies deep to the rectus femoris and is best seen when the rectus femoris muscle is removed.
Insertion: The tendons of quadriceps femoris muscles merge at the distal end of the thigh into one strong tendon that inserts into the base of the **patella**. The tendon then continues inferiorly and inserts into the tibial tuberosity as the **patellar ligament**.
Action: The entire quadriceps femoris extends the leg at the knee. The rectus femoris portion of the muscle also flexes the thigh at the hip.
Innervation: Femoral nerve.

Anterior Muscles of the Thigh—Identification

[Figure 8.1]

Examine the illustrations of the anterior, iliac, and thigh muscles and then color them. Then examine the thigh on a model or prepared human cadaver and identify the following iliac muscles:

1. iliacus
2. psoas major
3. tensor fasciae latae
4. sartorius

QUADRICEPS FEMORIS

1. rectus femoris
2. vastus medialis
3. vastus lateralis
4. vastus intermedius

Posterior Thigh Muscles (Gluteal, or Buttocks Region)—Description

The posterior muscles of the gluteal (buttocks) region that move the thigh include the **gluteus maximus**, **gluteus medius**, and **gluteus minimus**.

GLUTEUS MAXIMUS [Figures 8.1 and 8.2]

Description: Most of the buttock prominence and mass is formed by the gluteus maximus muscle, the largest and most superficial muscle in the gluteal region.
Origin: The ilium, lower part of the sacrum, and side of the coccyx.
Insertion: Gluteal tuberosity of the femur and iliotibial tract of the fascia lata.
Action: Extends and rotates the thigh laterally.
Innervation: Inferior gluteal nerve.

GLUTEUS MEDIUS [Figures 8.1 and 8.2]

Description: Superior and deep to the gluteus maximus. The inferior third of the muscle is covered by the gluteus maximus.
Origin: External surface of the ilium bone.
Insertion: Lateral surface of the greater trochanter of the femur.
Action: Abducts and rotates the thigh medially.
Innervation: Superior gluteal nerve.

GLUTEUS MINIMUS [Figure 8.2]

Description: The smallest of the three gluteal muscles, the gluteus minimus muscle is located immediately deep to the gluteus medius.
Origin: External surface of the ilium bone.
Insertion: Anterior surface of the greater trochanter of the femur.
Action: Abducts and rotates the thigh medially.
Innervation: Superior gluteal nerve.

Posterior Thigh Muscles (Gluteal Region)—Identification [Figures 8.1 and 8.2]

Examine the illustrations of the posterior thigh muscles of the gluteal region and color them. Then examine the posterior thigh in a model or prepared human cadaver and identify the following muscles:

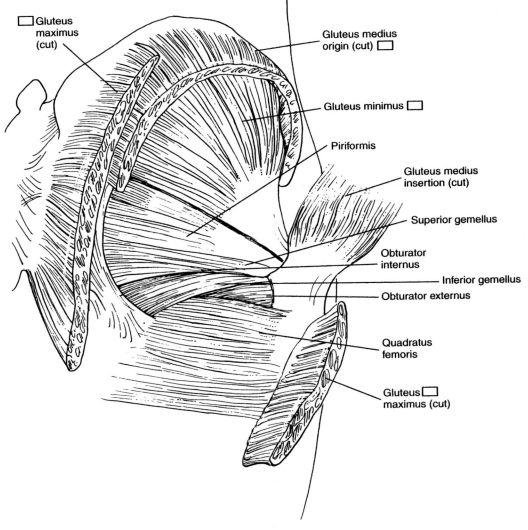

Figure 8.2 Superficial (cut) and deep gluteal muscles of the posterior thigh.

1. **gluteus maximus**
2. **gluteus medius**
3. **gluteus minimus**

Medial or Adductor Muscles of the Thigh— Description

The main function of the medial group of muscles in the thigh is adduction of the thigh. These muscles comprise the **adductor** group and include the **gracilis, pectineus, adductor longus, adductor brevis,** and **adductor magnus.**

GRACILIS [Figures 8.1 and 8.3]

> *Description:* A long, thin, straplike muscle on the medial side of the thigh and the knee, the gracilis is the most superficial adductor muscle and the only one that crosses the knee joint. The gracilis is a two-joint muscle in that it both adducts the thigh and flexes the leg.

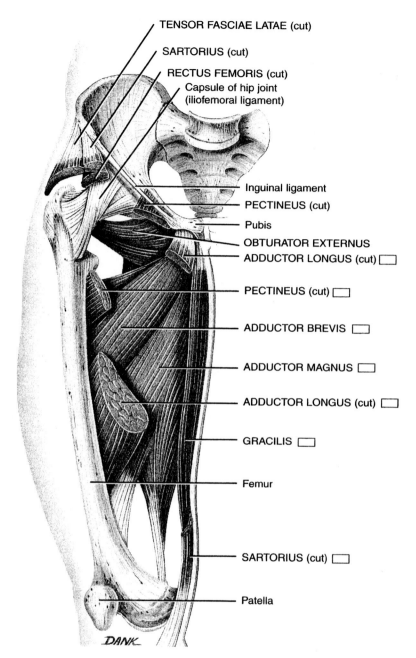

TENSOR FASCIAE LATAE (cut)

SARTORIUS (cut)

RECTUS FEMORIS (cut)
Capsule of hip joint
(iliofemoral ligament)

Inguinal ligament

PECTINEUS (cut)

Pubis

OBTURATOR EXTERNUS

ADDUCTOR LONGUS (cut) ☐

PECTINEUS (cut) ☐

ADDUCTOR BREVIS ☐

ADDUCTOR MAGNUS ☐

ADDUCTOR LONGUS (cut) ☐

GRACILIS ☐

Femur

SARTORIUS (cut) ☐

Patella

DANK

Figure 8.3 Anterior deep view of the major muscles of the medial (adductor) group of the thigh.

Origin: Inferior ramus of the pubic bone near the symphysis.
Insertion: Upper medial surface of the tibia.
Action: Adducts the thigh, flexes the leg at the knee, and rotates the leg medially.
Innervation: Obturator nerve.

PECTINEUS [Figures 8.1 and 8.3]

Description: A short, flat quadrangular muscle located on the anterior and medial side of the thigh.

Origin: Pubic bone (near pubic crest).
Insertion: On femur inferior to lesser trochanter.
Action: Flexes, adducts, and rotates the thigh medially.
Innervation: Femoral nerve.

ADDUCTOR LONGUS [Figures 8.1 and 8.3]

Description: Lying lateral to the gracilis muscle on the upper third of the thigh, the adductor longus muscle is the most anterior of the adductor muscles.
Origin: Anterior pubis.
Insertion: Middle third of the femur along the linea aspera.
Action: Adducts, flexes, and rotates the thigh medially.
Innervation: Obturator nerve.

ADDUCTOR BREVIS [Figures 8.1 and 8.3]

Description: A short muscle located deep to the pectineus and adductor longus.
Origin: Inferior ramus of pubis.
Insertion: Linea aspera of the femur.
Action: Adducts, flexes, and rotates the thigh medially.
Innervation: Obturator nerve.

ADDUCTOR MAGNUS [Figures 8.1 and 8.3]

Description: The largest adductor muscle which is located on the medial side of the thigh.
Origin: Inferior ramus of the pubis and ischium, and ischial tuberosity.
Insertion: By a broad aponeurosis into the linea aspera of the femur.
Action: Adducts the thigh.
Innervation: Obturator and sciatic nerves.

Medial or Adductor Muscles of the Thigh—Identification [Figures 8.1 and 8.3]

Examine the illustrations of the medial thigh muscles and color them. Then examine the medial thigh on a model or prepared human cadaver and identify the following muscles:

1. gracilis
2. adductor magnus
3. adductor longus
4. adductor brevis
5. pectineus

Posterior Muscles of the Thigh—Description

The posterior muscles of the thigh are called the hamstrings and include the **biceps femoris, semitendinosus,** and **semimembranosus** muscles. They span the hip and knee joints. They are **flexors** of the leg at the knee joint and **extensors** of the thigh at the hip joint.

BICEPS FEMORIS [Figure 8.1]

Description: The biceps femoris muscle has a long head and a short head of origin. This muscle is on the posterior side of the thigh and is the most lateral hamstring muscle.

Origin: The **long head** arises from the ischial tuberosity, and the **short head** arises from the linea aspera of the femur.

Insertion: Head of the fibula and lateral condyle of the tibia.

Action: Flexes the leg at the knee, rotates the leg laterally, and extends the thigh at the hip.

Innervation: Tibial portion of the sciatic nerve.

SEMITENDINOSUS [Figure 8.1]

Description: Lies medial to the biceps femoris and is the middle hamstring. The semitendinosus muscle has a long, slender tendon, which begins two-thirds of the way down the posterior thigh.

Origin: Ischial tuberosity by a common tendon with the long head of biceps femoris.

Insertion: Medial surface of superior portion of the tibia.

Action: Flexes of the leg at the knee, rotates the tibia medially, and extends the thigh at the hip.

Innervation: Tibial portion of the sciatic nerve.

SEMIMEMBRANOSUS [Figure 8.1]

Description: A flat muscle located deep to the semitendinosus. The semimembranosus is the most medial hamstring muscle.

Origin: Ischial tuberosity.

Insertion: Medial condyle of the tibia.

Action: Flexes the leg at the knee, rotates the tibia medially, and extends the thigh at the hip.

Innervation: Tibial portion of the sciatic nerve.

Posterior Muscles of the Thigh—Identification [Figure 8.1]

Examine the illustrations of the posterior thigh muscles. Then examine the posterior thigh in a model or prepared human cadaver and identify the following hamstring muscles:

1. **biceps femoris**
2. **semimembranosus**
3. **semitendinosus**

Muscles of the Lower Leg—Description

The muscles of the lower leg, the **crural muscles**, move the ankle, foot, and toes. They are divided by location into three groups, the **anterior**, **lateral**, and **posterior** crural muscles. The anteromedial surface of the leg along the shaft of the tibia lacks muscle attachments.

The tendons of the crural muscles that descend and cross the ankle joint are held tightly together close to the bone by a thick connective tissue fascia called the **retinaculum**. This fascia is similar to that of the wrist and prevents displacement of the tendons during muscle contraction.

Anterior Crural Muscles of the Leg—Description

The anterior crural muscles produce superior flexion of the foot (dorsiflexion), inversion of the foot, and extension of the toes. The anterior crural muscles include the **tibialis anterior**, **extensor digitorum longus**, **extensor hallicus longus**, and **peroneus tertius**.

TIBIALIS ANTERIOR [Figure 8.4]

Description: A long, thick muscle on the lateral surface of the tibia. The fibers of the tibialis anterior descend inferiorly along the tibia and its long tendon of insertion continues into the foot.
Origin: Lateral condyle and superior half of lateral surface of the tibia.
Insertion: First cuneiform (medial) and first metatarsal bone.
Action: Dorsiflexion and inversion of the foot.
Innervation: Deep peroneal nerve.

EXTENSOR DIGITORUM LONGUS [Figure 8.4]

Description: Lateral to the tibialis anterior on the anterolateral surface of the leg. The single tendon of the extensor digitorum longus passes to the metatarsals, where it divides into four tendons that continue to the four toes.
Origin: Lateral condyle of the tibia, anterior surface of the fibula, and the interosseous membrane.
Insertion: Middle and distal phalanges of toes 2 to 5.
Action: Extension of toes 2 to 5 and dorsiflexion of the foot.
Innervation: Deep peroneal nerve.

EXTENSOR HALLICUS LONGUS [Figure 8.4]

Description: A thin muscle. The extensor hallicus longus is partially covered by the tibialis anterior and extensor digitorum longus, but it emerges between the two muscles in the lower part of the leg.
Origin: Anterior surface of the fibula and interosseous membrane.
Insertion: Distal phalanx of the great toe.
Action: Extension of the great toe, dorsiflexion and inversion of the foot.
Innervation: Deep peroneal nerve.

PERONEUS TERTIUS [Figure 8.4]

Description: A small muscle that is usually a continuation or a partially separated inferior portion of the extensor digitorum longus muscle. Not present in all individuals.
Origin: Distal third of the anterior surface of the fibula and interosseous membrane.
Insertion: Dorsal surface of the fifth metatarsal bone.
Action: Dorsiflexion and eversion of the foot.
Innervation: Deep peroneal nerve.

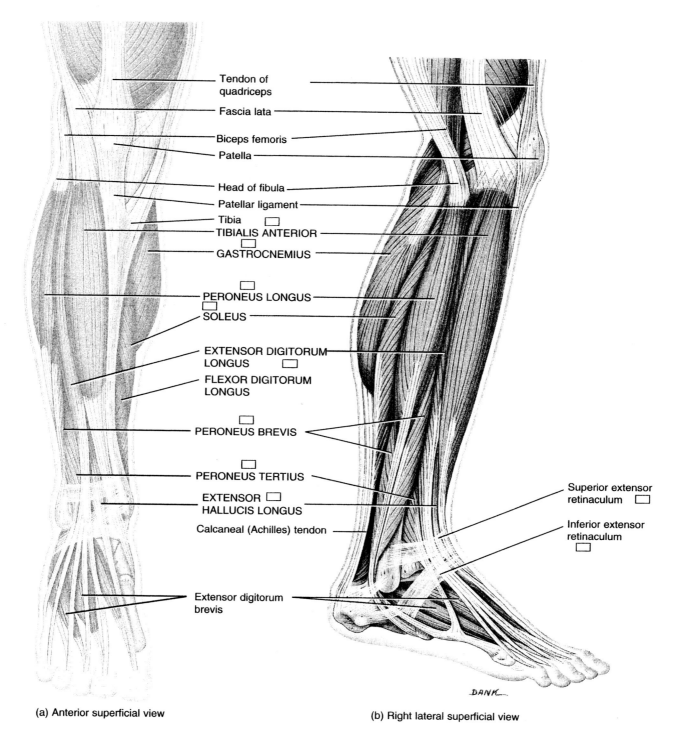

Tendon of quadriceps

Fascia lata

Biceps femoris

Patella

Head of fibula

Patellar ligament

Tibia ☐

TIBIALIS ANTERIOR ☐

GASTROCNEMIUS

PERONEUS LONGUS ☐

SOLEUS ☐

EXTENSOR DIGITORUM LONGUS ☐

FLEXOR DIGITORUM LONGUS

PERONEUS BREVIS ☐

PERONEUS TERTIUS ☐

EXTENSOR ☐ HALLUCIS LONGUS

Calcaneal (Achilles) tendon

Extensor digitorum brevis

Superior extensor retinaculum ☐

Inferior extensor retinaculum ☐

(a) Anterior superficial view

(b) Right lateral superficial view

Figure 8.4 (a) Anterior and (b) lateral views of the crural muscles of the leg.

Lateral Crural Muscles of the Leg—Description

There are two lateral crural muscles, the **peroneus longus** and **peroneus brevis**. They produce extension of the foot inferiorly (plantarflexion) and eversion of the foot. These muscles are on the lateral side of the leg and arise from the lateral surface of the fibula.

PERONEUS LONGUS [Figure 8.4]

Description: A long, straplike muscle on the lateral side of the leg, the peroneus longus is more superficial than the peroneus brevis and arises higher on the fibula. The peroneus longus terminates in a long tendon, which descends behind the lateral malleolus, where it is covered by the connective tissue sheath of the retinaculum. The muscle then crosses the sole of the foot to the medial side and attaches to the foot bones.

Origin: Head and proximal two-thirds of the lateral surface of the fibula.

Insertion: First metacarpal bone and medial cuneiform bone.

Action: Plantar flexion and eversion of the foot.

Innervation: Superficial peroneal nerve.

PERONEUS BREVIS [Figure 8.4]

Description: Lying deep to the peroneus longus and closer to the foot, the peroneus brevis muscle is shorter and smaller than the adjacent peroneus longus. The tendon of the peroneus brevis passes behind the lateral malleolus along with the tendon of peroneus longus.

Origin: Distal third of the lateral surface of the fibula.

Insertion: Fifth metatarsal bone.

Action: Plantar flexion and eversion of the foot.

Innervation: Superficial peroneal nerve.

Anterior and Lateral Crural Muscles of the Leg— Identification [Figure 8.4]

Examine the illustrations of the anterior and lateral crural muscles of the leg and color them. Then examine the leg on a model or prepared human cadaver and identify the following crural muscles:

1. **tibialis anterior**
2. **extensor hallicus longus**
3. **extensor digitorum longus**
4. **peroneus longus**
5. **peroneus brevis**
6. **peroneus tertius**

Posterior Crural Muscles of the Leg—Description

The posterior crural muscles produce plantar flexion of the foot and flexion of the toes. They are located posterior to the tibia and fibula and are divided into two groups, **superficial** and **deep**. The superficial muscles from the shape and contour of the calf in the leg.

The **posterior superficial crural muscles** include the **gastrocnemius, soleus,** and **plantaris**.

GASTROCNEMIUS [Figures 8.4 and 8.5]

Description: A large, and most superficial muscle of the posterior crural muscles, the gastrocnemius forms the proximal part of the calf of the leg. The muscle arises by two heads at the inferior margin of the popliteal fossa. The heads converge and form the mass of the gastrocnemius muscle.

Origin: Lateral and medial condyles of the femur.

Insertion: Via the calcaneal tendon (Achilles tendon) into the calcaneus bone of the foot.

Action: Plantar flexion of the foot and flexion of the leg.

Innervation: Tibial nerve.

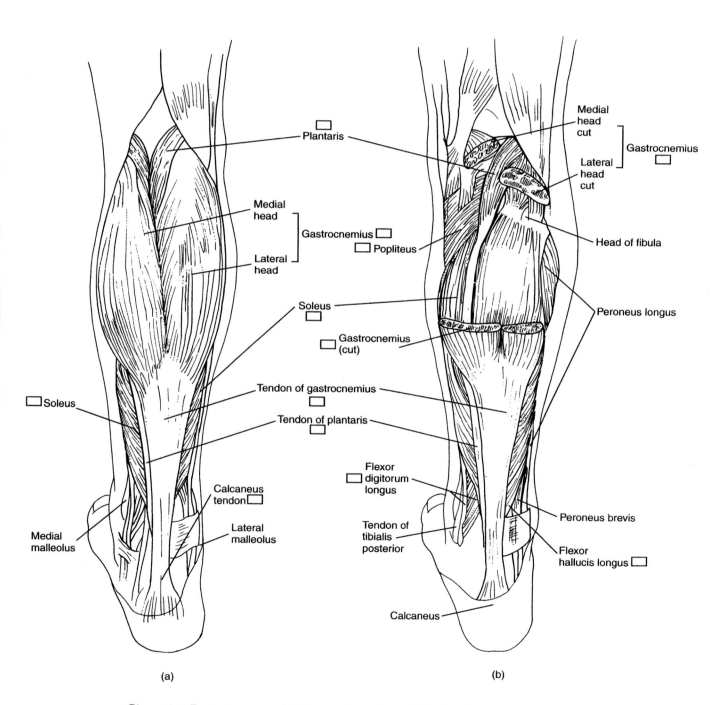

Figure 8.5 Posterior superficial crural muscles of the leg. (a) Superficial layer. (b) Gastrocnemius has been cut to reveal underlying muscles.

SOLEUS [Figure 8.5]

Description: A broad, flat muscle situated deep to the gastrocnemius muscle.
Origin: Head of the fibula and medial border of the tibia and interosseus membrane.
Insertion: Calcaneus bone via the calcaneal tendon, which is shared with the gastrocnemius.
Action: Plantar flexion of the foot.
Innervation: Tibial nerve.

The **posterior deep crural muscles** of the leg are the **popliteus, flexor digitorum longus, flexor hallicus longus,** and **tibialis posterior.** The popliteus muscle moves the knee joint, and the other muscles move the ankle and foot joints. The tendons of these latter muscles pass behind the medial malleolus and deep to the retinaculum in the foot.

POPLITEUS [Figures 8.5 and 8.6]

Description: Thin, triangular in shape, and located in the proximal leg deep to the heads of the gastrocnemius muscle. The popliteus muscle forms the distal floor of the popliteal fossa and its fibers descend medially across the upper leg to the tibial surface.
Origin: Lateral condyle of the femur.
Insertion: Posterior surface of proximal tibia.
Action: Flexion and medial rotation of the leg.
Innervation: Tibial nerve.

FLEXOR DIGITORUM LONGUS [Figures 8.5 and 8.6]

Description: A long, narrow muscle, the most medially located (tibial side of the leg) of the deep muscles. The flexor digitorum longus is also deep to the soleus muscle. Its long tendon passes posterior to the medial malleolus of the tibia and diagonally into the sole of the foot. In the middle of the sole, the tendon expands into four separate tendons that pass to bases of the lateral four toes.
Origin: Posterior surface of the tibia.
Insertion: Distal phalanges of the lateral four toes (toes 2 to 5).
Action: Flexion of the lateral four toes, plantar flexion and inversion of foot.
Innervation: Tibial nerve.

FLEXOR HALLICUS LONGUS [Figures 8.5 and 8.6]

Description: A powerful, and the largest of the deep muscles. The flexor hallicus longus is situated deep to the soleus muscle on the posterolateral, or fibula, side of the leg. Its long tendon passes obliquely downward and medial and then posterior to the medial malleolus. The tendon continues in the sole of the foot toward the great toe, passing deep to the tendon of the flexor digitorum longus.
Origin: Inferior two-thirds of the posterior fibula.
Insertion: Base of distal phalanx of the great toe.
Action: Flexion of the great toe, plantar flexion and eversion of the foot.
Innervation: Tibial nerve.

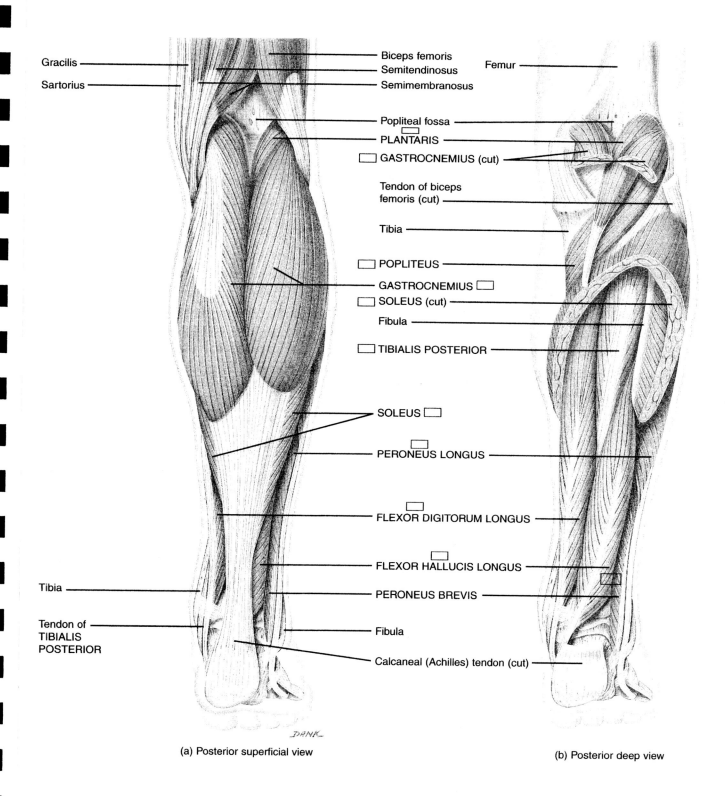

Gracilis

Sartorius

Biceps femoris

Semitendinosus

Semimembranosus

Femur

Popliteal fossa

PLANTARIS

GASTROCNEMIUS (cut)

Tendon of biceps
femoris (cut)

Tibia

POPLITEUS

GASTROCNEMIUS

SOLEUS (cut)

Fibula

TIBIALIS POSTERIOR

SOLEUS

PERONEUS LONGUS

FLEXOR DIGITORUM LONGUS

FLEXOR HALLUCIS LONGUS

Tibia

PERONEUS BREVIS

Tendon of
TIBIALIS
POSTERIOR

Fibula

Calcaneal (Achilles) tendon (cut)

DANK

(a) Posterior superficial view

(b) Posterior deep view

Figure 8.6 Deep posterior crural muscles of the leg. The gastrocnemius and soleus muscles have been cut to reveal the deep muscles.

TIBIALIS POSTERIOR [Figure 8.6]

Description: The deepest muscle of the posterior group, the tibialis posterior is located between the flexor digitorum longus and the flexor hallicus longus muscles in the upper half of the leg. The tendon of the tibialis posterior passes posterior to the medial malleolus to spread out in the sole of the foot.
Origin: Posterior surface of the tibia, interosseous membrane, and medial surface of the fibula.
Insertion: Plantar surfaces of the navicular, cuneiform, cuboidal, and second, third, and fourth metatarsal bones.
Action: Plantar flexion and inversion of the foot.
Insertion: Tibial nerve.

SUPERFICIAL AND DEEP POSTERIOR CRURAL MUSCLES OF THE LEG—IDENTIFICATION [Figures 8.5 and 8.6]

Examine the illustrations of the superficial and deep posterior crural muscles of the leg and color them. Then examine the leg in a model or human cadaver and identify the following muscles:

1. **gastrocnemius**
2. **soleus**
3. **popliteus**
4. **tibialis posterior**
5. **flexor digitorum longus**
6. **flexor hallicus longus**

Muscles of the Foot

The intrinsic muscles of the foot are comparable to those in the hand. While the muscles of the hand are specialized for highly skilled and precise movements, those of the foot are designed for strength, body support, and locomotion.

The muscles of the foot, like those of the hand, are in layers and in a complex arrangement. Like those in the hand, the muscles of the foot are difficult to prepare, preserve, and demonstrate.

CHAPTER SUMMARY AND CHECKLIST

I. MUSCLES OF THE PELVIC GIRDLE

   A. Description
     1. Muscles associated with pelvic girdle
     2. Muscles are massive

II. MUSCLES OF THE THIGH

   A. Description
     1. Muscles are attached to pelvic bones and thigh
     2. Muscles are divided into anterior, posterior, and medial groups
     3. Muscles are involved in moving the thigh

B. **Anterior Thigh Muscles of Iliac Region**
 1. Psoas major
 2. Iliacus

C. **Anterior Thigh Muscles**
 1. Tensor fasciae latae
 2. Sartorius
 3. Quadriceps femoris: four muscles
  a. Rectus femoris
  b. Vastus lateralis
  c. Vastus medialis
  d. Vastus intermedius

D. **Posterior Thigh Muscles of the Gluteal Region**
 1. Gluteus maximus
 2. Gluteus medius
 3. Gluteus minimus

E. **Medial or Adductor Muscles of the Thigh**
 1. Main function is adduction of thigh
 2. Muscles
  a. Gracilis
  b. Pectineus
  c. Adductor longus
  d. Adductor brevis
  e. Adductor magnus

F. **Posterior Muscles of the Thigh**
 1. Called hamstring muscles
 2. Extend between hip and knee
 3. Flex lower leg and extend thigh
 4. Muscles
  a. Biceps femoris
  b. Semitendinosus
  c. Semimembranosus

III. **MUSCLES OF THE LOWER LEG**

A. **Description**
 1. Move the ankle, foot, and toes
 2. Called crural muscles
 3. Crural muscles are anterior, lateral, and posterior

B. **Anterior Crural Muscles of the Leg**
 1. Dorsiflex and invert foot and extend toes
 2. Muscles
  a. Tibialis anterior
  b. Extensor digitorum longus
  c. Extensor hallicus longus
  d. Peroneus tertius

C. **Lateral Crural Muscles of the Leg**
 1. Plantar flex and invert the foot

2. Muscles occupy lateral region and arise from fibula
3. Muscles
   a. Peroneus longus
   b. Peroneus brevis

D. Posterior Crural Muscles of the Leg
1. Plantar flexion of foot and flexion of toes
2. Located posterior to tibia and fibula
3. Divided into superficial and deep groups
4. Posterior superficial crural muscles
   a. Gastrocnemius
   b. Soleus
5. Posterior deep crural muscles
   a. Popliteus
   b. Flexor digitorum longus
   c. Flexor hallicus longus
   d. Tibialis posterior

Laboratory Exercises 8

NAME _____

LAB SECTION _____ DATE _____

LABORATORY EXERCISE 8.1

Part 1

Muscles of the Thigh

1. Which muscles flex the thigh at hip joint?

    a. _____

    b. _____

2. Which muscle extends the leg at the knee joint and flexes the thigh at hip joint?

    _____

3. Which muscle flexes the thigh and leg and rotates the thigh laterally?

    _____

4. Which muscles adduct the thigh?

    a. _____

    b. _____

    c. _____

    d. _____

    e. _____

5. Which muscles abduct the thigh?

    a. _____

    b. _____

6. Which muscles flex the leg at knee joint, rotate the tibia medially, and extend the thigh at hip joint?

    a. _____

    b. _____

    c. _____

7. Which muscles comprise the quadratus femoris?

   a. _____

   b. _____

   c. _____

   d. _____

8. Which muscle crosses the anterior thigh obliquely from the lateral to the medial side?

   _____

9. Which muscle originates from the iliac fossa and iliac crest?

   _____

10. Which posterior muscles of the gluteal region move the thigh?

    a. _____

    b. _____

    c. _____

Muscles of the Lower Leg

1. Which muscles cause dorsiflexion of the foot?

   a. _____

   b. _____

   c. _____

   d. _____

2. Which muscles extend toes?

   a. _____

   b. _____

3. Which muscles cause inversion of the foot?

   a. _____

b. _____

4. Which muscles cause eversion and plantar flexion of the foot?

a. _____

b. _____

5. Which posterior crural muscles insert into the calcaneus bone of the foot?

a. _____

b. _____

6. Which muscles flex toes?

a. _____

b. _____

7. Which posterior muscle in the leg is the deepest of the group?

_____

8. Certain tendons of the posterior leg muscles pass posterior to what part of the tibia?

_____

9. What is the largest and most superficial posterior leg muscle?

_____

10. Which two posterior leg muscles insert jointly via the calcaneal tendon?

a. _____

b. _____

Part II

Using the listed terms, label the muscles on both sides of the thigh and then color them.

Psoas major	Sartorius	Vastus medialis
Tensor fascia latae	Adductor longus	Semitendinosus
Gracilis	Vastus lateralis	Iliacus
Rectus femoris	Gluteus medius	
Gluteus maximus	Semimembranosus	
Biceps femoris (short	Pectineus	
and long heads)	Adductor magnus	

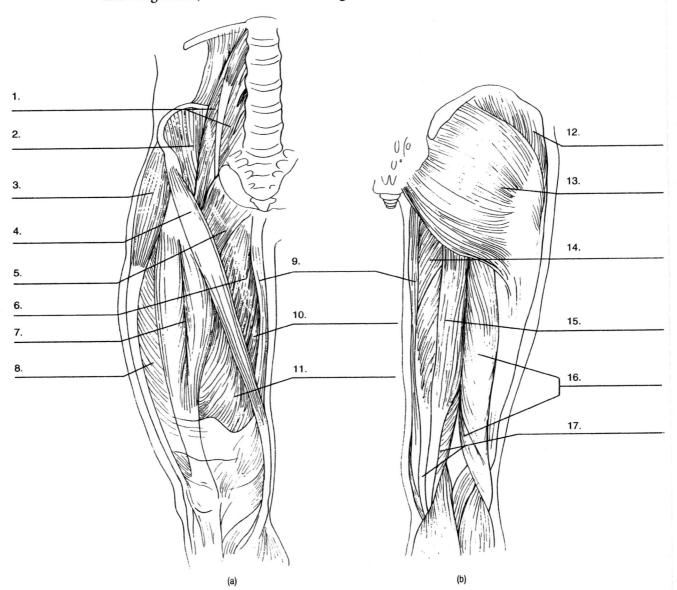

1.
2.
3.
4.
5.
6.
7.
8.
9.
10.
11.
12.
13.
14.
15.
16.
17.

(a)   (b)

Figure 8.7 (a) Superficial muscles of the thigh in the (a) anterior and (b) posterior views.

LABORATORY EXERCISE 8.2

Using the listed terms, label the muscles of the leg and then color them.

Tibialis anterior   Gastrocnemius
Peroneus longus   Extensor digitorum longus
Peroneus brevis   Extensor hallicus longus
Flexor hallicus longus   Soleus

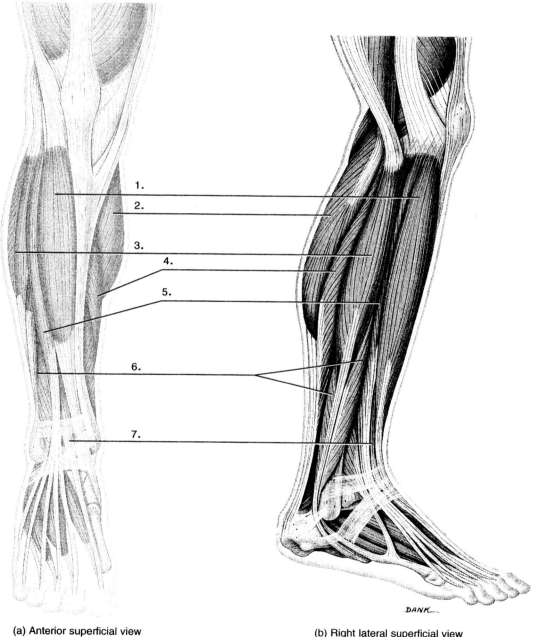

1.
2.
3.
4.
5.
6.
7.

(a) Anterior superficial view   (b) Right lateral superficial view

Figure 8.8 (a) Anterior and (b) lateral crural muscles of the leg.

LABORATORY EXERCISE 8.3

Using the listed terms, label the muscles on the leg and then color them.

Gastrocnemius (medial and lateral heads)   Peroneus brevis
Flexor digitorum longus   Soleus
Peroneus longus   Plantaris
Flexor hallicus longus   Popliteus
Tibialis posterior

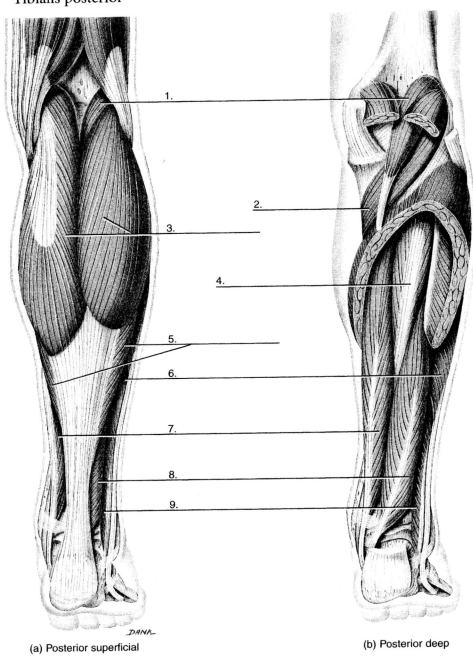

1.
2.
3.
4.
5.
6.
7.
8.
9.

(a) Posterior superficial   (b) Posterior deep

Figure 8.9 (a) Posterior superficial and (b) deep crural muscles of the leg.

9

The Muscles of the Axial Skeleton

MUSCLES OF THE FACE, HEAD, NECK, BACK, AND VERTEBRAL COLUMN

Objective

The Muscles of Facial Expression—Description

    The Frontalis and Occipitalis Muscles

The Muscles of Mastication—Description

Muscles of Facial Expression and Mastication—Identification

Muscles of Upper Back and Neck—Description

    Muscles of Upper Back

    Muscles of Upper Neck

Deep Muscles of the Vertebral Column and Back

    Erector Spinae

Muscles of the Upper Back, Neck, and Vertebral Column—Identification

Chapter Summary and Checklist

Laboratory Exercises

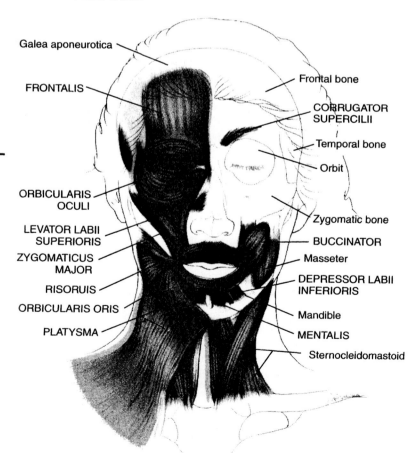

173

Objective

The objective of Chapter 9, "The Muscles of the Axial Skeleton," is for you to become acquainted with the muscles of:

1. **Facial expression**
2. **Mastication**
3. **The upper back and neck**
4. **The vertebral column and back**

The Muscles of Facial Expression—Description

The axial muscles both originate and insert on the bones of the axial skeleton. The muscles of facial expression tend to insert into the skin of the face.

The Frontalis and Occipitalis Muscles (Figures 9.1 and 9.2)

The superior and lateral sides of the skull are covered by a muscular and tendinous layer that contains two thin muscles: the **frontalis** and the **occipitalis**. These muscles are on the forehead and occipital region of the skull, respectively. There are no muscles on the top of the skull, only skin, on top of the broad aponeurosis called the **epicranial aponeurosis**, or **galea aponeurotica**. The frontalis and occipitalis muscles are connected by this cranial connective tissue tendon and use the galea aponeurotica as the region of either origin or insertion. Collectively, the occipitalis muscle, connective tissue galea aponeurotica, and frontalis muscle are called the **epicranius**.

FRONTALIS [Figure 9.1]

> *Origin:* Galea aponeurotica.
> *Insertion:* Skin and dense connective tissue surrounding the eyebrows.
> *Action:* Elevation of the eyebrows, as in surprise, and wrinkling of the forehead, as in worrying.
> *Innervation:* Facial nerve (cranial nerve VII).

OCCIPITALIS [Figure 9.2]

> *Origin:* Occipital bone of the skull.
> *Insertion:* Galea aponeurotica.
> *Action:* Pulls scalp posteriorly.
> *Innervation:* Facial nerve (cranial nerve VII).

ORBICULARIS OCULI [Figures 9.1 and 9.2]

> *Description:* Encircles the orbit of the eye.
> *Origin:* Maxillary and frontal bones of the skull and ligaments around the eye.
> *Insertion:* Skin around the eyelids.
> *Action:* Closing the eye, winking, blinking, squinting, and moving the eyebrows.
> *Innervation:* Facial nerve (cranial nerve VII).

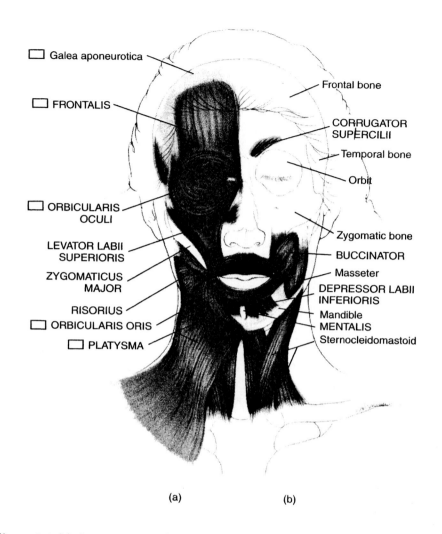

□ Galea aponeurotica

□ FRONTALIS

□ ORBICULARIS OCULI

LEVATOR LABII SUPERIORIS

ZYGOMATICUS MAJOR

RISORIUS

□ ORBICULARIS ORIS

□ PLATYSMA

Frontal bone

CORRUGATOR SUPERCILII

Temporal bone

Orbit

Zygomatic bone

BUCCINATOR

Masseter

DEPRESSOR LABII INFERIORIS

Mandible
MENTALIS

Sternocleidomastoid

(a)   (b)

Figure 9.1 (a) Anterior superficial and (b) anterior deep views of the muscles of facial expression.

ORBICULARIS ORIS [Figures 9.1 and 9.2]

Description: Surrounds the mouth.
Origin: Muscle fibers of facial muscles surrounding the mouth.
Insertion: Skin at base of lips and corners of mouth.
Action: Closes lips and mouth, puckers lips, protrudes lips for speech and kissing.
Innervation: Facial nerve (cranial nerve VII).

BUCCINATOR [Figures 9.1 and 9.2]

Description: A deep and principal muscle of the cheek. The buccinator forms the lateral wall of the oral cavity and lies lateral to the teeth.
Origin: Maxillary bone and mandible.
Insertion: Fibers of the orbicularis oris.
Action: Compresses cheeks and keeps food between the teeth during chewing by not allowing it to accumulate in the cheek area.
Innervation: Facial nerve (cranial nerve VII).

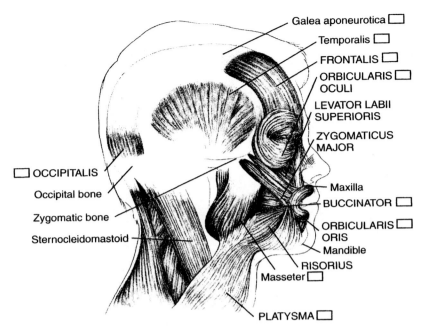

Figure 9.2 Muscles of facial expression. Right lateral superficial view.

PLATYSMA [Figures 9.1 and 9.2]

Description: A thin, sheetlike superficial neck muscle that extends from the upper chest over the neck to the face.
Origin: Connective tissue (fascia) of pectoralis major and deltoid muscles.
Insertion: Lower border of mandible and the skin and muscles of the mouth.
Action: Depresses the mandible, draws corners of the mouth downward and backward, tightens the skin of the neck.
Innervation: Facial nerve (cranial nerve VII).

The Muscles of Mastication—Description

The muscles of mastication move the mandible during biting and chewing. The major muscles of mastication are the **masseter** and **temporalis**.

MASSETER [Figures 9.1 and 9.2]

Description: The most lateral, superficial muscle of mastication, the masseter is located on the ramus of the mandible.
Origin: Zygomatic process of the maxillary bone and inferior border of the zygomatic arch.
Insertion: Angle and ramus of the mandible.
Action: Closes the jaw by elevating the mandible.
Innervation: Mandibular division of trigeminal nerve (cranial nerve V).

TEMPORALIS [Figures 9.1 and 9.2]

Description: The largest muscle of mastication, located on the lateral side of the head over the temporal bone of the skull. The temporalis passes deep to the zygomatic arch to insert into the mandible. Contraction of this muscle can be felt on the temple of the head when the teeth are clenched.

Origin: Temporal fossa of the skull.

Insertion: Coronoid process of the mandible.

Action: Elevates the jaw and closes the mouth.

Innervation: Mandibular division of trigeminal nerve (cranial nerve V).

Muscles of Facial Expression and Mastication—Identification *[Figures 9.1 and 9.2]*

Examine the illustrations and color the muscles of facial expression and mastication. Then examine the face and head of a model or prepared human cadaver and identify as many of the following muscles as you can:

1. frontalis
2. occipitalis
3. orbicularis oculi
4. orbicularis oris
5. buccinator
6. platysma
7. masseter
8. temporalis

Muscles of the Upper Back and Neck—Description

The head is moved by muscles located in the neck and upper back regions. These muscles include those that attach to the skull or thorax and those that originate or insert in the neck. Some back muscles are connected to the **ligamentum nuchae,** a strong, elastic connective tissue ligament located in the posterior neck region. This ligament extends from the occipital bone of the skull over the spinous processes of the cervical vertebrae.

The upper back and neck muscles include the **splenius capitis, semispinalis capitis,** and **longissimus capitis,** which extend the head and neck. The term **capitis** appears in the name of the muscle if the muscle reaches and inserts into the skull.

Muscles of the Upper Back

SPLENIUS CAPITIS [Figures 9.3 and 9.4]

Description: In back of the neck deep to the trapezius and sternocleidomastoid muscles. The splenius capitis is a wide muscle that connects the base of the skull to the vertebrae in the neck and upper thorax.

Origin: Ligamentum nuchae and spinous processes of the seventh cervical vertebra and the first three or four thoracic vertebrae.

Insertion: Occipital bone of the skull under the sternocleidomastoid muscle and into the mastoid process of the temporal bone.

Action: Extends and rotates the head and neck.

Innervation: Cervical nerves.

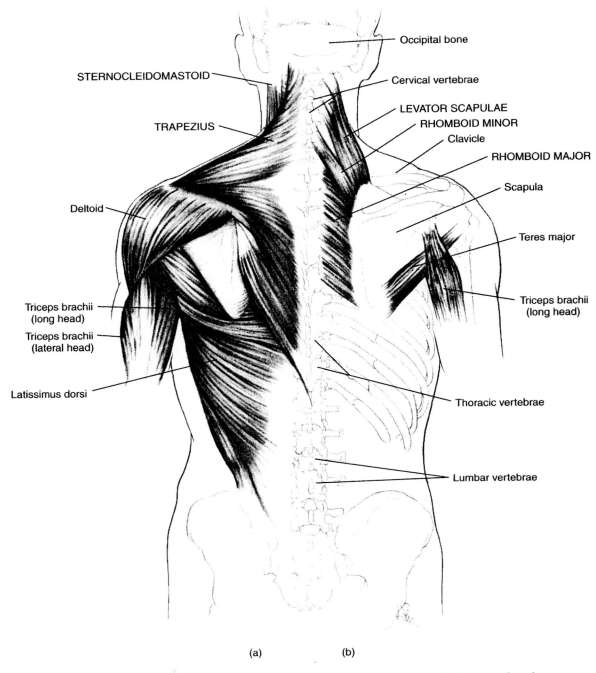

Occipital bone

STERNOCLEIDOMASTOID

Cervical vertebrae

LEVATOR SCAPULAE
RHOMBOID MINOR

TRAPEZIUS

Clavicle

RHOMBOID MAJOR

Scapula

Deltoid

Teres major

Triceps brachii
(long head)

Triceps brachii
(long head)

Triceps brachii
(lateral head)

Latissimus dorsi

Thoracic vertebrae

Lumbar vertebrae

(a)   (b)

Figure 9.3 (a) Posterior superficial and (b) posterior deep views of the muscles that move the pectoral (shoulder) girdle.

SEMISPINALIS CAPITIS [Figure 9.4]

Description: A wide, sheetlike muscle located deep to the splenius capitis muscle. The semispinalis capitis extends superiorly from the cervical and thoracic vertebrae toward the skull.

Origin: Transverse processes of the first six or seven thoracic vertebrae and the seventh cervical vertebra.

Insertion: Occipital bone of the skull.

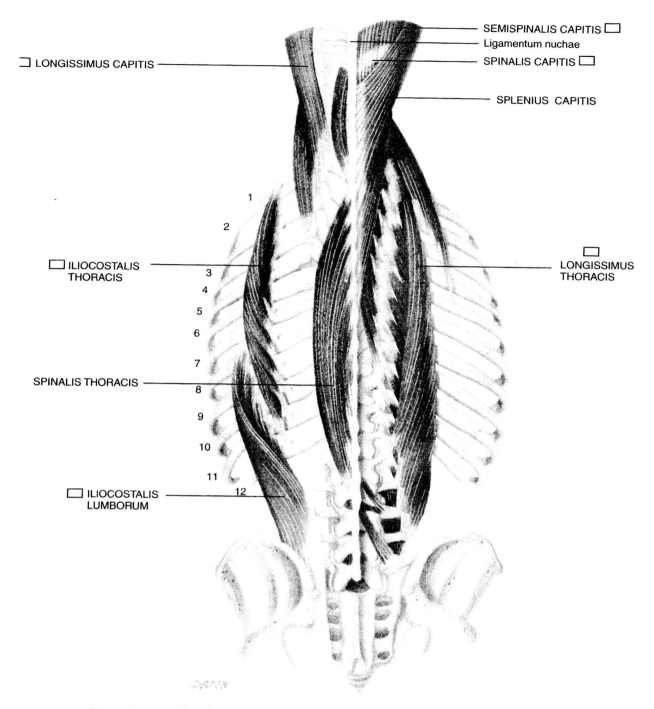

LONGISSIMUS CAPITIS ☐ ——————

SEMISPINALIS CAPITIS ☐
Ligamentum nuchae
SPINALIS CAPITIS ☐

SPLENIUS CAPITIS

1
2

☐ ILIOCOSTALIS
THORACIS

3
4
5
6
7

☐
LONGISSIMUS
THORACIS

SPINALIS THORACIS ——————
8
9
10
11

☐ ILIOCOSTALIS
LUMBORUM
12

Figure 9.4 Muscles that move the vertebral column. Posterior view of deep muscles of the neck and back illustrates the different groups of erector spinae muscles.

Action: Extends the head and rotates it to the opposite side.
Innervation: Cervical nerves.

LONGISSIMUS CAPITIS [Figure 9.4]

Description: A narrow strip of muscle that ascends from the cervical and thoracic vertebrae toward the temporal bone of the skull, deep to the splenius capitis and sternocleidomastoid.

Origin: Transverse processes of upper four or five thoracic vertebrae.
Insertion: Posterior region of the mastoid process of the temporal bone.
Action: Extends the head and rotates it laterally.
Innervation: Cervical nerves.

Muscles of the Upper Neck [Figures 9.5 and 9.6]

Two groups of muscles are associated with the hyoid bone (see Chapter 2). The **suprahyoid** muscles are situated superior and the **infrahyoid** muscles inferior to the hyoid bone. The suprahyoid muscles are the **digastric, mylohyoid, stylohyoid,** and **hyoglossus.** The infrahyoid muscles are named on the basis of their origin and insertion and include the **sternohyoid, sternothyroid, thyrohyoid,** and **omohyoid.**

THE SUPRAHYOID GROUP

The suprahyoid muscles assist in swallowing (deglutition) by elevating the hyoid bone, the floor of the mouth, and the tongue. The infrahyoid muscles fix and/or depress the hyoid bone during mastication, swallowing, and speech.

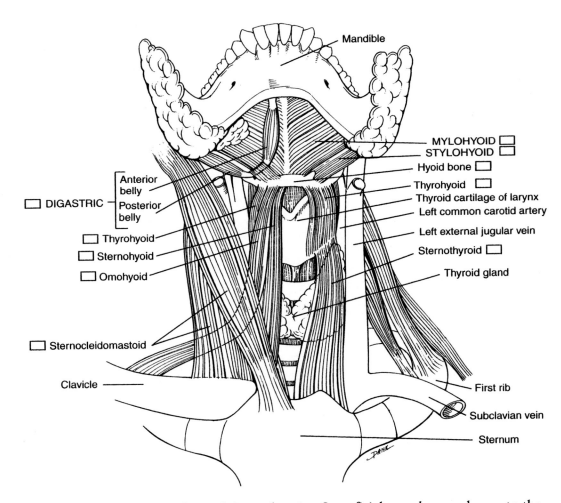

Figure 9.5 Muscles of the floor of the oral cavity. Superficial muscles are shown to the left; deep muscles are shown to the right.

DIGASTRIC [Figure 9.5]

Description: The most superficial muscle of the upper neck and floor of the mouth. The digastric consists of two parts, a posterior belly and an anterior belly, connected by a tendon.
Origin: Lower portion of inner mandible (anterior belly) and mastoid process of the temporal bone (posterior belly).
Insertion: Hyoid bone via a tendon intermediate to both bellies.
Action: Elevates the hyoid bone and opens the mouth.
Innervation: Trigeminal nerve (anterior belly) and facial nerve (posterior belly).

MYLOHYOID [Figure 9.5]

Description: Located superior to the anterior belly of the digastric muscle, the mylohyoid forms the floor of the oral cavity.
Origin: Entire length of the inner surface of the mandible.
Insertion: Hyoid bone.
Action: Raises the hyoid bone and the floor of the mouth during swallowing.
Innervation: Mandibular division of trigeminal nerve (cranial nerve V).

STYLOHYOID [Figure 9.5]

Description: The stylohyoid lies anterior, superior, and parallel to the posterior belly of the digastric.
Origin: Styloid process of the temporal bone.
Insertion: Body of the hyoid bone.
Action: Elevates and retracts the hyoid bone and elongates the floor of the mouth.
Innervation: Facial nerve (cranial nerve VII).

THE INFRAHYOID GROUP

STERNOHYOID [Figures 9.5 and 9.6]

Description: A thin, most medially located muscle in the neck.

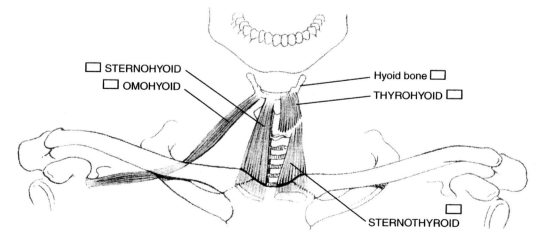

Figure 9.6 Neck muscles of the infrahyoid group.

Origin: Posterior region of the manubrium of the sternum and medial end of the clavicle.
Insertion: Inferior border of the hyoid bone.
Action: Draws the hyoid bone inferiorly.
Innervation: Cervical spinal nerves.

STERNOTHYROID [Figures 9.5 and 9.6]

Description: Located deep to the sternohyoid.
Origin: Posterior surface of the manubrium of the sternum and the first costal cartilage.
Insertion: Thyroid cartilage.
Action: Draws the thyroid cartilage and larynx inferiorly.
Innervation: Cervical nerves.

THYROHYOID [Figures 9.5 and 9.6]

Description: A small, short muscle that appears as a superior continuation of the sternothyroid muscle. The thyrohyoid lies deep to the sternohyoid muscle.
Origin: Thyroid cartilage.
Insertion: Inferior border of the hyoid bone.
Action: Depresses the hyoid bone and elevates the thyroid cartilage.
Innervation: Cervical nerves.

OMOHYOID [Figures 9.5 and 9.6]

Description: Long and thin, the omohyoid muscle consists of two parts or bellies separated by an intermediate tendon. It is located lateral to the sternohyoid muscle.
Origin: Cranial border of the scapula.
Insertion: Hyoid bone.
Action: Depresses the hyoid bone.
Innervation: Cervical nerves.

STERNOCLEIDOMASTOID [Figure 9.5]

Description: A long muscle that passes at an oblique angle across the lateral and anterior side of the neck. The sternocleidomastoid extends superiorly from the thorax to the base of the skull. When contracted this muscle forms a prominent surface landmark on the lateral surface of the neck.
Origin: Anterior surface of the sternum (manubrium) and upper surface of the clavicle.
Insertion: Mastoid process of the temporal bone.
Action: When the right or left muscle contracts individually, the head flexes toward the opposite side. When both muscles contract simultaneously, the neck and head flex ventrally toward the chest.
Innervation: Spinal accessory nerve (cranial nerve XI).

Deep Muscles of the Vertebral Column and Back

Erector Spinae [Figure 9.5]

The erector spinae are large muscles of the back that consist of three vertically oriented columns. The principal origin sites are the sacrum, iliac crest, and spines of the lumbar and lower thoracic vertebrae. As erector spinae extend superiorly, they split into lateral, intermediate, and medial columns of parallel muscles. Each muscle group extends over several vertebrae.

The lateral column represents the **iliocostalis** muscles, which insert onto the ribs. The **iliocostalis lumborum** is the largest muscle of this column. The intermediate column represents the **longissimus** muscles, which insert onto the ribs and the transverse processes of the vertebrae. It forms the bulk of the erector spinae muscles. The medial column represents the **spinalis** muscles, which originate from the spines of the vertebrae and insert onto the vertebrae spines higher in the vertebral column.

All muscles (erector spinae) extend the trunk, maintain erect posture, and oppose forces of gravity when the vertebral column is flexed. Lateral flexing is due to contractions of the iliocostalis muscles on one side only. During walking, vertical muscle groups contract on alternate sides of the body, which prevents the trunk from falling to the unsupported side.

Muscles of the Upper Back, Neck, and Vertebral Column—Identification *[Figures 9.5 and 9.6]*

Examine the illustrations of the muscles in the upper back and upper neck and color them. Then examine the back and neck of a model or prepared human cadaver and identify the following muscles:

1. **splenius capitis**
2. **semispinalis capitis**
3. **erector spinae (iliocostalis, longissimus, spinalis)**
4. **spinalis capitis**
5. **longissimus capitis**
6. **digastric**
7. **mylohyoid**
8. **stylohyoid**
9. **sternohyoid**
10. **sternothyroid**
11. **thyrohyoid**
12. **omohyoid**
13. **sternocleidomastoid**

CHAPTER SUMMARY AND CHECKLIST

I. MUSCLES OF FACIAL EXPRESSION

A. Description
1. Origin and insertion on skull bones or fascia
2. Delicate strands of muscle below skin
3. Muscles shape skin of face
4. Muscles interconnected around eyes, mouth, and lips

B. The Muscles
1. Frontalis
2. Occipitalis
3. Orbicularis oculi
4. Orbicularis oris
5. Buccinator
6. Platysma

II. MUSCLES OF MASTICATION

A. Description
1. Muscles move mandible during chewing and biting

B. Muscles
1. Masseter
2. Temporalis

III. MUSCLES OF THE UPPER BACK

A. Description
1. Back muscles attach to skull or thorax
2. Some back muscles connect to ligamentum nuchae

B. Muscles
1. Splenius capitis
2. Semispinalis capitis
3. Longissimus capitis

IV. MUSCLES OF UPPER NECK

A. Description
1. Neck muscles located within neck region
2. Neck muscles associated with larynx or hyoid bone
3. Suprahyoid muscles situated superior to hyoid bone
4. Infrahyoid muscles situated inferior to hyoid bone

B. Suprahyoid Group
1. Digastric
2. Mylohyoid
3. Stylohyoid

C. Infrahyoid Group
1. Sternohyoid
2. Sternothyroid
3. Thyrohyoid
4. Omohyoid

D. Sternocleidomastoid

V. DEEP MUSCLES OF VERTEBRAL COLUMN AND BACK

A. Erector Spinae
1. Muscles are large and powerful
2. Consist of three vertical columns
3. Originate in lower back region and ascend

Laboratory Exercises 9

NAME _____

LAB SECTION _____ DATE _____

LABORATORY EXERCISE 9.1

Part I

Muscles of Facial Expression, Mastication, and the Upper Back

1. Which muscles originate from the connective tissue of skull bones?

    a. _____

    b. _____

2. Which muscle causes blinking and squinting of the eyes?

    _____

3. Which muscle causes movement of the lips and mouth?

    _____

4. Which muscle lines the cheeks?

    _____

5. Which muscles are used for mastication?

    a. _____

    b. _____

6. Which muscles of the upper back insert into skull bones (occipital or temporal)?

    a. _____

    b. _____

    c. _____

Muscles of the Upper Neck

1. Which muscles insert into the hyoid bone?

    a. _____

    b. _____

    c. _____

    d. _____

e. _____

f. _____

2. Which muscle in the upper neck originates on the scapula?

_____

3. Which muscles elevate and retract the hyoid bone?

a. _____

b. _____

4. Which muscle originates from the inner surface of the mandible?

_____

5. Which upper neck muscle originates in part from the sternum?

_____

6. Which muscle draws the thyroid cartilage and larynx inferiorly?

_____

7. Which muscle originates on the thyroid cartilage?

_____

8. Which two muscles, when contracting simultaneously, flex the head and neck ventrally?

_____

9. Which upper neck muscles depress the hyoid bone?

a. _____

b. _____

c. _____

10. What is the origin of the stylohyoid muscle?

_____

Part II

Using the listed terms, label the muscles and then color them.

Galea aponeurotica   Platysma
Orbicularis oris   Occipitalis
Masseter   Buccinator
Frontalis   Temporalis
Orbicularis oculi

1.

2.

3.

4.

5.

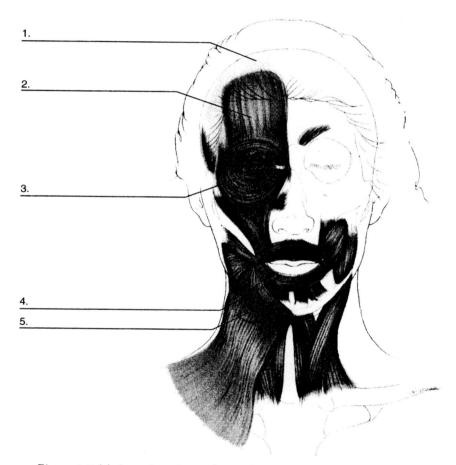

Figure 9.7 (a) Anterior view of muscles of facial expression and mastication.

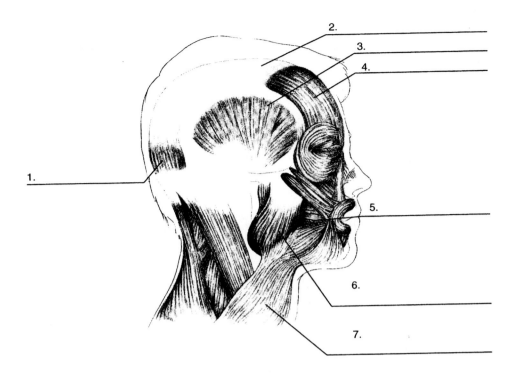

Figure 9.7 (b) Lateral view of muscles of facial expression and mastication.

LABORATORY EXERCISE 9.2

Using the listed terms, label the muscles and then color them.

Digastric
Sternohyoid
Sternocleidomastoid
Mylohyoid

Omohyoid
Thyrohyoid
Stylohyoid
Sternothyroid

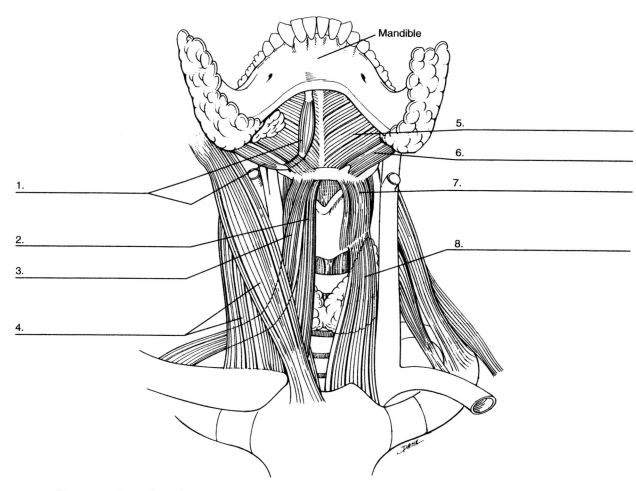

Figure 9.8 Muscles of the upper neck.

10

The Muscles of the Axial Skeleton

MUSCLES OF THE THORAX, ABDOMEN, AND PELVIC FLOOR

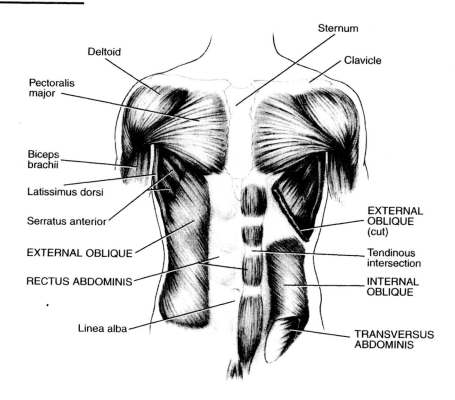

Objective

The objective of Chapter 10, "The Muscles of the Axial Skeleton," is to familiarize you with the muscles of:

1. **Respiration**
2. **Thoracic cage**
3. **Abdomen**
4. **Pelvic floor and perineum**

The Muscles of Respiration—Description

The muscles of normal, quiet respiration are attached to the rib cage and the ribs. Their primary function is to produce rhythmic expansion and contraction of the thorax. The muscles that attach to the thorax include the **intercostal muscles** and the **diaphragm.** Their contractions cause the lateral and vertical dimensions of the thorax to increase.

EXTERNAL INTERCOSTAL MUSCLES [Figure 10.1]

Description: Eleven external intercostal muscles lie between the ribs. Their fibers run at an oblique angle inferiorly and medially toward the anterior side of the thorax. Most external intercostal muscles extend from the verte-

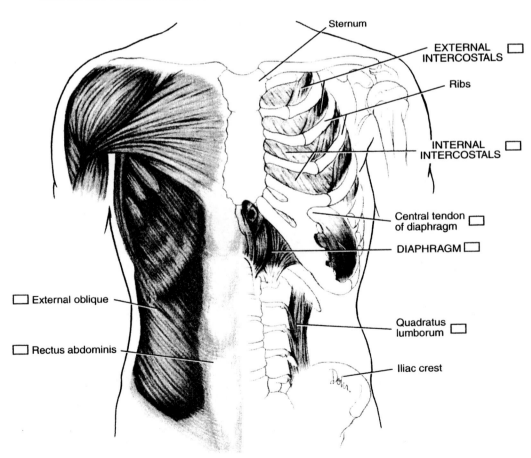

Figure 10.1 Muscles of respiration (right side). The left side shows superficial muscles of the thorax and abdomen.

brae to the costochondral (bone-cartilage) junction of the ribs. Anterior to this junction, a connective tissue fascia replaces the muscles.

Origin: Inferior border of the rib above.

Insertion: Superior border of the rib below.

Action: Elevate the ribs and expand the thorax laterally during inspiration.

Innervation: Intercostal nerves.

INTERNAL INTERCOSTAL MUSCLES [Figures 10.1 and 10.2]

Description: Eleven internal intercostal muscles lie on each side of the thorax deep to the external intercostal muscles. Their fibers run at an oblique angle at right angles to the external intercostal muscles. The internal intercostal muscles start at the sternum and continue to the angles of the ribs. Posterior to the angle of the rib the muscles are replaced by a connective tissue fascia.

Origin: Superior border of the rib and the corresponding costal cartilage of the lower rib.

Insertion: Inferior border of the upper rib.

Action: Respiratory movement of the thorax.

Innervation: Costal nerves.

TRANSVERSUS THORACIS [Figure 10.2]

Description: The transversus thoracis is located on the inner surface of the ventral thorax.

Origin: Lower third of the inner surface of the body of the sternum, from dorsal surface of the xiphoid process, and from sternal ends of costal cartilage of the last three or four true ribs.

Insertion: Into the inner surface of the costal cartilages of the second through sixth ribs.

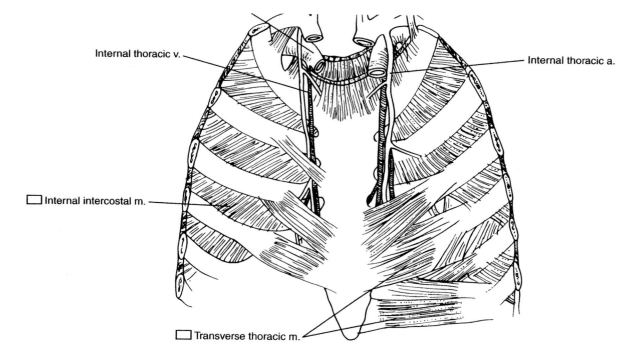

Internal thoracic v.

Internal thoracic a.

Internal intercostal m.

Transverse thoracic m.

Figure 10.2 Anterior thoracic wall and the internal muscles.

Action: Depresses the ribs.
Innervation: Costal nerves.

THE DIAPHRAGM [Figure 10.3]

Description: A dome-shaped musculofibrous sheet that separates the thoracic and abdominal cavities. The peripheral muscles of the diaphragm originate around the lower margin of the bony thorax (thoracic cage) and pass radially toward the **central tendon**, a thin but strong connective tissue aponeurosis situated caudal to the pericardium of the heart.
Origin: Dorsal xiphoid process, the inner surface of lower six ribs of the thorax, and the right and left tendinous **crura** (**crus**, singular) that attach along the lateral bodies of the upper lumbar vertebrae and the twelfth rib.
Insertion: Central tendon.
Action: The muscle of inspiration, the diaphragm increases the thoracic cavity, volume, and abdominal pressure.
Innervation: Phrenic nerves.

The Muscles of Respiration—Identification

[Figures 10.1–10.3]

Examine and color the illustrations of the thorax and muscles of respiration. Then examine the musculature of the external and internal thorax of a model or prepared human cadaver and identify the following:

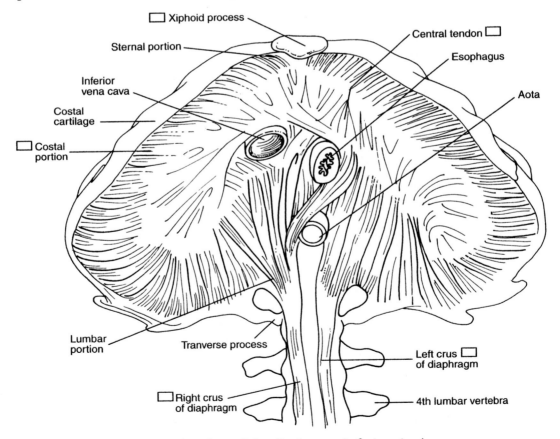

Figure 10.3 The abdominal surface of the diaphragm (inferior view).

1. external intercostal muscle
2. internal intercostal muscle
3. transversus thoracis muscle
4. diaphragm
5. central tendon of the diaphragm

The Muscles of the Abdominal Wall— Description

In humans, the abdominal wall is almost entirely muscular (no ribs). It consists of anterior muscles (**rectus abdominis**) and three lateral muscles (**external oblique**, **internal oblique**, and **transversus abdominis**). These muscles are broad, flat, and constructed in layers. The **quadratus lumborum** muscle forms part of the posterior abdominal wall.

Each layer of fibers in abdominal muscles bears a different orientation. The thin aponeuroses of the three lateral abdominal muscles merge to form the **rectus sheath**. This sheath encloses each rectus abdominis muscle and then fuses in the midline to form the **linea alba**, a band of connective tissue that extends vertically from the xiphoid process of the sternum to the symphysis pubis of the pelvic bone.

The Anterolateral Group of Abdominal Muscles

RECTUS ABDOMINIS [Figures 10.1 and 10.4]

Description: The rectus abdominis muscles are long, narrow bands of paired muscle strips that extend vertically the length of the anterior abdominal wall on either side of the midline. The **linea alba** separates these muscles in the midline. Each rectus abdominis muscle is crossed by three connective tissue bands called **tendinous intersections** (insertions or inscriptions). The first intersection is located immediately below the xiphoid process, the second halfway to the umbilicus, and the third at the level of the umbilicus.
Origin: Pubic crest and symphysis pubis of the pelvis.
Insertion: Cartilages of the fifth through seventh ribs.
Action: Compresses the abdominal contents and flexes the vertebral column.
Innervation: Intercostal nerves.

EXTERNAL OBLIQUE [Figures 10.1 and 10.4]

Description: The most superficial abdominal wall muscle, the external oblique is located in the lateral and anterior regions of the abdomen. Its fibers, like those of the external intercostal muscles of the thorax, run inferiorly and medially over the anterior abdominal wall. The inferior, thickened aponeurosis of the external oblique muscle forms the **inguinal ligament** that extends from the anterior superior iliac spine to the pubic tubercle.
Origin: The outer surfaces of the lower eight ribs.
Insertion: Linea alba, pubic tubercle, and iliac crest of the pelvic bone.
Action: Flexes the vertebral column when both sides contract and compresses the abdominal contents.
Innervation: Intercostal nerves and iliohypogastric nerves.

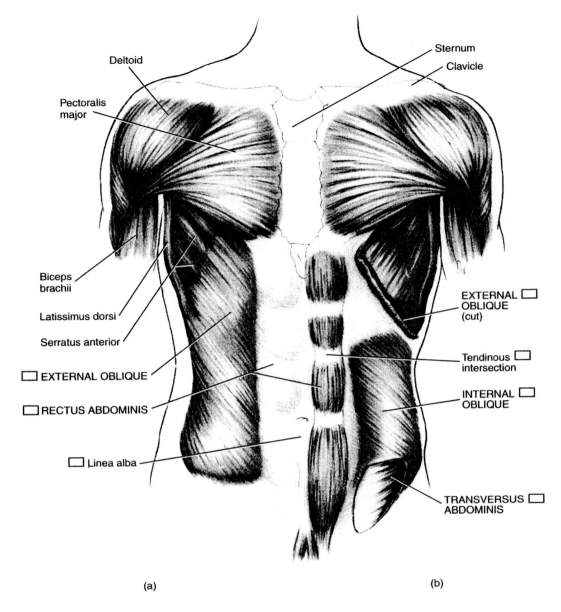

Deltoid

Sternum

Clavicle

Pectoralis
major

Biceps
brachii

Latissimus dorsi

Serratus anterior

☐ EXTERNAL OBLIQUE

☐ RECTUS ABDOMINIS

☐ Linea alba

EXTERNAL ☐
OBLIQUE
(cut)

Tendinous ☐
intersection

INTERNAL ☐
OBLIQUE

TRANSVERSUS ☐
ABDOMINIS

(a)

(b)

Figure 10.4 (a) Superficial and (b) deep views of the muscles of the anterior abdominal wall.

INTERNAL OBLIQUE [Figure 10.4]

Description: Internal to the external oblique muscle, the fibers of the internal oblique muscle pass anteriorly and superiorly at right angles to those of the external oblique muscle. In contrast to the external oblique muscle, the aponeurosis of the internal oblique muscle splits into two layers that enclose the rectus abdominis muscle.

Origin: Inguinal ligament, iliac crest, and thoracolumbar fascia.

Insertion: Costal cartilages of the lower three or four ribs, linea alba, and pubic crest.

Action: Similar to that of the external oblique.

Innervation: Intercostal nerves and iliohypogastric nerves.

TRANSVERSUS ABDOMINIS [Figure 10.4]

Description: The deepest anterolateral abdominal wall muscle, the fibers of the transversus abdominis muscle pass horizontally or transversely across the abdominal wall.
Origin: Inguinal ligament, iliac crest, and cartilages of the last six ribs.
Insertion: Linea alba and pubic crest.
Action: Compresses the abdominal contents.
Innervation: Intercostal, iliohypogastric, and ilioinguinal nerves.

The Posterior Group of Abdominal Muscles

QUADRATUS LUMBORUM [Figure 10.1]

Description: A thick, quadrilateral muscle in the posterior abdominal wall. The quadratus lumborum is situated adjacent to the transverse processes of the lumbar vertebrae and extends from the twelfth rib and lumbar transverse processes to the iliac crest.
Origin: Iliac crest and transverse processes of the lower lumbar vertebrae.
Insertion: Anterior surface of the twelfth rib and transverse processes of upper four lumbar vertebrae.
Action: Stabilizes the twelfth rib and flexes the trunk laterally.

The Muscles of the Abdominal Wall—
Identification *[Figures 10.1 and 10.4]*

Examine and color the illustrations of the abdominal musculature. Then examine the anterior and posterior regions of the abdominal wall of a model or prepared human cadaver and identify the following:

1. **rectus abdominis**
2. **tendinous intersections**
3. **linea alba**
4. **rectus sheath**
5. **external oblique muscle**
6. **inguinal ligament**
7. **internal oblique muscle**
8. **transversus abdominis**
9. **quadratus lumborum**

Muscles of the Pelvic Floor—
Description *[Figures 10.5–10.7]*

The pelvic floor is formed by two muscle sheets—the **pelvic diaphragm** and **urogenital diaphragm**—and their connective tissue.

The pelvic diaphragm is funnel shaped and located superior in the pelvis (Figure 10-7a). It forms the floor of the pelvic cavity and consists of the **levator ani** anteriorly and the **coccygeus** muscles posteriorly. The muscles of the pelvic diaphragm are pierced by the anal canal and urethra in both males and females and by the vagina in females.

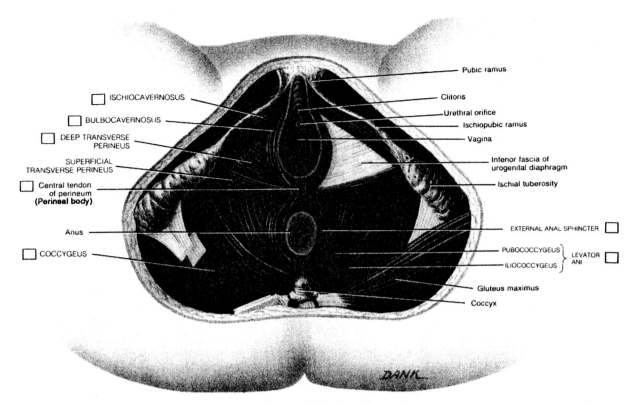

Figure 10.5 Muscles in the pelvic floor of the female perineum.

The urogenital diaphragm (Figure 10.7b) is located more superficially than the pelvic diaphragm, between the two ischiopubic rami. Note the thin **deep transverse perinei** muscles and **spincter urethrae** muscle that surrounds the urethra in both genders (Figure 10.7b).

LEVATOR ANI [Figures 10.5–10.7]

Description: The levator ani forms the major portion of the pelvic floor. It is a thin muscle sheet that attaches to the inner surface of the true pelvis.
Origin: Anteriorly from the inner surface of the pubic ramus and posteriorly from the spine of the ischium bone.
Insertion: Inner surface of the coccyx, external sphincter, and perineal body.
Action: Supports the pelvic floor; produces sphincter action on the anal canal and vagina and raises the pelvic floor during increased intraabdominal pressure.
Innervation: Pudendal nerves of the sacral plexus.

COCCYGEUS [Figures 10.5–10.7]

Description: A small triangular muscle located posterior and superior to the levator ani muscle, the coccygeus forms the posterior portion of the pelvic floor.
Origin: Spine of the ischium.
Insertion: Sacrum and coccyx bones.
Action: Supports and draws the coccyx ventrally after defecation.
Innervation: Sacral spinal nerves.

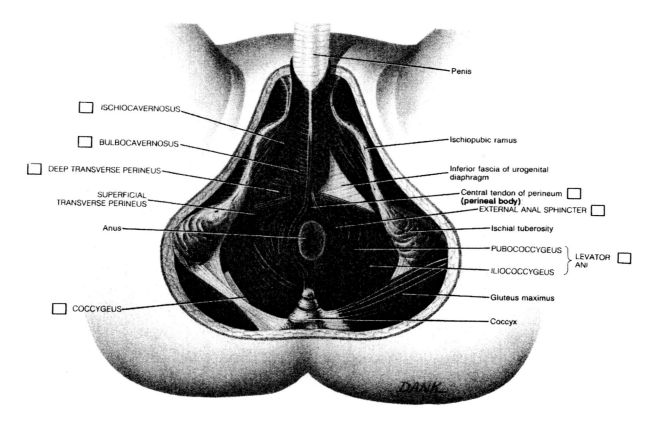

Figure 10.6 Muscles of the male perineum.

Muscles of the Perineum—Description

The perineum lies over the pelvic floor. It is the diamond-shaped region located beneath the pelvic diaphragm (the levator ani and coccygeus muscles) between the thighs and buttocks.

The perineum is bounded anteriorly by the pubic arch, laterally by the ischiopubic rami and ischial tuberosities, and posteriorly by the coccyx. The **central tendon** of the perineum, called the **perineal body**, attaches numerous perineal muscles.

The perineal muscles are located inferior to the pelvic diaphragm and are associated with the external genital organs and deep muscles of the urogenital diaphragm. The superficial perineal muscles include the **ischiocavernosus, bulbospongiosus (bulbocavernosus),** and **superficial transverse perinei** muscles. The deep perineal muscles are the **deep transverse perinei** muscles.

The anal region of the perineum contains the anal orifice and **external anal sphincter.**

ISCHIOCAVERNOSUS [Figures 10.5 and 10.6]

Description: An elongated, paired muscle located laterally in the perineum. In the male, each ischiocavernosus muscle covers one of the crura of the penis. In the female, the ischiocavernosus muscle is smaller, also located laterally, and covers the crura of the clitoris.

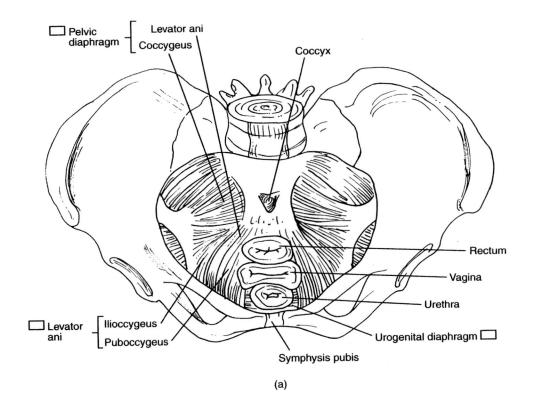

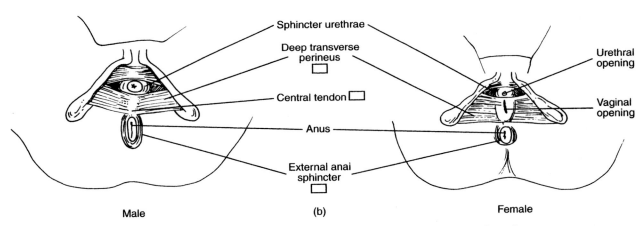

Figure 10.7 (a) Muscles of the pelvic floor viewed from above in the female. (b) Muscles of the urogenital diaphragm (sphincter urethrae and deep transverse perinei) of the perineum are present in both genders.

Origin: Ischial tuberosity and pubic rami.
Insertion: In the male, the crura of the corpora cavernosa of the penis. In the female, the sides and inferior surface of the crura of the clitoris.
Action: Compression of the crura of the clitoris or penis decreases venous blood return and thereby maintains an erection during sexual arousal.
Innervation: Pudendal nerve.

BULBOSPONGIOSUS (BULBOCAVERNOSUS)
[Figures 10.5 and 10.6]

Description: Surrounds the bulb of the penis and the adjacent corpus spongiosum. In the female, the bulbospongiosus (bulbocavernosus) muscle surrounds the opening of the vaginal canal and covers the vestibular bulb.
Origin: Perineal body of the perineum.
Insertion: In the male, the corpus spongiosum of the penis. In the female, the corpora cavernosa of the clitoris.
Action: In males, assists in emptying the urethra after micturition, propels semen during ejaculation, and assists in maintaining erection of the penis. In females, constricts the vaginal orifice and assists in maintaining erection of the clitoris.
Innervation: Pudendal nerve.

SUPERFICIAL TRANSVERSE PERINEI [Figures 10.5 and 10.6]

Description: Narrow slips of muscle that run transversely across the perineum anterior to the anus.
Origin: Ischial tuberosity.
Insertion: Perineal body.
Action: Stabilizes the perineal body.
Innervation: Pudendal nerve.

DEEP TRANSVERSE PERINEI [Figures 10.6 and 10.7]

Description: Those from one side of the body interlace with those from the other side.
Origin: Inferior rami of the ischium bone.
Insertion: Perineal body.
Action: Support the perineal body and perineal region.
Innervation: Pudendal nerve.

EXTERNAL ANAL SPHINCTER [Figures 10.6 and 10.7]

Description: Surrounds the anus.
Origin: Annococcygeal raphe.
Insertion: Perineal body.
Action: Closes anal canal and draws it forward.
Innervation: Branches from the pudendal nerve.

Muscles of the Pelvic Floor—Identification
[Figures 10.5–10.7]

Examine and color the illustrations of the pelvic floor, perineum, and associated muscles. Then examine the pelvic floor and perineum of a model or, if possible, prepared female and male cadavers. Try to identify all of the following:

1. **levator ani**
2. **coccygeus**

3. perineal body
4. ischiocavernosus (in males and females)
5. bulbospongiosus (in males and females)
6. deep transverse perinei
7. external anal sphincter

CHAPTER SUMMARY AND CHECKLIST

I. MUSCLES OF RESPIRATION
  A. Description
    1. Attached to rib cage and ribs
    2. Enlarge thoracic cavity for inspiration
  B. Muscles
    1. External intercostal
    2. Internal intercostal
    3. Transversus thoracis
    4. Diaphragm

II. MUSCLES OF THE ABDOMINAL WALL
  A. Description
    1. Abdominal wall is almost entirely muscular
    2. Muscles are broad, flat, and in layers
    3. Aponeuroses of muscles merge to form linea alba
    4. Linea alba lies vertically between rectus abdominis muscles
    5. Muscles are anterolateral and posterior
  B. Anterolateral Abdominal Muscles
    1. Rectus abdominis muscle
    2. External oblique muscle
    3. Internal oblique muscle
    4. Transverse abdominis
  C. Posterior Abdominal Muscles
    1. Quadratus abdominis

III. MUSCLES OF THE PELVIC FLOOR
  A. Description
    1. Support abdominopelvic organs
    2. Form pelvic and urogenital diaphragms
    3. Pierced by urethra, anal canal, and vagina
  B. Muscles
    1. Levator ani
    2. Coccygeus

IV. MUSCLES OF THE PERINEUM
  A. Description
    1. Muscles are inferior to pelvic diaphragm
    2. Structure for muscle attachments is perineal body

3. Perineal muscles associated with external genitalia

B. Muscles
   1. Ischiocavernosus
   2. Bulbospongiosus
   3. Deep and superficial transverse perinei
   4. External anal sphincter

NAME _____

LAB SECTION _____ DATE _____

LABORATORY EXERCISE 10.1

Muscles of the Thorax and Abdomen

Part I

1. Which three muscles on the thoracic cage are involved in normal, quiet respiration?

    a. _____

    b. _____

    c. _____

2. Which muscle inserts into a central tendon?

    _____

3. Which abdominal muscle is separated in the midline by the linea alba and crossed by tendinous intersections?

    _____

4. Which abdominal muscles have fibers that run in the same direction as those of the external intercostal muscles?

    _____

5. Which abdominal muscles have fibers that run in the same direction as those of the internal intercostal muscles?

    _____

6. Which large muscle is located in the posterior region of the abdomen and allows lateral flexion of the trunk?

    _____

Muscles of Pelvic Floor and Perineum

1. Which muscles form the pelvic floor?

    a. _____

    b. _____

2. Which muscle forms the major portion of the pelvic floor?

    _____

3. What structure in the perineum serves as an important site of attachment for perineal muscles?

   _____

4. Which muscle inserts into the crura of the corpora cavernosa of the penis and the crura of the clitoris?

   _____

5. Which muscle inserts into the corpus spongiosum of the penis?

   _____

6. Which muscle closes the anal canal and draws it forward?

   _____

Part II

Using the listed terms, label the muscles on the thorax and abdomen and then color them.

External intercostals   External oblique
Central tendon of Diaphragm   Diaphragm
Quadratus lumborum   Rectus abdomini
Internal intercostals

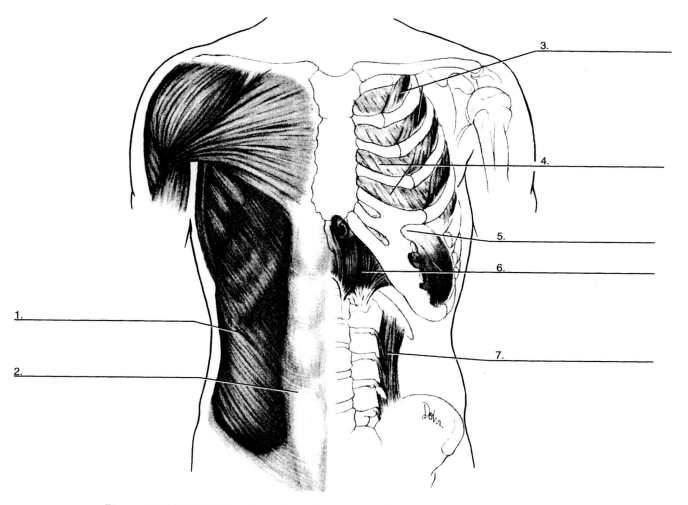

Figure 10.8 (Right) Muscles of respiration and (left) anterior and posterior abdominal muscles.

LABORATORY EXERCISE 10.2

Using the listed terms, label the muscles of the anterior abdominal wall and then color them.

External oblique
Tendinous intersection
Internal oblique

Transverse abdominis
Rectus abdominis
Linea alba

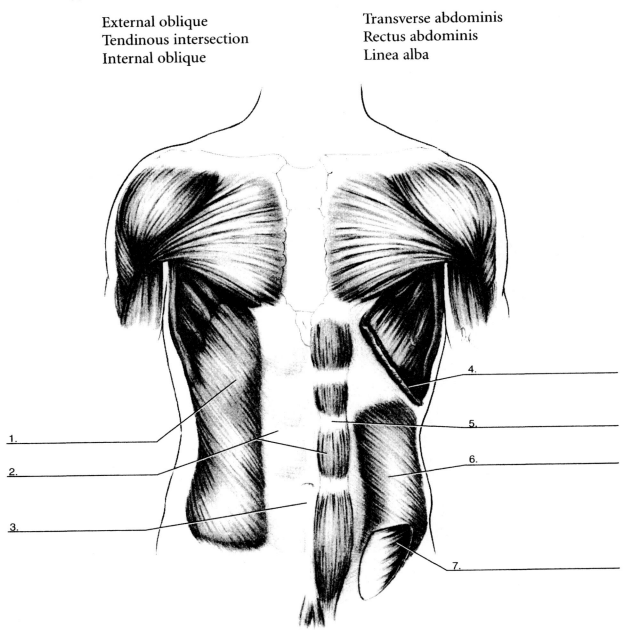

1.

2.

3.

4.

5.

6.

7.

Figure 10.9 Superficial and deep views of the muscles of the anterior abdominal wall.

LABORATORY EXERCISE 10.3

Using the listed terms, label the muscles of the pelvic floor and then color them.

Ischiocavernosus
Bulbocavernosus
Coccygeus
Perineal body

Deep transverse perineus
Superficial transverse perineus
Levator ani
External anal sphincter

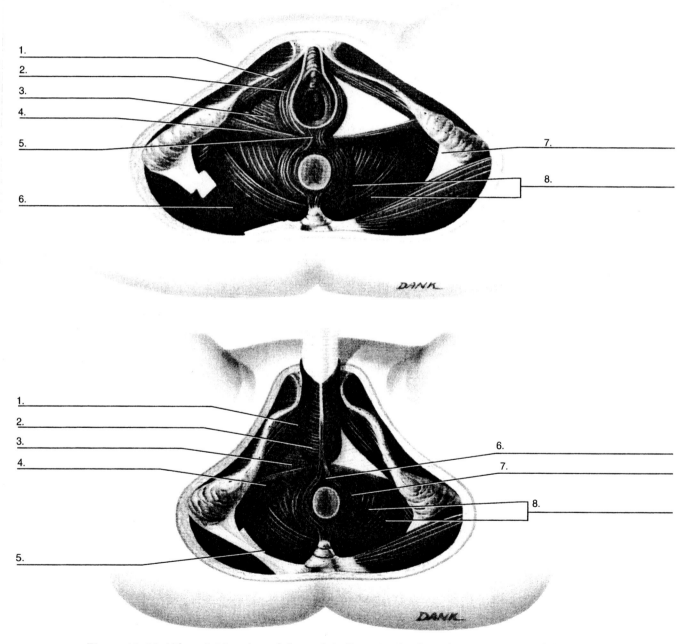

1. _____
2. _____
3. _____
4. _____
5. _____
6. _____
7. _____
8. _____

1. _____
2. _____
3. _____
4. _____
5. _____
6. _____
7. _____
8. _____

Figure 10.10 (Above) Muscles of the pelvic floor in the female perineum and (below) male perineum.

Part Four

Integration and Control Systems of the Human Body

11

The Central Nervous System

THE BRAIN

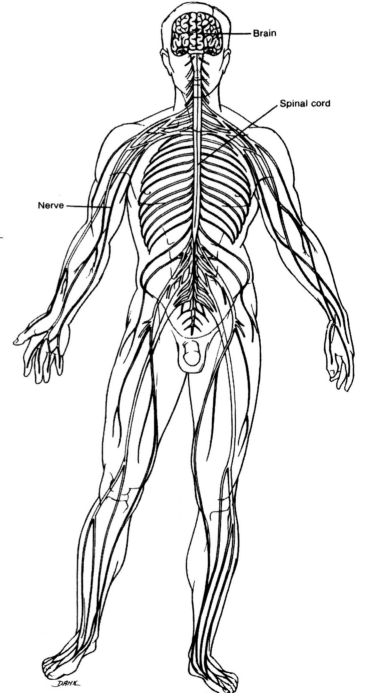

Objective

The objective of Chapter 11, "The Central Nervous System," is for you to become familiar with the following parts of the human brain:

1. Its surface anatomy and that of the cerebellum
2. Lobes
3. Midsagittal and coronal sections
4. Protective coverings and their extensions
5. Ventricles

The Nervous System

The human nervous system is divided into two major components, the **central nervous system** and the **peripheral nervous system**. The central nervous system consists of the **brain** and **spinal cord**. The peripheral nervous system consists of the **cranial nerves** that originate in the brain and the **peripheral nerves** that arise from the spinal cord and their ganglia.

The Brain and Its Embryological Development—A Brief Description

The brain develops from the cephalic end of a hollow neural tube. By four weeks of gestation **three** primary, fluid-filled brain swellings, or **vesicles**, are recognizable: the **prosencephalon (forebrain)**, **mesencephalon (midbrain)**, and **rhombencephalon (hindbrain)**.

Five secondary vesicles are visible by five weeks of development. The prosencephalon divides into **diencephalon** and **telencephalon**. The rhombencephalon divides into the **myelencephalon** and **metencephalon**. Posterior to the brain, the neural tube develops into the spinal cord. Internal changes occur in the **mesencephalon (midbrain)** as well. Eventually the following transformations take place:

a. The **telencephalon** becomes the **cerebrum** (cerebral hemispheres)
b. The **diencephalon** becomes the **hypothalamus, thalamus, subthalamus**, and **epithalamus**
c. The **mesencephalon** becomes the **midbrain**
d. The **metencephalon** becomes the **pons** and **cerebellum**
e. The **myelencephalon** becomes the **medulla**

As the brain matures, the cavities in the vesicles develop into **brain ventricles** and become filled with **cerebrospinal fluid**. The first two ventricles develop in the cerebral hemispheres and become the **lateral ventricles**. The **third ventricle** forms in the diencephalon and the **fourth ventricle** in the hindbrain. A narrow **cerebral aqueduct** in the midbrain connects the third and fourth ventricles. All the ventricles are connected with each other and with the central canal of the spinal cord.

The Cerebrum: Its Surface and Lobes

The adult human brain is commonly separated into four principal parts: **cerebrum (cerebral hemispheres)**, **brain stem**, **diencephalon**, and **cerebellum**. The cerebrum and the cerebellum are best be examined in whole brains. Other regions of the brain, the brain stem, and diencephalon are best be examined either when the brain or brain model is cut in midsagittal plane or when the cerebral and cerebellar hemispheres are removed.

The Gyri, Major Sulci, and Fissures—Description
[Figures 11.1–11.3]

The cerebrum is the largest and most rostral region of the brain. It consists of two layers. The outer layer is the **cerebral cortex**, composed of gray matter and containing mostly nerve cell bodies. Deep to the cerebral cortex is the thick second layer of **white matter**, which largely consists of **myelinated axons** coursing to various brain regions. The surface of the cerebrum is composed of elevated folds or convolutions called **gyri** (singular, **gyrus**), which are separated from each other by grooves. The deep grooves are called **fissures**, and the shallow grooves are called **sulci** (singular, **sulcus**).

The cerebrum consists of **right** and **left hemispheres**, incompletely separated superiorly by a deep **longitudinal fissure** that runs in an anterior-posterior direction. Inferiorly, the right and left hemispheres are connected by a thick **corpus callosum** (best seen when the brain is sectioned in a midsagittal plane, as in Figure 11.3).

Passing at right angles to the longitudinal fissure is the **central sulcus,** an important functional landmark in the brain. The gyrus immediately **anterior** to the central sulcus is the **precentral gyrus**, the motor region. The gyrus immediately **posterior** to the central sulcus is the **postcentral gyrus**, the sensory region. On the lateral side of each cerebral hemisphere and inferior to the central sulcus is the **lateral sulcus.** A **transverse fissure** separates the **cerebellum** from the overlying cerebrum.

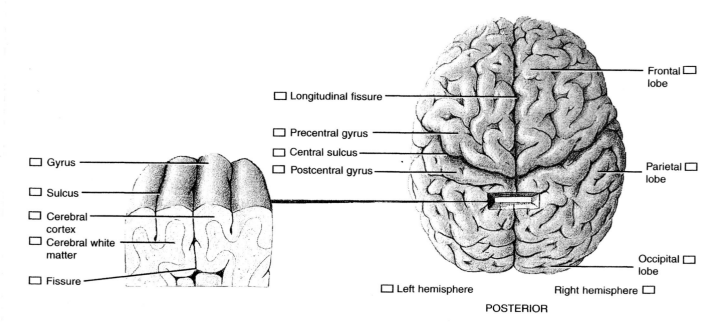

Figure 11.1 Lobes and fissures of the cerebrum as seen from the superior aspect.

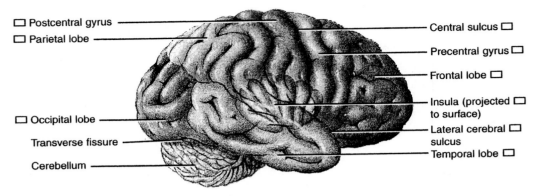

□ Postcentral gyrus
□ Parietal lobe

Central sulcus □
Precentral gyrus □
Frontal lobe □
Insula (projected □ to surface)
Lateral cerebral □ sulcus
Temporal lobe □

□ Occipital lobe
Transverse fissure
Cerebellum

Figure 11.2 Lobes and fissures of the cerebrum as seen from the lateral aspect.

Other fissures and sulci divide the cerebrum into lobes named for the cranial bones that are adjacent to them.

The Lobes [Figures 11.1–11.3]

The cerebral hemispheres are divided into four main lobes: the **frontal, parietal, temporal,** and **occipital lobes.**

The **frontal lobes** are the largest and form the anterior part of the cerebrum. They lie anterior to the **central sulcus** and superior to the **lateral sulcus.** Posterior to the central sulcus are the **parietal lobes,** each bounded anteriorly by the central sulcus and posteriorly by the upper part of the parieto-occipital sulcus. The **parieto-occipital sulcus** separates the parietal lobes from the occipital lobes and can be seen only on the medial side of the brain (Figure 11.3). The **occipital lobes** are small and form the posterior portion of the cerebral hemispheres. The **calcarine fissure,** also seen only on the medial side of the brain, bisects the occipital lobe in a transverse plane (Figure 11.3). The **temporal lobes** are separated from both the frontal and parietal lobes by the **lateral sulcus.**

The **insula** is often called the fifth lobe of the brain. It lies deep within the lateral sulcus and deep to the parietal, frontal, and temporal lobes. It can be seen only when the lateral sulcus is widened by pulling back the temporal lobe (Figure 11.2).

The Gyri, Fissures, Sulci, and Lobes— Identification [*Figures 11.1–11.3*]

Examine a model, diagrams, or prepared human brain and color the parts in the illustrations. Then identify the following structures:

Lobes

1. **frontal lobe**
2. **parietal lobe**
3. **occipital lobe**
4. **temporal lobe**
5. **insula**

Gyri

1. **precentral gyrus**
2. **postcentral gyrus**

Sulci/fissures

1. **longitudinal fissure**
2. **central sulcus**
3. **lateral sulcus**
4. **parieto-occipital sulcus**
5. **calcarine sulcus**

The Other Principal Parts of the Brain— Description [Figures 11.3 and 11.4]

Brain Stem [Figures 11.3 and 11.4]

The brain stem consists of the **medulla, pons,** and **midbrain**. The medulla is the most inferior region of the brain and merges with the spinal cord at the level of the foramen magnum.

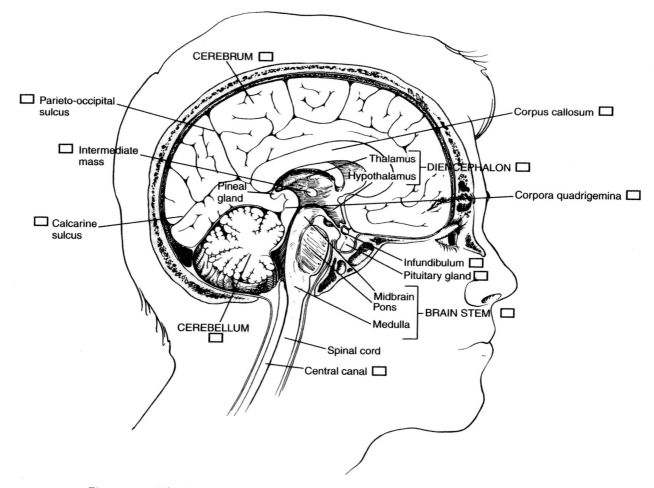

Figure 11.3 The brain and its principal parts seen in sagittal section.

MEDULLA [Figures 11.3, 11.4, and 11.8]

Although the medulla is connected to the spinal cord, it is distinguished by several identifiable features. On its anterior (ventral) side, the medulla exhibits a **median fissure**. Lateral to the fissure are elevated, triangular structures called **pyramids**. The pyramids contain nerve tracts that carry motor information from the cerebral cortex to the spinal cord. Near the medulla–spinal cord junction, most of the fibers in the pyramids decussate (cross over) to the opposite side. This crossover is called the **pyramidal decussation**. Lateral to the pyramids are oval enlargements called **olives** that contain important relay centers for information destined for the cerebellum. The posterior surface of the medulla forms the floor of the **fourth ventricle**, which continues into the spinal cord as a small **central canal** (Figure 11.3).

PONS [Figures 11.3 and 11.4]

The pons is a prominent, rounded structure located superior to the medulla and anterior to the cerebellum. The internal structure of the pons is composed of nerve fibers that serve as a connection between the cerebellum, spinal cord, and cerebrum.

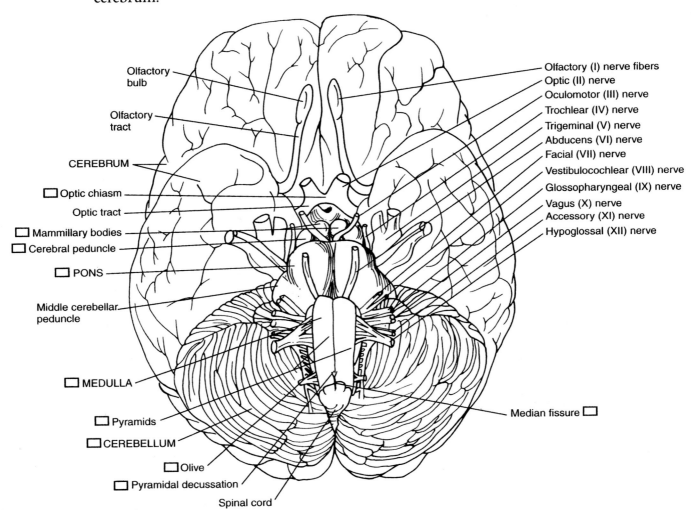

Figure 11.4 Brain stem: Ventral surface of the brain shows the structure of the brain in relation to the cranial nerves and associated structures.

MIDBRAIN [Figures 11.3, 11.4, 11.6, and 11.10]

The midbrain of the brain stem is small and short. Within it is a small **cerebral aqueduct,** a canal that allows for the flow of cerebrospinal fluid from the third to the fourth ventricle. The cerebral aqueduct also divides the midbrain into anterior and posterior regions.

On the anterior (ventral) side of the midbrain are two prominent fiber tracts called the right and left **cerebral peduncles** (Figure 11.4). They contain fiber tracts that convey motor information from the cerebral cortex to the pons and spinal cord.

The posterior (dorsal) region of the midbrain forms a roof over the cerebral aqueduct called the **tectum.** The tectum consists of four rounded elevations or bodies called the **corpora quadrigemina** (Figure 11.3). The pair of superior elevations are called the **superior colliculi;** they coordinate the visual reflexes. The pair of inferior elevations are called the **inferior colliculi;** they coordinate the auditory reflexes (Figures 11.3 and 11.6).

Diencephalon [Figures 11.3–11.5]

The diencephalon is located in the central region of the brain and is surrounded by the **cerebrum.** It is not visible except for a portion that lies in the inferior ventral region. This narrow region is bound posteriorly by the cerebral peduncles and anteriorly by the **optic chiasm** (decussating optic nerve fibers). The **hypophysis (pituitary gland)** is attached by a slender **infundibular (pituitary) stalk** to the ventral region of the hypothalamus (Figure 11.5). In brains removed from cadavers, the pituitary gland is usually torn and only the infundibular stalk remains. The opening in the stalk is the inferior portion of the third ventricle.

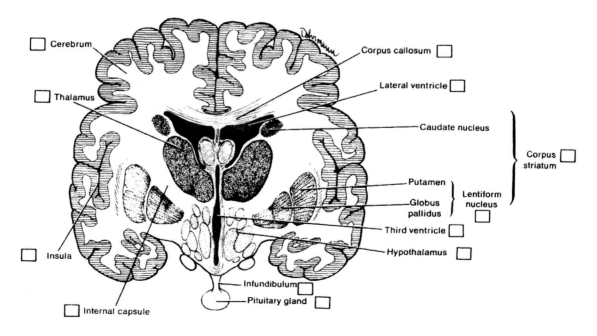

Figure 11.5 Coronal section of the brain illustrates internal structures of the cerebral hemisphere such as the ventricles, thalamus, basal ganglia, hypothalamus, and other structures.

The diencephalon consists of the **thalamus, subthalamus, hypothalamus**, and **epithalamus** (pineal gland). These structures enclose and form the boundaries of the **third ventricle**.

The **thalamus** forms the superior and lateral boundaries of the third ventricle. The thalamus is composed of two oval masses of gray matter located on each side of the third ventricle and joined in the middle across the ventricle by the **intermediate mass** (Figure 11.3). The lateral boundary of the thalamus is the tract of white fibers called the **internal capsule**. This capsule contains fibers or axons of major motor and sensory tracts that pass to and from the cerebral cortex (Figure 11.5).

The **subthalamus** is a small area immediately inferior to the thalamus. It contains subthalamic nuclei associated with the **basal ganglia (cerebral nuclei)** of the cerebral hemispheres. The basal ganglia consist of several large nuclei deep within the base of the cerebral hemispheres. The largest basal ganglion is the **corpus striatum**, which consists of the **caudate nucleus** and **lentiform nucleus**. The lentiform nucleus is further subdivided into a lateral **putamen** and a medial **globus pallidus**. The internal capsule passes between the lentiform nucleus and caudate nucleus (Figure 11.5). The basal ganglia are important in coordinating motor movement and body posture. Their major effect is to inhibit unwanted muscular activity.

The portion of the diencephalon located inferior to the thalamus is the **hypothalamus**. It forms the floor and part of the lower lateral walls of the third ventricle. The hypothalamus extends from the optic chiasm to the posterior region of the mammillary bodies. The **mammillary bodies** are small, paired, oval swellings on the inferior portion of the hypothalamus posterior to the infundibular stalk of the pituitary gland (Figures 11.4 and 11.5). The **hypothalamus** performs numerous vital integrative functions, most of which control visceral activities via the autonomic nervous system.

The **epithalamus** is the dorsal portion of the diencephalon that forms a roof over the third ventricle. A small, cone-shaped mass extends outward from the epithalamus, the **pineal body**, or **pineal gland** (Figures 11.3 and 11.6).

The **thalamus, hypothalamus, epithalamus,** and **basal ganglia** can be seen best when the brain is prepared in the coronal and midsagittal planes (Figures 11.3 and 11.5).

Cerebellum [Figure 11.6]

The **cerebellum** overlies the superior regions of the pons and medulla and is located in the inferior and posterior region of the skull. It is separated from the overlying cerebrum by the **transverse fissure.**

The cerebellum consists of two lateral **cerebellar hemispheres** connected medially by the **vermis**. The cerebellar surface is highly convoluted into parallel gyri called **folia**; these are separated from one another by **transverse fissures**. Similar to the cerebrum, the cerebellum has a thin, outer layer of gray matter called the **cerebellar cortex**. Deep to the cerebellar cortex is the **white matter**. Its branching patterns, called the **arbor vitae** (tree of life), resemble tree branches.

The cerebellum connects to the brain stem via three **cerebellar peduncles (inferior, middle,** and **superior)**. The superior peduncles connect the cerebellum to the midbrain, the middle peduncles connect the cerebellum to the pons, and the inferior peduncles connect the cerebellum to the medulla.

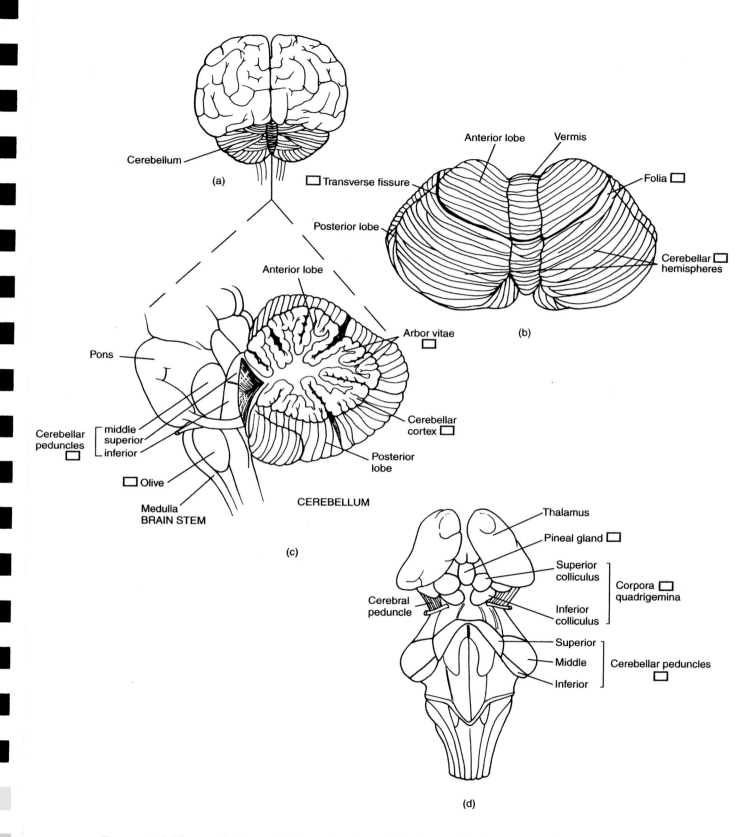

Cerebellum

(a)

Anterior lobe   Vermis

Transverse fissure   Folia ☐

Posterior lobe   Cerebellar ☐ hemispheres

(b)

Anterior lobe

Arbor vitae ☐

Pons

middle
superior
inferior

Cerebellar peduncles ☐

☐ Olive

Medulla
BRAIN STEM

Cerebellar cortex ☐

Posterior lobe

CEREBELLUM

(c)

Thalamus

Pineal gland ☐

Superior colliculus

Inferior colliculus

Corpora ☐ quadrigemina

Cerebral peduncle

Superior

Middle   Cerebellar peduncles ☐

Inferior

(d)

Figure 11.6 The cerebellum. (a) Posterior view of the brain. (b) Superior surface of the cerebellum shows major anatomical landmarks. (c) Sagittal section of the cerebellum and brain stem. (d) Posterior view of the brain stem shows cerebellar peduncles, corpora quadrigemina, and pineal gland.

The Brain and Its Protective Cover—Description

The Meninges [Figure 11.7]

The brain and spinal cord are protected by **bones**, connective tissue **membranes** called **meninges**, and **cerebrospinal fluid (CSF)**. The meninges are the outer **dura mater**, a middle **arachnoid**, and the inner **pia mater**.

The **dura mater** is a thick, double-layered membrane composed of tough connective tissue fibers. Its outer layer adheres to the inner surface of the cranial bones as the **periosteal layer**. The inner layer of the dura mater is the **meningeal layer**, which covers the brain and spinal cord.

The **arachnoid** membrane lies underneath the dura mater and forms a loose cover over the brain surface. It is separated from the dura by a narrow **subdural space**. Delicate weblike projections extend from the arachnoid to the pia mater, forming a **subarachnoid space** between the two membranes. In life, the subarachnoid space is filled with circulating CSF. The arachnoid does not extend into the grooves (sulci) on the brain surface.

The **pia mater** is a thin, delicate connective tissue and vascular membrane that attaches to the surface of the brain; it follows the surface convolutions into the grooves.

Dural Reflections [Figures 11.7 and 11.8]

The two layers of the dura mater are fused over most of the brain surface. In certain regions of the brain, however, the inner or meningeal dural layers reflect inwardly between the cerebral or cerebellar hemispheres to form rigid membrane **folds**, or **septa** (singular, **septum**). The septa partition the cranial cavity into communicating compartments and stabilize the brain in the skull. Some septa contain

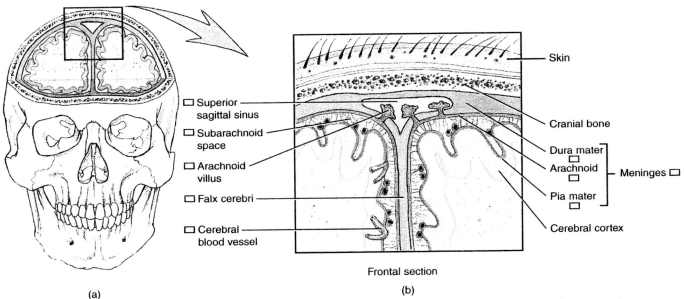

(a)

(b)

Frontal section

Figure 11.7 (a) Coronal section through the skull and cerebral hemispheres. (b) Coronal section of the area within the box shows internal structures of the meninges.

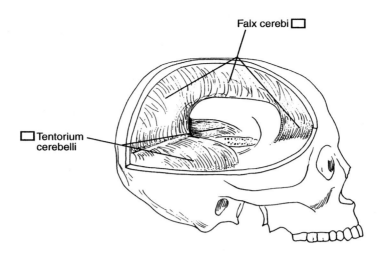

Figure 11.8 The skull with a portion of the cranium removed to show the falx cerebri and tentorium cerebelli.

sinuses, which are noncollapsible veins that drain venous blood from different regions of the brain.

The **flax cerebri** is a large, crescent-shaped, midsagittal septum located between and partially separating the left and right cerebral hemispheres. Its narrow end attaches anteriorly to the **crista galli** of the ethmoid bone. Posteriorly, the falx cerebri attaches to the **tentorium cerebelli**, a wide, horizontal septum that separates the posterior (occipital) lobes of the cerebrum from the cerebellum. A venous sinus called the **superior sagittal sinus** is located in the upper margin of the falx cerebri.

The Ventricles and Cerebrospinal Fluid (CSF)
[Figures 11.3, 11.5, 11.7, 11.9, and 11.10]

The human brain contains numerous fluid-filled cavities called **ventricles**. They are numbered by flow of the CSF. The ventricles communicate with each other, with the central canal of the spinal cord, and with the subarachnoid space via various foramina. The roof of each ventricle contains the **choroid plexus** and capillaries; the choroid plexus continually produces CSF.

The first two brain ventricles are the **lateral ventricles** (Figure 11.9), the U-shaped cavities that occupy the medial region of the two cerebral hemispheres inferior to the **corpus callosum**. The **third ventricle** is located in the diencephalon, between the right and left halves of the thalamus and hypothalamus. The lateral ventricles communicate with the third ventricle by narrow channels, the interventricular foramina. The **fourth ventricle** is located inferior to the cerebellum in the pons and medulla. It communicates with the third ventricle via the **cerebral aqueduct**, which passes through the midbrain. The roof of the fourth ventricle contains three small openings through which CSF exits into the subarachnoid space. From here, the fluid circulates around the brain, spinal cord, cerebrum, and cerebellum. The brain and spinal cord "float" in the CSF, a protective cushion against jarring injuries (Figures 11.9 and 11.10).

The CSF is reabsorbed from the subarachnoid space into the venous blood through the **arachnoid villi** (singular, **villus**), which project from the arachnoid

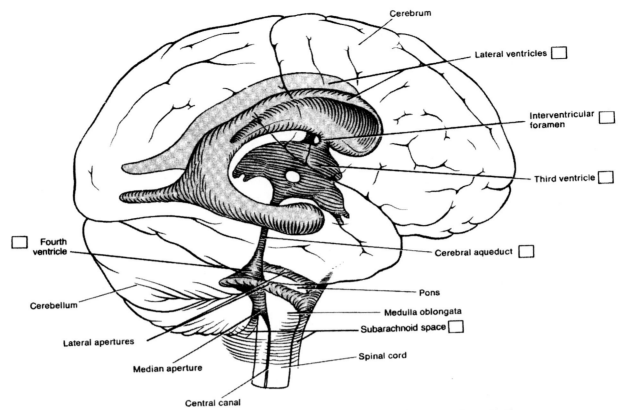

Cerebrum

Lateral ventricles ☐

Interventricular ☐
foramen

Third ventricle ☐

Cerebral aqueduct ☐

Pons

Medulla oblongata

Subarachnoid space ☐

Spinal cord

Fourth ☐
ventricle

Cerebellum

Lateral apertures

Median aperture

Central canal

Figure 11.9 Ventricles of the brain superimposed on the cerebrum and cerebellum.

layer into the venous blood sinuses in the dura mater. Most of the CSF is absorbed into the **superior sagittal sinus** (Figures 11.7 and 11.10).

The ventricles are best studied when the brain is sectioned in either the midsagittal or coronal plane or prepared as a ventricular cast. Ventricles can also be seen with x-rays and other medical imaging techniques.

The Principal Parts of the Brain—Identification

[Figures 11.3, 11.4 and 11.10]

Brain Stem

Examine the inferior region of a model of the brain or human brain and the diagrams, and color the parts. Identify the following structures:

1. **midbrain**
2. **corpora quadrigemina**
3. **cerebral peduncles**
4. **pons**
5. **cerebral aqueduct**
6. **medulla**
7. **medullary pyramids**
8. **pyramidal decussation**
9. **median fissure in the medulla**
10. **medullary olives**

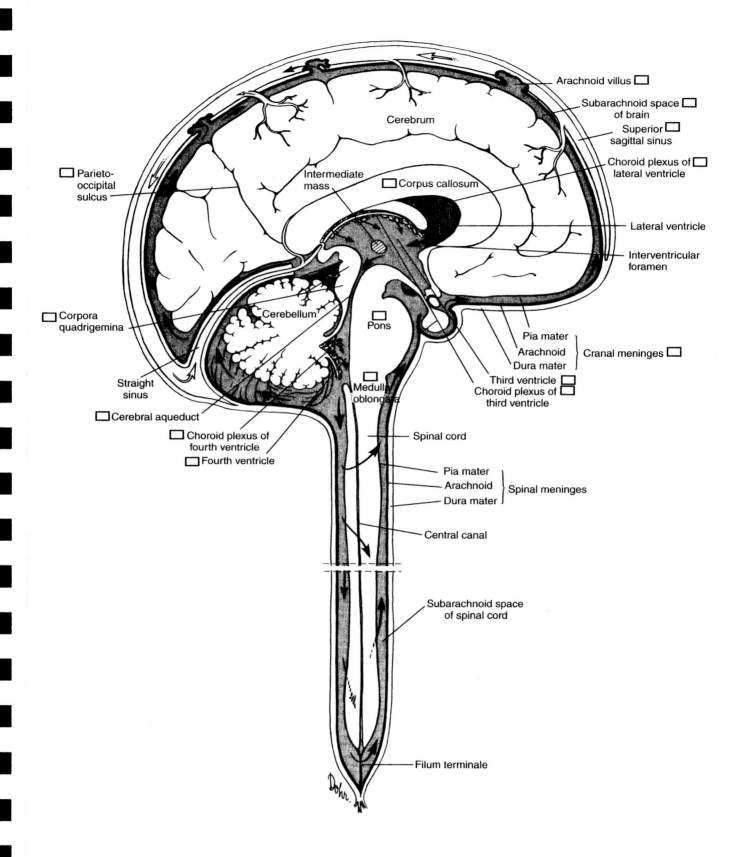

□ Parieto-
occipital
sulcus

Cerebrum

Arachnoid villus □

Subarachnoid space □
of brain

Superior □
sagittal sinus

Choroid plexus of □
lateral ventricle

Intermediate
mass

□ Corpus callosum

Lateral ventricle

Interventricular
foramen

□ Corpora
quadrigemina

Cerebellum

Pons □

Pia mater ⎫
Arachnoid ⎬ Cranal meninges □
Dura mater ⎭

Third ventricle □
Choroid plexus of □
third ventricle

Straight
sinus

Medulla
oblongata

□ Cerebral aqueduct

Spinal cord

□ Choroid plexus of
fourth ventricle

Pia mater ⎫
Arachnoid ⎬ Spinal meninges
Dura mater ⎭

□ Fourth ventricle

Central canal

Subarachnoid space
of spinal cord

Filum terminale

Dohr.

Figure 11.10 Meninges and ventricles of the brain in sagittal section. Arrows indicate the flow of cerebrospinal fluid.

Diencephalon [Figures 11.3, 11.4, 11.5, and 11.10]

Examine the midsagittal and coronal sections of a model of the brain or human brain and the diagrams, and color the parts. Identify the following structures in the diencephalon:

Midsagittal Plane (Section)

1. thalamus
2. pineal body
3. intermediate mass
4. hypothalamus
5. infundibular stalk
6. mammillary bodies

Coronal Plane (Section)

1. lentiform nucleus
2. putamen
3. globus pallidus
4. corpus striatum
5. caudate nucleus
6. internal capsule

Cerebellum [Figure 11.6]

Examine a model of the brain or the human brain and an illustration of the cerebellum, and color the parts. If possible, examine a brain cut in the midsagittal plane. Identify the following structures in the cerebellum:

1. cerebellar hemispheres
2. folia
3. vermis
4. arbor vitae
5. cerebellar cortex
6. cerebellar peduncles
7. transverse fissure

The Brain, Meninges, and Dural Reflections—Identification [Figures 11.6 and 11.7]

Examine a model of the brain or human brain and the diagram of the brain, and color the parts. If possible, examine a detached dura mater as well and identify the following structures:

1. dura mater
2. falx cerebri
3. tentorium cerebelli
4. arachnoid
5. arachnoid villi
6. superior sagittal sinus
7. pia mater

Brain Ventricles—Identification [Figure 11.9]

Examine the diagram of the brain ventricles and a prepared cast of the human brain ventricles. Examine a prepared brain or model sectioned in the midsagittal and coronal planes and identify the following structures:

1. lateral ventricles
2. interventricular foramen
3. third ventricle
4. cerebral aqueduct
5. fourth ventricle
6. ventricular choroid plexuses

CHAPTER SUMMARY AND CHECKLIST

I. ORGANIZATION OF THE NERVOUS SYSTEM

   A. **Composition of the Central Nervous System**
      1. Brain and spinal cord

   B. **Composition of the Peripheral Nervous System**
      1. Cranial and peripheral nerves

II. THE BRAIN AND ITS EMBRYOLOGICAL DEVELOPMENT

   A. **Initial Formation of Three Brain Vesicles**
      1. Prosencephalon—forebrain
      2. Mesencephalon—midbrain
      3. Rhombencephalon—hindbrain

   B. **During Brain Development**
      1. Prosencephalon divides into diencephalon and telencephalon
      2. Rhombencephalon divides into myelencephalon and metencephalon
      3. Mesencephalon does not divide further

   C. **Transformations in the Adult Brain**
      1. Diencephalon forms the hypothalamus, thalamus, and epithalamus
      2. Telencephalon forms the cerebrum (cerebral hemispheres)
      3. Mesencephalon forms the midbrain
      4. Myelencephalon forms the medulla
      5. Metencephalon forms the pons and cerebellum

   D. **Formation of the Brain Ventricles**
      1. Lateral ventricles from in the cerebral hemispheres
      2. Third ventricle forms in the midbrain
      3. Fourth ventricle forms in the hindbrain
      4. Cerebral aqueduct connects third and fourth ventricles

III. THE CEREBRUM, ITS SURFACE, AND LOBES

   A. **Gyri, Sulci, and Fissures**
      1. Right and left cerebral hemispheres separated by longitudinal fissure
         a. most prominent fissure
         b. cerebral hemispheres connected by corpus callosum
      2. Outer brain surface is gray matter (cerebral cortex)
      3. Inner layer is white matter
      4. Gyri-folds or convolutions in brain surface
      5. Deep grooves are fissures; shallow grooves are sulci
      6. Central sulcus separates precentral from postcentral gyrus
      7. Transverse cerebral fissure separates cerebrum from cerebellum

B. Lobes

1. Frontal lobes are largest and most anterior
   a. lie anterior to central sulcus
   b. lie superior to lateral sulcus
2. Parietal lobes lie behind central sulcus
   a. separated from occipital lobe by parieto-occipital sulcus
3. Occipital lobes from posterior portion of cerebrum
   a. form posterior region of cerebral hemispheres
4. Temporal lobes lie inferior to parietal and frontal lobes
   a. separated by lateral sulcus

IV. THE OTHER PRINCIPAL PARTS OF THE BRAIN

A. Brain Stem
1. Consists of medulla, pons, and midbrain

B. Medulla
1. Most inferior region of brain
2. Contains pyramids and decussation on its anterior side
3. Lateral oval enlargements are olives
4. Posterior surface forms floor for fourth ventricle
5. Contains vital reflex centers and nuclei of cranial nerves

C. Pons
1. Superior to medulla and anterior to cerebellum
2. Contains vital centers for respiratory regulation
3. Contains nuclei of nerves that innervate face

D. Midbrain
1. Between pons and diencephalon
2. Cerebral aqueduct passes through it
3. Cerebral peduncles on ventral side
4. Corpora quadrigemina involved in visual and auditory reflexes

E. Diencephalon
1. Located centrally in brain and surrounded by cerebrum
2. Contains thalamus, hypothalamus, and epithalamus
3. Thalamus and hypothalamus form sides of third ventricle
4. Intermediate mass connects both sides of thalamus
5. Epithalamus (pineal gland) forms roof over third ventricle
6. Thalamus is a relay center for sensory information
7. Subthalamus is located inferior to thalamus
   a. nuclei associated with basal ganglia
   b. largest basal ganglion is corpus striatum
   c. corpus striatum consists of caudate nucleus and lentiform nucleus
   d. lentiform nucleus divided into putamen and globus pallidus
8. Hypothalamus controls visceral functions

V. CEREBELLUM

A. Structural Features
1. Overlies pons and medulla
2. Separated from cerebrum by transverse fissure

3. Cerebellar hemispheres connected by vermis
4. Gray matter is cerebellar cortex
5. White matter exhibits arbor vitae branching pattern
6. Coordinates muscular activities, equilibrium, and posture

VI. THE BRAIN AND ITS PROTECTIVE COVER

A. The Meninges
1. Dura mater-thick, outer membrane over brain surface
2. Arachnoid-thinner, middle membrane
3. Pia mater-thin, delicate inner membrane attached to brain surface
4. Space under arachnoid is subarachnoid space
5. Subarachnoid space contains CSF

B. Dural Reflections (Septa)
1. Falx cerebri-septum between cerebral hemispheres
2. Tentorium cerebelli-septum between posterior cerebral hemispheres and cerebellum

VII. BRAIN VENTRICLES AND CEREBROSPINAL FLUID

A. Location and Structure of Ventricles
1. Four fluid-filled cavities
2. Lateral ventricles in cerebral hemispheres
3. Third ventricle located in hypothalamus
4. Fourth ventricle in pons and medulla
5. Cerebral aqueduct connects third and fourth ventricles

B. Production and Circulation of Cerebrospinal Fluid
1. Produced in choroid plexus capillaries of brain ventricles
2. CSF circulates over brain and spinal cord in subarachnoid space
3. CSF reabsorbed from subarachnoid space through arachnoid granulations into superior sagittal sinus

Laboratory Exercises 11

NAME _____

LAB SECTION _____ DATE _____

LABORATORY EXERCISE 11.1

The Brain

Part I

1. The brain is separated into four parts. They are:

   a. _____

   b. _____

   c. _____

   d. _____

2. Nerve cell bodies are located in _____ matter. Myelinated nerve fibers, or axons, are located in _____matter.

3. What structure connects the left and right hemispheres?

   _____

4. Supply the correct sulcus or fissure.

   a. separates left and right hemispheres _____

   b. located between precentral and postcentral gyrus _____

   c. separates the cerebellum from the cerebrum _____

5. What are the major lobes of the brain?

   a. _____

   b. _____

   c. _____

   d. _____

6. What part of the brain lies deep within the lateral sulcus?

   _____

7. What structures constitute the brain stem?

   a. _____

b. _____

c. _____

8. In which part of the brain stem is the fourth ventricle located?

_____

9. The connection between the third and fourth ventricle is through the

_____.

10. In what region of the brain are found the thalamus, hypothalamus, basal

ganglia, and third ventricle? _____

11. Intermediate mass in the third ventricle connects what region of the brain?

_____

12. What three structures attach the cerebellum to the brain?

a. _____

b. _____

c. _____

13. a. What is the thick outer connective tissue cover of the brain?

_____

b. What is the delicate connective tissue that attaches to the brain?

_____

c. What is the space between the arachnoid membrane and the pia mater?

_____

d. What are the connective tissue membranes that surround the brain?

_____

14. What dural reflections separate the

a. cerebral hemispheres _____

b. cerebrum from the cerebellum _____

15. What structures form the cerebrospinal fluid and where are they located?

    a. _____

    b. _____

16. Between which ventricles is the cerebral aqueduct found?

    a. _____

    b. _____

17. What structure absorbs the CSF from the subarachnoid space?

    _____

18. Between which meningeal layers is CSF found?

    _____

19. How many ventricles are there in the brain?

    _____

20. Into which structure is most of CSF absorbed?

    _____

Part II

Using the listed terms, label the following structures in the brain and then color them.

Frontal lobe
Temporal lobe
Precentral gyrus
Longitudinal fissure
Occipital lobe

Central sulcus
Postcentral gyrus
Parietal lobe
Insula
Lateral sulcus

ANTERIOR

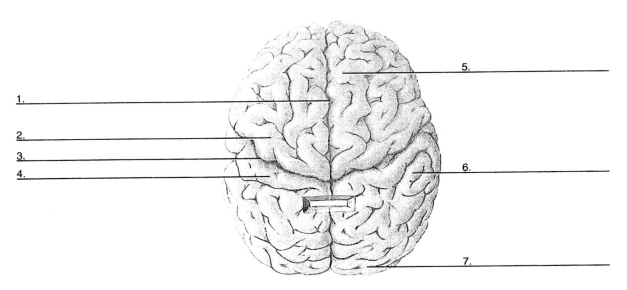

1. _____
2. _____
3. _____
4. _____

5. _____
6. _____
7. _____

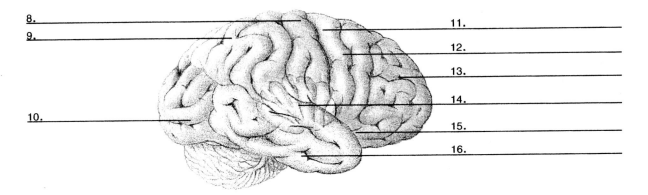

8. _____
9. _____

10. _____

11. _____
12. _____
13. _____
14. _____
15. _____
16. _____

Figure 11.11 Superior (above) and lateral (below) views of the brain, brain lobes, fissures, and sulci.

LABORATORY EXERCISE 11.2

Using the listed terms, label the following structures in the sagittal section of the brain and then color them.

Corpus callosum
Parieto-occipital sulcus
Cerebellum
Pituitary gland
Hypothalamus
Corpora quadrigemina
Pons

Intermediate mass
Midbrain
Medulla
Cerebrum
Thalamus
Calcarine sulcus

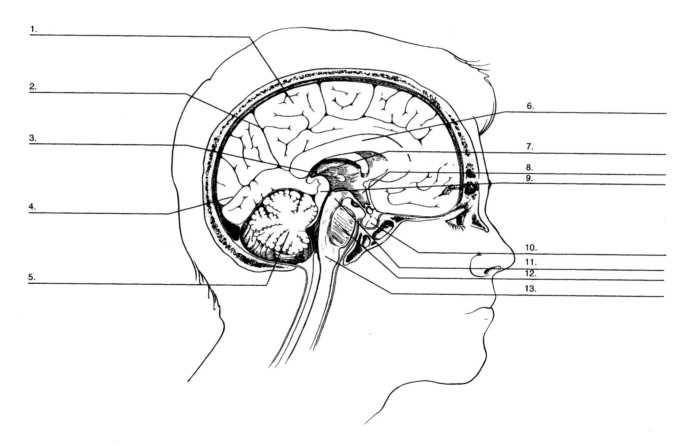

Figure 11.12 The brain in sagittal section.

LABORATORY EXERCISE 11.3

Using the listed terms, label the following structures in the coronal section of the brain and then color them.

Thalamus
Lateral ventricle
Third ventricle
Basal ganglia
Hypothalamus

Internal capsule
Pituitary gland
Corpus callosum
Insula
Infundibulum

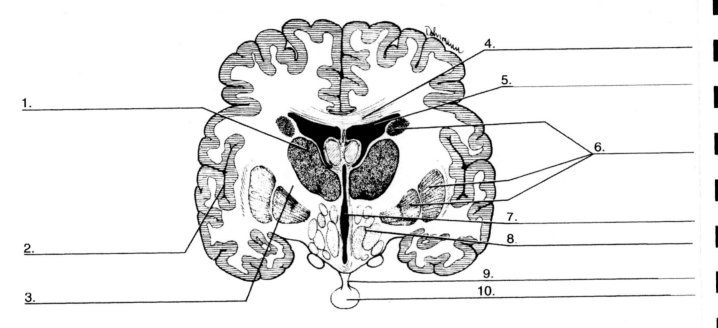

Figure 11.13 The brain in coronal section.

LABORATORY EXERCISE 11.4

Using the listed terms, label the following structures in the meninges and then color them.

Dura mater   Superior sagittal sinus
Cranial bone   Falx cerebri
Arachnoid villus   Arachnoid membrane
Pia mater   Subarachnoid space

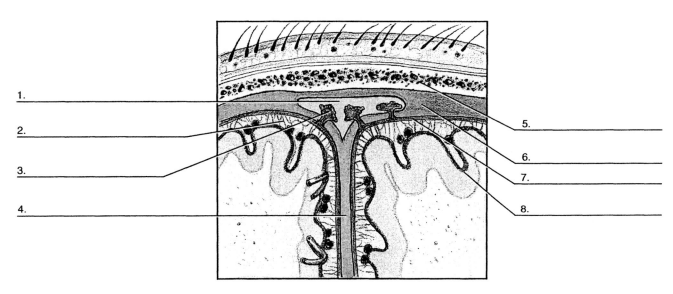

Frontal section

Figure 11.14 Coronal section of the brain shows the internal structures of the meninges.

12

The Central Nervous System

THE SPINAL CORD

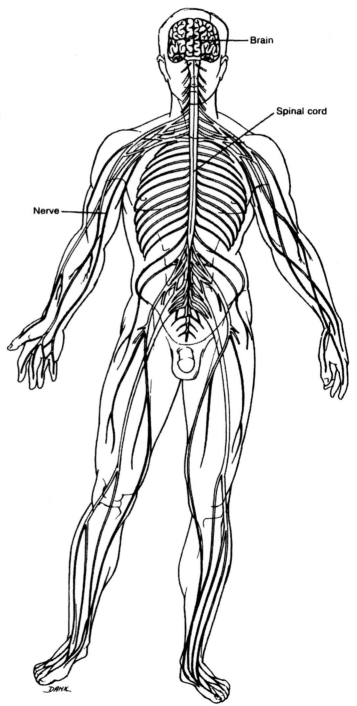

Brain

Spinal cord

Nerve

DANK

Objective

The objective of Chapter 12, "The Central Nervous System: The Spinal Cord," is for you to learn about the:

1. **General characteristics of the spinal cord**
2. **Spinal cord meninges and their attachments**
3. **Cross-sectional anatomy of the spinal cord**
4. **Composition of spinal nerves**

Functions of the Spinal Cord

The spinal cord consists of nerve cells, nerve fibers, and nerve tracts that enable it to function as a major pathway for conducting sensory information to different brain centers. Ascending nerve tracts bring information to the centers and deliver motor impulses from them via the descending nerve tracts to the blood vessels, glands, and muscles. The spinal cord is also an integrating center for spinal reflexes. Reflexes, which are rapid responses to changes in the environment, are initiated and completed at the spinal cord level.

General Characteristics of the Spinal Cord— Description *[Figures 12.1–12.3]*

The spinal cord is cylindrical and exhibits slight anterior and posterior flattening. In cross section, the cord is oval. A continuation of the brain stem, the spinal cord is contained within the vertebral canal of the vertebrae. It starts at the level of the medulla and extends in the vertebral canal from the foramen magnum of the skull to the first or second lumbar vertebra. Inferiorly, the connective tissue meninges extend in the vertebral canal beyond the terminal end of the spinal cord to approximately the second sacral vertebra.

The spinal cord is divided into five regions: **cervical, thoracic, lumbar, sacral,** and **coccygeal** regions. It is not uniform in diameter throughout its length. Superiorly, the cord exhibits the **cervical enlargement**; the nerves from this region innervate the skin and muscles of the upper extremity. Inferiorly, the cord exhibits a **lumbar enlargement**; the nerves from this region innervate the lower extremity. The thoracic spinal cord segment is uniform and has no enlargement.

Inferior to the lumbar enlargement, the spinal cord narrows to a cone-shaped terminal portion called the **conus medullaris**. Arising from the conus medullaris is the **filum terminale**, a connective tissue extension of the pia mater that extends inferiorly and anchors the spinal cord to the coccyx bone.

The adult spinal cord is shorter than the vertebral canal and does not extend beyond the first lumbar vertebra. As a result, the spinal nerve roots that arise from the caudal region of the spinal cord become elongated and do not exit from the vertebral column at the same level that they exit from the spinal cord. Consequently, the lumbar and sacral nerve roots travel inferiorly in the vertebral canal before exiting through their respective intervertebral foramina as spinal nerves. Because these nerve bundles at the inferior end of the vertebral canal resem-

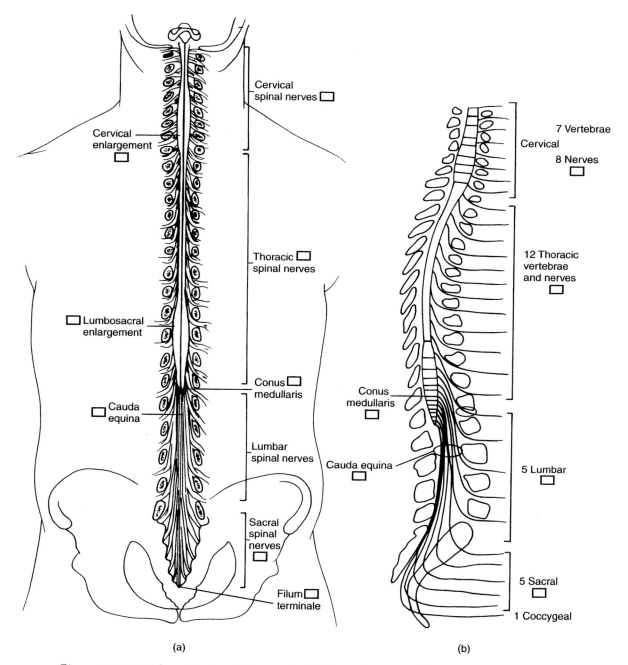

Cervical
spinal nerves □

Cervical
enlargement
□

Thoracic □
spinal nerves

Lumbosacral □
enlargement

Conus □
medullaris

Cauda □
equina

Lumbar
spinal nerves

Sacral
spinal
nerves
□

Filum □
terminale

(a)

7 Vertebrae
Cervical

8 Nerves
□

12 Thoracic
vertebrae
and nerves
□

Conus
medullaris
□

Cauda equina
□

5 Lumbar
□

5 Sacral
□

1 Coccygeal

(b)

Figure 12.1 (a) The spinal cord, from the skull to the coccyx bone. The vertebral arches have been removed to show different regions. (b) Lateral view illustrates the relation of the spinal cord segments and spinal nerves to the vertebral column. Note the length of the spinal cord and the length of nerves in the cauda equina.

ble a horse's tail, they are called **cauda equina**. The cauda equina float in the cerebrospinal fluid.

The anterior surface of the spinal cord has a deep vertical groove, the **anterior median fissure**, and a more shallow posterior median groove, the **posterior median sulcus**. These grooves extend the length of the spinal cord and separate it into right and left sides (Figure 12.3).

The Spinal Cord and Its Protective Covering

[Figures 12.2 and 12.3]

The spinal cord is protected by bones of the vertebrae, cerebrospinal fluid, and the connective tissue meninges. The **spinal cord meninges** enclose the spinal cord from the sacrum to the foramen magnum. The outer covering, the **spinal dura mater**, is continuous at the foramen magnum with the inner layer (meningeal) of the cranial dura mater that covers the brain. In contrast to the cranial dura, the spinal dura mater does not attach to the bones of the vertebral column. As a result, an **epidural space** is formed between the spinal dura and surrounding vertebrae. This space is filled with fat, loose connective tissue and blood vessels and serves as protective padding around the spinal cord.

The middle spinal meningeal layer is the **arachnoid**, which is continuous with the arachnoid of the brain. Between the dura and the arachnoid is the subdural space, which contains a small amount of fluid.

The innermost meningeal layer is the delicate **pia mater**, which adheres to the surface of the spinal cord and roots of the spinal nerves. The three spinal meningeal

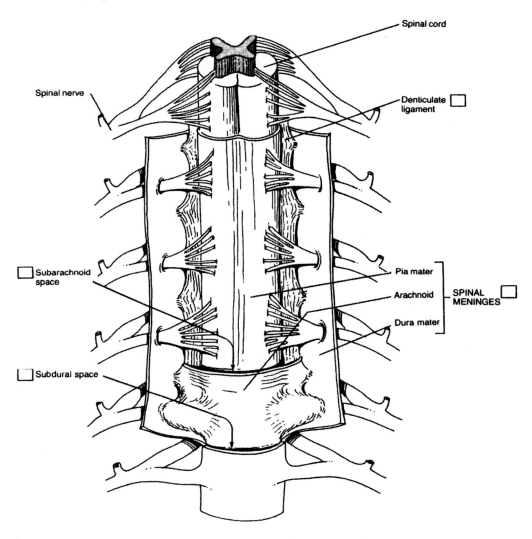

Figure 12.2 The spinal meninges as seen in sections of the spinal cord.

layers also cover the spinal nerves and blend with the connective tissue as the nerves exit the spinal cord.

Along the length of the cord, the pia mater exhibits lateral extensions between the spinal nerve roots on each side. These extensions look like small teeth and are called the **denticulate ligaments**; they pass laterally to the spinal cord and attach to the arachnoid layer and dura mater. The denticulate ligaments suspend and secure the spinal cord in the middle of the dural sheath along its entire length. The spinal cord also attaches inferiorly to the coccyx bone by the **filum terminale** and superiorly to the meninges of the brain.

As it does in the brain, the cerebrospinal fluid (CSF) fills the **subarachnoid space** of the spinal cord to the lower border of the second sacral vertebra.

The Spinal Cord in Cross Section— Description *[Figure 12.3]*

Similar to the cerebrum, the spinal cord consists of **gray matter** and **white matter**. The white matter contains primarily myelinated nerve fibers, and the gray matter nerve cell bodies. In the cerebrum and cerebellum, the gray matter is located peripherally in the cortex. By contrast, the gray matter in the spinal cord is centrally located and forms a darker H-shaped area surrounded by the white matter. The gray matter on each side of the spinal cord is connected by the **gray commissure**. In the center of the gray commissure is a narrow, CSF-filled **central canal**. This

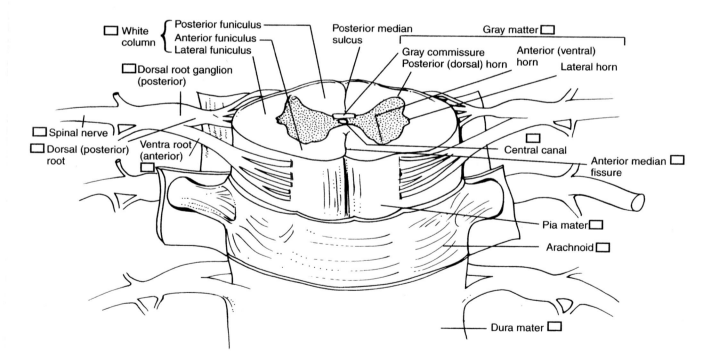

Figure 12.3 The thoracic region of the spinal cord; internal structures and meninges.

canal is continuous with the fourth ventricle of the brain and extends the length of the spinal cord.

Extensions of gray matter toward the outer edges of the spinal cord, as viewed in cross section, are called **horns**. The horns that extend toward the front or anterior region of the cord are called the **anterior (ventral) gray horns**. The narrower horns that extend toward the back or posterior region of the cord are called the **posterior (dorsal) gray horns**.

The anterior horns contain cell bodies of the **motor neurons**; these transmit motor impulses to the periphery of the spinal nerves. The posterior horns contain axons from **sensory neurons** of the spinal nerves and interneurons of the spinal cord. The thoracic, lumbar, and sacral regions of the spinal cord contain **lateral horns** located between the two gray horns. In the lateral horns are motor neurons of the **autonomic nervous system** that innervate the visceral organs of the body.

The horns partition the white matter on each side of the spinal cord into three **white columns** or **funiculi** (singular, **funiculus**): the **anterior (ventral) white column**, **posterior (dorsal) white column**, and **lateral white column**.

The size and shape of the gray and white matter vary in different regions of the spinal cord. White matter increases in the upper, cervical region as the spinal cord and nerve tracts increase in size. Gray matter is more pronounced in the cervical and lumbar regions, indicating increased neuronal connections needed for the innervation of upper and lower limbs.

The Spinal Nerves [Figure 12.3]

The spinal nerves comprise part of the peripheral nervous system. Afferent (sensory) fibers from the peripheral receptors form the **posterior roots** on each side of the spinal cord and enter the spinal cord in the **posterior gray horn**. The nerve cell bodies (neurons) of each posterior root lie outside the spinal cord in a swelling or enlargement called the **dorsal root ganglion**, or **spinal ganglion**. The **anterior roots** that leave the spinal cord are motor (efferent) axons of neurons located in the **anterior** and **lateral gray horns** of the spinal cord.

The anterior and posterior roots on each side of the spinal cord unite laterally to form the **spinal nerve**, which is now a mixture of both sensory and motor fibers. The 31 pairs of spinal nerves are named and numbered according to the region and level of the spinal cord from which they emerge.

General Characteristics—Identification

[Figures 12.1, 12.2, and 12.3]

Examine a model and diagrams of the spinal cord and color them. Then examine the isolated spinal cord and identify the following structures:

1. **cervical spinal cord**
2. **thoracic spinal cord**
3. **lumbar spinal cord**
4. **cervical enlargement**
5. **lumbar enlargement**
6. **conus medullaris**

7. filum terminale
8. cauda equina
9. epidural space
10. spinal dura mater
11. spinal arachnoid
12. subarachnoid space
13. spinal pia mater
14. denticulate ligaments
15. anterior median fissure
16. posterior median sulcus

The Spinal Cord in Cross Section—
Identification *[Figure 12.3]*

Examine a model and illustrations of the cross sections of the spinal cord and color the parts. Then examine cross sections from the cervical, thoracic, and lumbar regions and identify the following structures:

1. central canal
2. anterior gray horns
3. posterior gray horns
4. gray commissure
5. anterior white columns
6. lateral white columns
7. posterior white columns
8. anterior roots
9. posterior roots
10. dorsal root ganglia

CHAPTER SUMMARY AND CHECKLIST

I. THE SPINAL CORD

A. **General Characteristics**
1. Contained within the vertebrae
2. Starts at the medulla and ends at second lumbar vertebra
3. Named according to cervical, thoracic, lumbar, and sacral regions
4. Contains cervical and lumbar enlargements
5. Terminates as a cone-shaped conus medullaris
6. Filum terminale attaches the cord to coccyx bone
7. Peripheral nerve root bundles at the end are cauda equina
8. Anterior surface bears a deep anterior median fissure
9. Posterior surface bears a posterior median sulcus

B. **The Spinal Cord and Its Protective Cover**
1. Spinal meninges enclose the cord from foramen magnum to sacrum
2. Outer membrane is the spinal dura mater
   a. continuous with meningeal dura mater of the brain
   b. does not attach to vertebral bones
3. Epidural space between spinal dura and vertebrae

4. Middle meningeal layer is arachnoid
   a. continuous with the arachnoid in the brain
   b. Underneath it is the subarachnoid space, filled with CSF
5. Innermost layer is pia mater
   a. lateral extensions from pia mater, the denticulate ligaments, attach the cord to the arachnoid membrane

II. THE SPINAL CORD IN CROSS SECTION

A. Internal Anatomy
1. Gray matter is composed of cell bodies and is central in cord, H-shaped
2. White matter contains axons and is peripheral in cord
3. Central canal is in the gray commissure
4. Gray matter exhibits anterior and posterior horns
   a. anterior horns contain motor neurons
   b. posterior horns contain sensory association neurons
   c. lateral horns contain autonomic motor neurons
5. Peripheral white matter is organized into
   a. anterior, posterior, and lateral white columns

III. THE SPINAL NERVES

A. Attachments
1. Anterior (ventral) and posterior (dorsal) roots attach to cord
2. Afferent fibers enter spinal cord through posterior roots
3. Efferent fibers exit spinal cord through anterior roots
4. Dorsal root ganglia contain sensory neuron cell bodies
5. Anterior and posterior roots merge laterally to form spinal nerve

Laboratory Exercises 12

NAME _____

LAB SECTION _____ DATE _____

LABORATORY EXERCISE 12.1
The Spinal Cord
Part I

1. Into what three regions is the spinal cord is divided?

   a. _____

   b. _____

   c. _____

2. What spinal cord enlargement supplies nerves to the upper extremities?

   _____

3. What spinal cord enlargement supplies nerves to the lower extremities?

   _____

4. What is the name of the terminal portion of the spinal cord?

   _____

5. What structure anchors the spinal cord to the coccyx bone?

   _____

6. The nerve bundles at the inferior end of the spinal cord are called

   _____. They are surrounded by

   _____ fluid.

7. What three connective tissue layers surround the spinal cord?

   a. _____

   b. _____

   c. _____

8. What structures secure the spinal cord along its length to the dural sheath?

   _____

9. What structure in the spinal cord contains CSF and is continuous with the fourth ventricle?

   _____

10. Where is the gray matter in the spinal cord located?

    _____

Part II

Match the description of the spinal cord on the left with the terms on the right.

Anterior horns contain _____

Posterior horns contain _____

Nerve cells in dorsal root ganglion _____

Gray matter connected by _____

Afferent nerve fibers found in _____

Anterior and posterior roots form _____

Motor neurons found in _____

Myelinated axons found in _____

Opening in gray commissure _____

Lateral horns found in _____

A. sensory axons

B. white matter

C. motor neurons

D. posterior roots

E. spinal nerve

F. gray commissure

G. thoracic spinal cord

H. central canal

I. sensory neurons

K. anterior gray horn

Part III

Using the listed terms, label the spinal cord.

Cervical enlargement
Cervical spinal nerves
Conus medullaris
Cauda equina
Filum terminale

Thoracic spinal nerves
Lumbosacral enlargement
Lumbar spinal nerves
Sacral spinal nerves

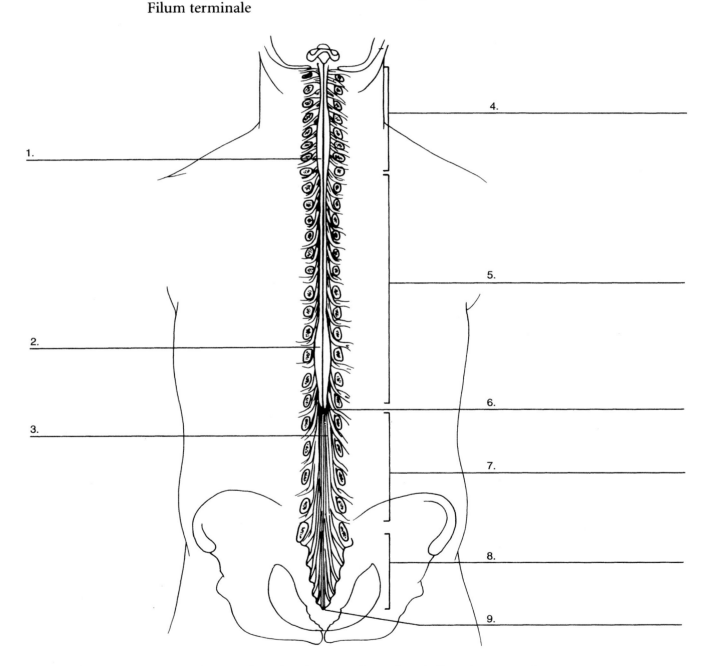

Figure 12.4 The spinal cord and nerves, from the skull to the coccyx bone.

LABORATORY EXERCISE 12.2

Using the listed terms, label the spinal cord.

White columns (posterior, anterior, and lateral funiculi)
Gray matter (anterior, posterior, and lateral horns)
Gray commissure
Dorsal root ganglion
Dorsal root
Arachnoid

Posterior median sulcus
Central canal
Anterior median fissure
Dura mater
Spinal nerve
Ventral root
Pia mater

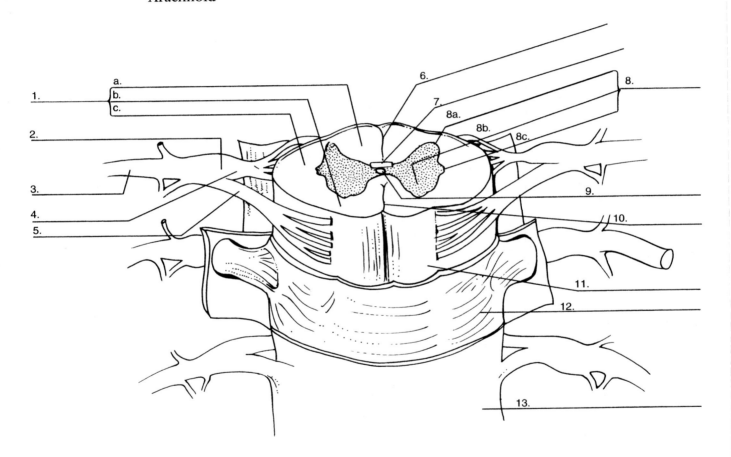

Figure 12.5 Thoracic region of spinal cord in cross section.

13

The Peripheral Nervous System

THE CRANIAL AND SPINAL NERVES

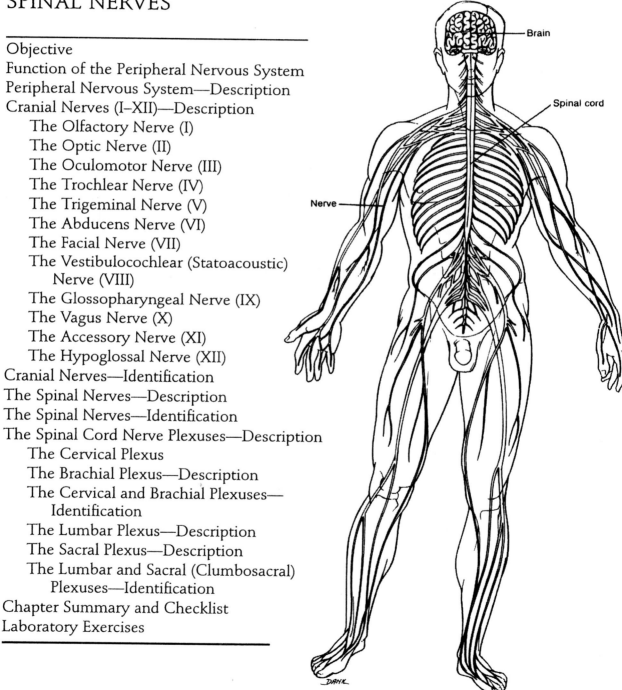

Brain

Spinal cord

Nerve

Objective

The objective of Chapter 13, "The Peripheral Nervous System: The Cranial and Spinal Nerves," is for you to know the location, distribution, and functions of the major:

1. **Cranial nerves**
2. **Spinal cord plexuses and their nerves**

Function of the Peripheral Nervous System

The peripheral nervous system conveys sensory information from the body and external environment from the peripheral receptors to the **afferent** fibers and into the central nervous system. Motor responses are carried via the **efferent** fibers to the effector organs in the blood vessels, glands, and muscles.

The Peripheral Nervous System—Description

Nerve cells and fibers external to the brain and spinal cord form the **peripheral nervous system**. Nerves that arise from the brain are called **cranial nerves**, and nerves that arise from the spinal cord are called **spinal nerves**. Most peripheral nerves are **mixed**; that is, they are both sensory and motor. Some peripheral nerves are responsible for highly specialized senses such as vision, hearing, smell, and balance; such nerves are **afferent**, or **sensory**, only.

Cranial Nerves I–XII—Description [Figure 13.1]

Twelve pairs of cranial nerves originate from the inferior surface of the brain. They pass to the periphery through various foramina in the skull and innervate structures in head and neck. The exception is the **vagus nerve**, which innervates visceral organs in the thoracic and abdominal cavities.

Each cranial nerve is designated by a Roman numeral and a name. The Roman numerals refer to the order, from rostral to caudal, in which the cranial nerves arise from the brain. The names indicate their major function, distribution, or structures that they innervate.

The Olfactory Nerve (I) [Figures 13.1 and 13.2]

The olfactory nerve originates from the olfactory cells in the upper nasal cavity (superior nasal concha and adjacent nasal septum). The afferent axons or nerve fibers from the olfactory cells pass through the **cribriform plate** or **ethmoid bone** of the skull, enter the cranial cavity, and synapse with neurons in the **olfactory bulb**, which lies directly over the foramina of the cribriform plate. The nerve fibers from the olfactory bulb then extend to the brain as the **olfactory tract**. The olfactory nerve is purely sensory; it conveys **olfaction**, or smell, from the receptors in the nose to the brain.

The Optic Nerve (II) [Figures 13.1 and 13.3]

The optic nerve originates from the light-sensitive receptor cells in the **retina** of the eye. From the posterior surface of each eyeball, the optic nerve exits the orbital cav-

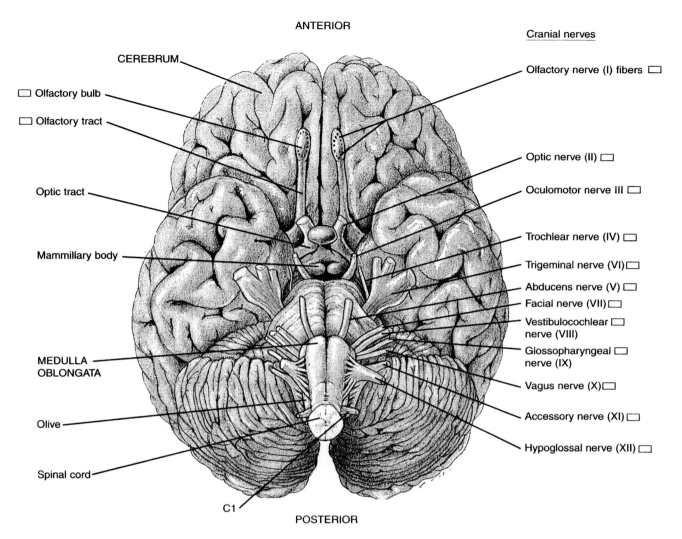

ANTERIOR

Cranial nerves

CEREBRUM

Olfactory bulb

Olfactory tract

Optic tract

Mammillary body

MEDULLA
OBLONGATA

Olive

Spinal cord

C1

POSTERIOR

Olfactory nerve (I) fibers ☐

Optic nerve (II) ☐

Oculomotor nerve III ☐

Trochlear nerve (IV) ☐

Trigeminal nerve (VI) ☐

Abducens nerve (V) ☐

Facial nerve (VII) ☐

Vestibulocochlear ☐
nerve (VIII)

Glossopharyngeal ☐
nerve (IX)

Vagus nerve (X) ☐

Accessory nerve (XI) ☐

Hypoglossal nerve (XII) ☐

Figure 13.1 Brain stem. Ventral surface of the brain shows the brain stem in relation to the cranial nerves.

ity, passes through the **optic foramina** in the sphenoid bone, and enters the cranial cavity. In the cranial cavity, the right and left optic nerves meet in an X-shaped **optic chiasm** situated anterior to the **infundibulum** (stalk) of the **hypophysis** (pituitary gland). In the optic chiasm, the nerve fibers from the medial half of each retina cross to the opposite side, while the fibers from the lateral half of each retina remain uncrossed. The fibers then continue posteriorly into the brain as the **optic tract**. After they synapse in the thalamic nucleus (lateral geniculate), the fibers terminate in the occipital (visual) cortex. The optic nerve is sensory for vision only.

The Oculomotor Nerve (III) [Figures 13.1 and 13.4]

The oculomotor nerve originates in the **midbrain**; its fibers pass into the orbit through the **superior orbital fissure**. Although this nerve is mixed sensory and motor, it is primarily motor. The motor portion innervates the muscles of the upper eyelid and the extrinsic eye muscles (the superior, inferior, and medial rectus and the inferior oblique). These muscles raise the eyelid and move the eyeball. In

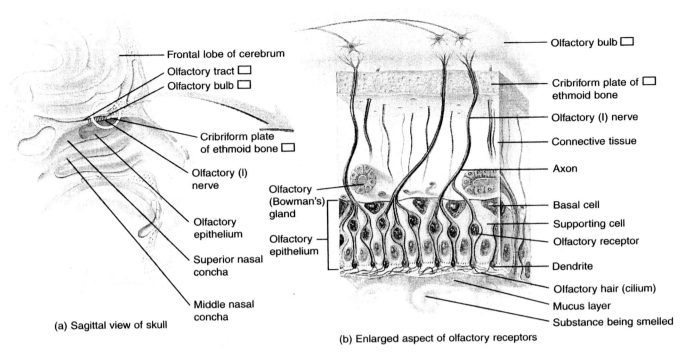

Labels for figure (a) Sagittal view of skull:
- Frontal lobe of cerebrum
- Olfactory tract ☐
- Olfactory bulb ☐
- Cribriform plate of ethmoid bone ☐
- Olfactory (I) nerve
- Olfactory epithelium
- Superior nasal concha
- Middle nasal concha

Labels for figure (b) Enlarged aspect of olfactory receptors:
- Olfactory bulb ☐
- Cribriform plate of ethmoid bone ☐
- Olfactory (I) nerve
- Connective tissue
- Axon
- Olfactory (Bowman's) gland
- Olfactory epithelium
- Basal cell
- Supporting cell
- Olfactory receptor
- Dendrite
- Olfactory hair (cilium)
- Mucus layer
- Substance being smelled

(a) Sagittal view of skull

(b) Enlarged aspect of olfactory receptors

Figure 13.2 (a) Location of olfactory receptors in nasal cavity and (b) enlarged aspect.

addition, autonomic nerve fibers run alongside the oculomotor nerve to innervate the smooth muscles of the iris and lens. The sensory portion of this nerve conveys impulses from the stretch receptors in the extrinsic eye muscles.

The Trochlear Nerve (IV) [Figures 13.1 and 13.4]

The smallest cranial nerve in diameter, the trochlear nerve originates in the midbrain. The fibers pass with the oculomotor nerve through the **superior orbital fissure** into the orbit. The trochlear nerve is mixed sensory and motor but primarily motor; it innervates the superior oblique muscles of the eyeball. The sensory impulses originate from the stretch receptors of the extrinsic muscles.

The Trigeminal Nerve (V) [Figures 13.1 and 13.5]

The trigeminal nerve is the largest cranial nerve and the major sensory nerve of the face. It originates on the ventrolateral side of the pons and has three major subdivisions: the **ophthalmic nerve, maxillary nerve,** and **mandibular nerve.** The trigeminal nerve carries sensory information from the head and face and innervates the muscles of mastication (chewing). A large **semilunar (trigeminal) ganglion** contains the cell bodies of the sensory neurons for the three branches of the trigeminal nerve. The three branches arise and diverge distal to the semilunar ganglion. A ganglion is a structure outside the central nervous system that contains nerve cell bodies.

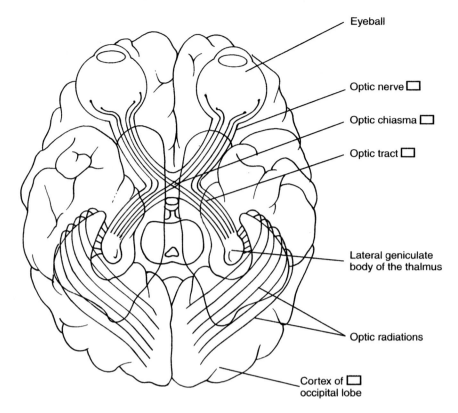

Eyeball

Optic nerve □

Optic chiasma □

Optic tract □

Lateral geniculate
body of the thalmus

Optic radiations

Cortex of □
occipital lobe

Figure 13.3 Optic nerve (II) fibers cross over from the medial retina and extend to the thalamus and occipital lobe of the brain.

THE OPHTHALMIC NERVE

The first branch of the trigeminal nerve enters the orbit through the **superior orbital fissure**. Its sensory fibers innervate the upper head and face (skin of the forehead, anterior scalp, upper eyelid, nose, nasal mucosa, and lacrimal gland).

THE MAXILLARY NERVE

The second branch of the trigeminal nerve leaves the cranium through the **foramen rotundum**. Its sensory fibers innervate the maxillary region of the face (upper teeth, upper lip, gums, mucosa of the nose, lower eyelid, skin of the cheek, and roof of the mouth).

THE MANDIBULAR NERVE

The third and largest branch of the trigeminal nerve is both sensory and motor. It leaves the cranium through the **foramen ovale**. Its sensory fibers innervate the lower region of the face (lower teeth, gums, skin over the mandible, lateral face anterior to the ear, anterior two-thirds of the tongue [except taste], and floor of the mouth).

All motor fibers of the trigeminal nerve are located in the mandibular nerve, exit the cranium through the foramen ovale, and innervate the muscles of mastication.

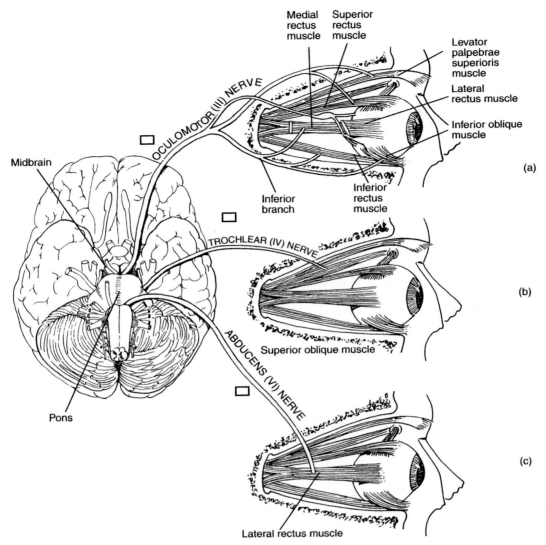

Figure 13.4 (a) Oculomotor nerve (III). (b) Trochlear nerve (IV). (c) Abducens nerve (VI).

The Abducens Nerve (VI) [Figures 13.1 and 13.4]

This small nerve originates near the posterior region of the pons and medulla. The nerve enters the orbit through the **superior orbital fissure** along with the oculomotor (III) and trochlear (IV) nerves. The abducens nerve is mixed sensory and motor but primarily motor. The motor fibers innervate the **extrinsic muscles** of the eyeball, the lateral rectus muscles. The sensory impulses originate in the stretch receptors of the lateral rectus muscles.

The Facial Nerve (VII) [Figures 13.1 and 13.6]

This nerve originates at the lower region of the **pons**, lateral to the abducens nerve (VI), enters the temporal bone through the internal auditory meatus, transverses the petrous portion of the temporal bone, and emerges from the cranium via the **stylomastoid foramina**. The facial nerve is mixed sensory and motor. The motor fibers innervate the muscles of facial expression, muscles in the scalp and platysma,

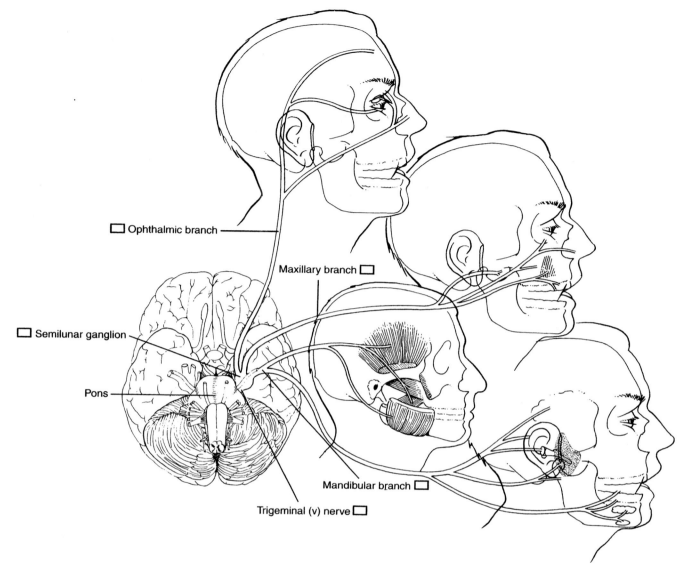

Figure 13.5 The trigeminal nerve (V) and its distribution in the face.

□ Ophthalmic branch

Maxillary branch □

□ Semilunar ganglion

Pons

Mandibular branch □

Trigeminal (v) nerve □

posterior belly of the digastric muscle, and stylohyoid muscles. Also, various secretory glands (lacrimal, submandibular, sublingual, nasal, and palatine) are innervated by motor fibers of the autonomic nerves in the facial nerve. The sensory portion of the facial nerve conveys taste sensations from the anterior two-thirds of the tongue and information to move facial muscles (from stretch receptors) of the face and scalp.

The Vestibulocochlear (Statoacoustic) Nerve (VIII)
[Figures 13.1 and 13.7]

The vestibulocochlear nerve is sensory and consists of the **cochlear** and **vestibular nerves**. The sensory fibers originate from the receptors located in the auditory and equilibrium apparatus; these are located within the inner ear of the temporal bone. The two nerves merge to form a single cranial nerve, the vestibulocochlear nerve. This nerve leaves the inner ear through the **internal auditory meatus** of the tem-

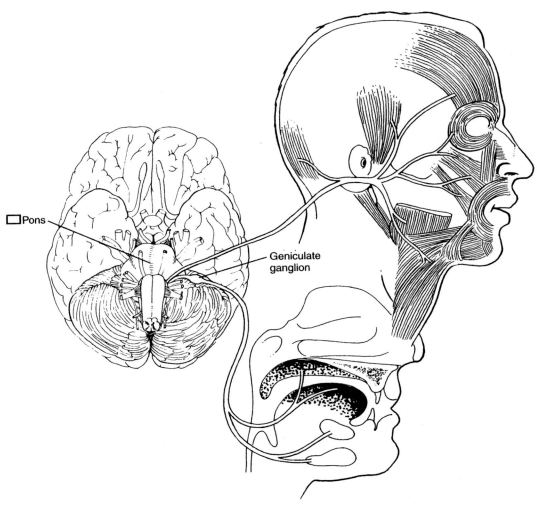

Pons

Geniculate
ganglion

Figure 13.6 The facial nerve (VII) and its distribution.

poral bone and enters the brain stem at the pons-medulla border. The vestibulo-cochlear nerve conveys impulses to the brain for hearing, equilibrium, and balance.

The Glossopharyngeal Nerve (IX)

[Figures 13.1 and 13.8]

The glossopharyngeal nerve originates in the lateral medulla, leaves the cranium through the **jugular foramen**, and innervates the tongue and throat. It is mixed sensory and motor. The motor fibers innervate some muscles of the tongue and pharynx. The glossopharyngeal nerve also contains the autonomic nerve fibers that innervate the parotid salivary glands. As a result, this nerve controls the swallowing reflex and saliva secretion. The sensory fibers convey sensations from the pharynx, soft palate, and taste buds from the posterior third of the tongue.

The Vagus Nerve (X) [Figures 13.1 and 13.9]

The vagus nerve originates in the lateral medulla, leaves the cranium through the **jugular foramina**, and descends in the neck to innervate viscera in the thorax and

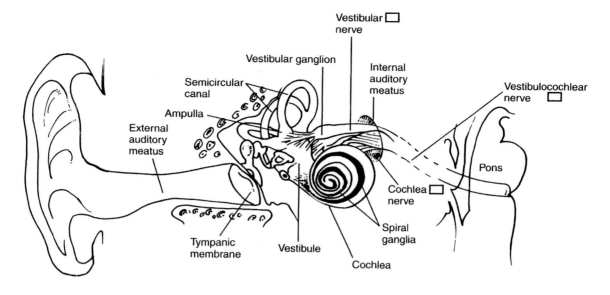

Figure 13.7 Vestibulocochlear nerve (VIII) and its distribution in the inner ear.

abdomen. The vagus nerve is mixed sensory and motor. The motor fibers, including autonomic nerve fibers, innervate the muscles of the pharynx, larynx, respiratory tract, lungs, heart, esophagus, and most of the abdominal viscera except the lower part of the large intestine. Sensory information is conveyed from the same organs innervated by the motor fibers. Sensory stretch fibers also arise from the muscles innervated by the motor fibers.

The Accessory Nerve (XI) [Figures 13.1 and 13.10]

The accessory nerve has a dual origin in that each accessory nerve is formed by two nerves. The cranial portion arises from the medulla, and the spinal portion from the first five or six segments in the cervical region of the spinal cord. The spinal portion of the accessory nerve ascends along the spinal cord and enters the cranium through the foramen magnum. The spinal and cranial portions merge in the cranium and leave the cranium through the jugular foramina. The accessory nerve is primarily motor. The cranial portion joins the vagus nerve (X) and innervates the pharynx, larynx, and soft palate. The spinal portion innervates the sternocleidomastoid and trapezius muscles, which move the head and neck. The sensory fibers pass from the sense organs in the innervated muscles.

The Hypoglossal Nerve (XII) [Figures 13.1 and 13.11]

The hypoglossal nerve originates in the medulla, passes through the **hypoglossal canals** in the occipital bones of the cranium, and innervates the tongue muscles. This nerve is mixed sensory and motor but is primarily motor to the tongue muscles. The coordinated contractions of the tongue muscles allow for chewing, swallowing, and speech. The sensory portion of the nerve consists of nerve fibers arising from sensory receptors in tongue muscles.

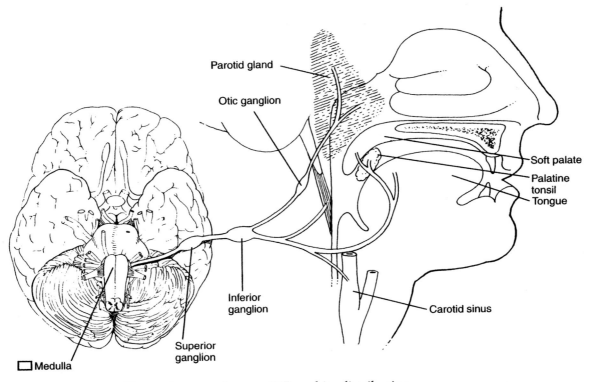

Parotid gland

Otic ganglion

Soft palate

Palatine tonsil

Tongue

Inferior ganglion

Carotid sinus

Superior ganglion

☐ Medulla

Figure 13.8 Glossopharyngeal nerve (IX) and its distribution.

Cranial Nerves—Identification *[Figures 13.1–13.11]*

To learn the location of various cranial nerves in the brain, examine brain models and illustrations carefully before you proceed to prepared cadaver brains. Identify as many of the following cranial nerves and associated parts as you can. Color the nerves and structures they innervate.

1. **olfactory nerve (I)**
   a. **olfactory bulb**
   b. **olfactory tract**
2. **optic nerve (II)**
   a. **optic chiasm**
3. **oculomotor nerve (III)**
4. **trochlear nerve (IV)**
5. **trigeminal nerve (V)**
6. **abducens nerve (VI)**
7. **facial nerve (VII)**
8. **vestibulocochlear nerve (VIII)**
9. **glossopharyngeal nerve (IX)**
10. **vagus nerve (X)**
11. **accessory nerve (XI)**
12. **hypoglossal nerve (XII)**

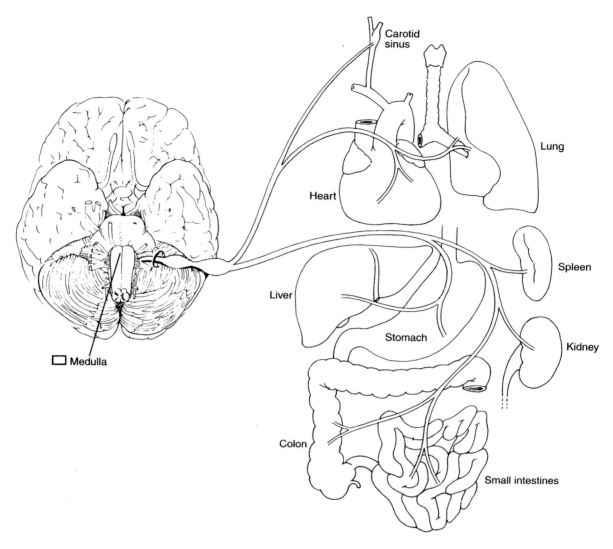

Figure 13.9 Vagus nerve (X) and its distribution in the thorax and abdomen.

The Spinal Nerves—Description *[Figures 13.12–13.14]*

The spinal nerves attach to the spinal cord and lie in the intervertebral foramina between the adjoining vertebrae; their branches pass to peripheral regions of the body. All spinal nerves contain both afferent and efferent fibers. The spinal nerves are named and numbered according to the spinal cord region and segment from which they originate. There are 31 pairs of spinal nerves:

1. Eight pairs of cervical nerves (C1–C8)
2. Twelve pairs of thoracic nerves (T1–T12)
3. Five pairs of lumbar nerves (L1–L5)
4. Five pairs of sacral nerves (S1–S5)
5. One pair of coccygeal nerves

The first pair of cervical nerves lies above the first cervical vertebra between the occipital bone and atlas. Cervical nerves 2 to 7 lie above the vertebrae for which

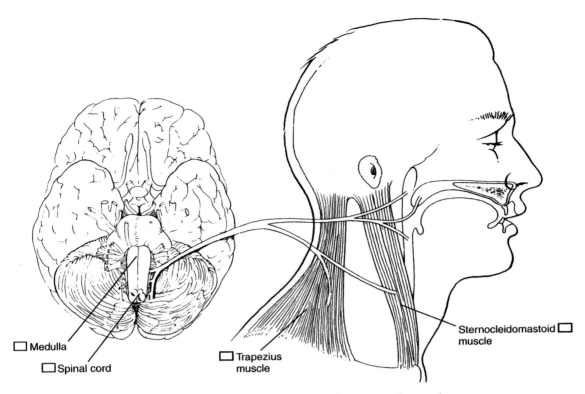

☐ Medulla

☐ Spinal cord

☐ Trapezius
muscle

Sternocleidomastoid ☐
muscle

Figure 13.10 Accessory nerve (XI) and its distribution in the neck.

they are named. The eighth pair of cervical nerves lies between the seventh cervical and first thoracic vertebra. Below this level, all pairs of spinal nerves leave the spinal cord inferior to the vertebrae for which they are named (Figure 13.12).

Each spinal nerve is connected to the spinal cord by a **posterior (dorsal) root** composed of sensory fibers and a **anterior (ventral) root** composed of motor fibers. The posterior (dorsal) root bears an enlargement called a **posterior (dorsal) root ganglion**, which contains the cell bodies of the sensory neurons. The posterior and anterior roots pass laterally from the spinal cord and merge to form a single, mixed **spinal nerve** in the region of the **intervertebral foramen**. After passing through the intervertebral foramen, each spinal nerve divides into two major branches: a large **ventral (anterior) ramus** and a smaller **dorsal (posterior) ramus**. (The primary rami or branches of the spinal nerves are also called **primary divisions.**)

The thin dorsal (posterior) ramus of each spinal nerve passes posteriorly to innervate the skin and muscles of the posterior neck and trunk (back). The ventral (anterior) ramus of each spinal nerve passes laterally to innervate the muscles and skin on the anterolateral side of the body and limbs. Because the ventral rami innervate a larger body area than the dorsal rami, they are larger and thicker (Figures 13.13 and 13.14).

The Spinal Nerves—Identification

[Figures 13.12–13.14]

Examine the illustrations and a model of a typical spinal nerve attached to the spinal cord, and color the parts. If possible, examine a prepared spinal cord and dis-

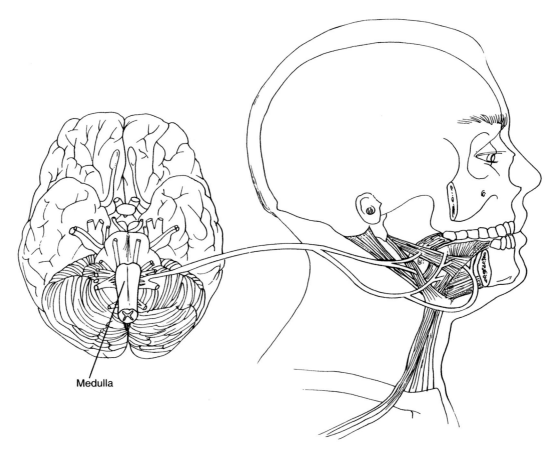

Figure 13.11 Hypoglossal nerve (XII).

sected vertebral column. Using the model, illustrations, and spinal cord from the cadaver, identify the following structures:

1. **cervical spinal nerves**
2. **thoracic spinal nerves**
3. **lumbar spinal nerves**
4. **sacral spinal nerves**
5. **anterior (ventral) root**
6. **ventral ramus**
7. **posterior (dorsal) root**
8. **dorsal root ganglion**
9. **dorsal ramus**

The Spinal Cord Nerve Plexuses—Description

[Figures 13.12 and 13.13]

In the cervical, brachial, lumbar, and sacral regions of the spinal cord, the ventral rami of spinal nerves merge to form nerve networks called **plexuses**. The nerves from these plexuses then supply the skin, muscles, and joints of the upper and lower extremities. Within such plexuses are nerves from different nerve roots of the spinal

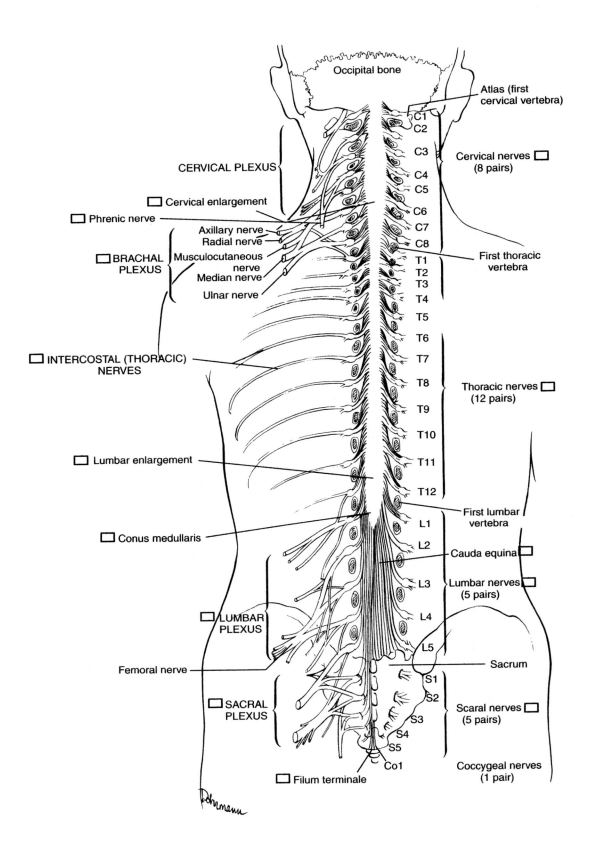

Occipital bone

Atlas (first cervical vertebra)

C1
C2

Cervical nerves ☐
(8 pairs)

C3

CERVICAL PLEXUS

C4
C5

☐ Cervical enlargement

C6

☐ Phrenic nerve

C7

Axillary nerve
Radial nerve

C8

☐ BRACHAL PLEXUS

Musculocutaneous nerve

T1

First thoracic vertebra

Median nerve

T2

Ulnar nerve

T3
T4
T5
T6

☐ INTERCOSTAL (THORACIC) NERVES

T7
T8

Thoracic nerves ☐
(12 pairs)

T9
T10

☐ Lumbar enlargement

T11
T12

First lumbar vertebra

☐ Conus medullaris

L1
L2

Cauda equina ☐

L3

Lumbar nerves ☐
(5 pairs)

☐ LUMBAR PLEXUS

L4
L5

Femoral nerve

Sacrum

S1
S2

☐ SACRAL PLEXUS

Scaral nerves ☐
(5 pairs)

S3
S4
S5
Co1

Coccygeal nerves
(1 pair)

☐ Filum terminale

Figure 13.12 Posterior view of the spinal cord and spinal nerves.

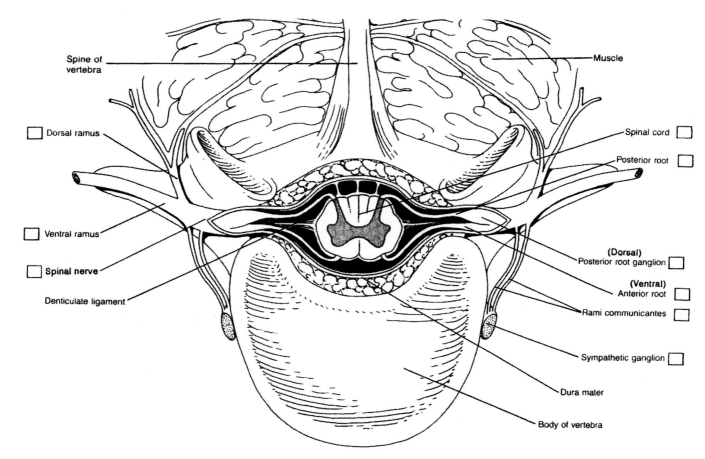

Spine of vertebra

Muscle

Dorsal ramus

Spinal cord

Posterior root

Ventral ramus

(Dorsal)
Posterior root ganglion

Spinal nerve

(Ventral)
Anterior root

Denticulate ligament

Rami communicantes

Sympathetic ganglion

Dura mater

Body of vertebra

Figure 13.13 Branches of a typical spinal nerve in cross section through the vertebra.

cord. The major plexuses in the human body are the **cervical, brachial, lumbar,** and **sacral or lumbosarcal plexuses**.

The thoracic nerves do not form plexuses. Instead, the ventral rami of the thoracic nerves (T1–T12) extend in the intercostal spaces between the ribs to innervate the skin and muscles of the lateral and anterior thorax.

The Cervical Plexus

The cervical plexus is a small plexus deep in the neck. It is formed from the ventral rami of the first four cervical nerves (C1–C4), with contribution from the fifth cervical nerve (C5). The nerves from this plexus innervate most of the skin and muscles of the neck, portions of the head, and shoulders.

The most important nerve from the cervical plexus is the **phrenic nerve,** which descends along each side of the neck and enters the thorax. Here it passes in front of the lung root, descends inferiorly on the surface of the pericardium of the heart, and innervates the chief muscle of respiration, the diaphragm (Figure 13.12).

The Brachial Plexus—Description [Figures 13.12 and 13.15]

The brachial plexus is located in the neck and axilla of the arm. It extends inferiorly and laterally from the spinal cord, passes posteriorly to the clavicle, and enters the

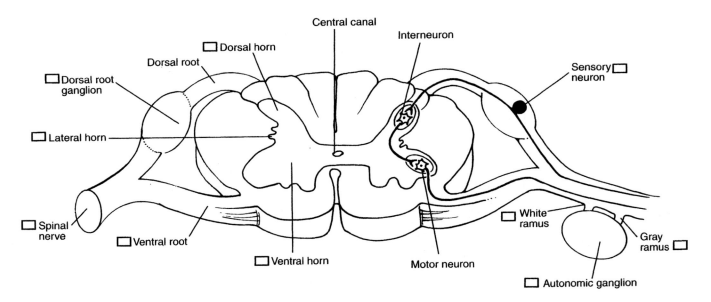

Figure 13.14 Cross section through thoracic segment of the spinal cord. (Left) the morphology of the cord. (Right) the location of neurons in the gray horn.

axilla. The brachial plexus supplies all major nerves to the upper extremity and shoulder.

The brachial plexus is formed by the ventral rami of spinal nerves C5–C8 and T1. It is a complex plexus composed of different **roots, trunks, divisions**, and **cords**. Five major nerves arise from the brachial plexus as the terminal branches of the cords. These nerves are the **axillary, radial, musculocutaneous, median**, and **ulnar**. Following are their origins and distributions:

Nerve	*Origin*	*Distribution*
Axillary	C5 and C6	Skin of shoulder; deltoid and teres minor muscles
Radial	C5–C8, T1	Skin of posterolateral surface of upper limb; all extensor muscles of upper arm, forearm, and hand
Musculocutaneous	C5–C7	Skin on lateral surface of forearm; flexor muscles of arm
Median	C5–C8, T1	Skin of lateral two-thirds of ventral part of hand; most flexors of anterior forearm, lateral palm, and first two fingers

Nerve	Origin	Distribution
Ulna	C8 and T1	Skin of medial third of hand; some flexor muscles of anterior forearm, medial palm, and intrinsic muscles of hand

Following are other smaller nerves of the brachial plexus and their distributions:

Nerve	Distribution
Dorsal scapular	Levator scapulae and rhomboid muscles
Suprascapular	Supraspinatus and infraspinatus muscles
Long thoracic	Serratus anterior muscle
Subscapular	Teres major and subscapular muscles
Thoracodorsal	Lattisimus dorsal muscle
Pectoral (medial and lateral)	Pectoralis minor and major muscles

The Cervical and Brachial Plexuses—Identification

[Figures 13.12 and 13.15]

Examine models and the illustrations of the cervical and brachial plexuses and color the parts. If possible, examine the prepared cervical plexus in a cadaver and identify the phrenic nerves. Examine the brachial plexus of the axillary region and identify the following:

Cords	Nerves
lateral cord	musculocutaneous nerve
posterior cord	median nerve
medial cord	ulnar nerve
	axillary nerve
	radial nerve

The following smaller nerves of the brachial plexus are difficult to prepare for demonstration, but try to identify them:

1. long thoracic
2. dorsal scapular
3. suprascapular
4. superior subscapular
5. thoracodorsal
6. inferior subscapular
7. lateral pectoral
8. medial pectoral

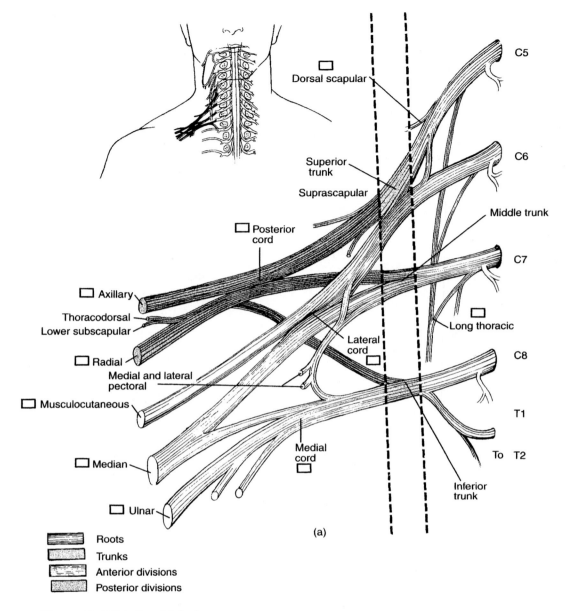

Labels in figure:

Dorsal scapular

C5

Superior trunk

Suprascapular

C6

Middle trunk

Posterior cord

C7

Axillary

Long thoracic

Thoracodorsal

Lower subscapular

Lateral cord

C8

Radial

Medial and lateral pectoral

Musculocutaneous

T1

Median

Medial cord

To T2

Ulnar

Inferior trunk

(a)

Roots
Trunks
Anterior divisions
Posterior divisions

Figure 13.15(a) Brachial plexus.

The Lumbar Plexus—Description [Figures 13.12 and 13.16]

The lumbar plexus originates from the ventral rami of spinal nerves L1–L4 and some fibers from the T12 nerve. The plexus is located laterally to the first four lumbar vertebrae. The nerves from the lumbar plexus innervate structures of the **anterolateral abdominal wall**, **external genitalia**, and **anterior** and **medial** regions of the **lower limb**. Because some nerves of the lumbar plexus intermix and contribute to the sacral plexus, the two plexuses are often described as the **lumbosacral plexus**.

The lumbar plexus is located within the **psoas major muscles**. The peripheral nerves of the plexus emerge from the lateral and medial borders of the muscle and

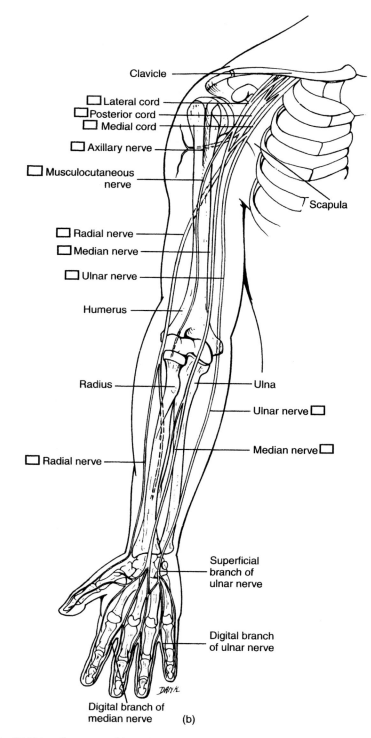

Clavicle

☐ Lateral cord
☐ Posterior cord
☐ Medial cord

☐ Axillary nerve

☐ Musculocutaneous
nerve

Scapula

☐ Radial nerve

☐ Median nerve

☐ Ulnar nerve

Humerus

Radius

Ulna

Ulnar nerve ☐

Median nerve ☐

☐ Radial nerve

Superficial
branch of
ulnar nerve

Digital branch
of ulnar nerve

Digital branch of
median nerve   (b)

Figure 13.15(b) Distribution of brachial plexus nerves in the arm.

its anterior surface. The major peripheral nerves of the lumbar plexus are the **femoral** and **obturator nerves**.

The **femoral nerve**, the largest branch of the lumbar plexus, passes inferior to the inguinal ligament to supply the thigh flexor and knee extensor muscles (iliopsoas, sartorius, and quadriceps femoris). The femoral nerve also innervates the skin of anterior and medial thigh and the medial leg and foot.

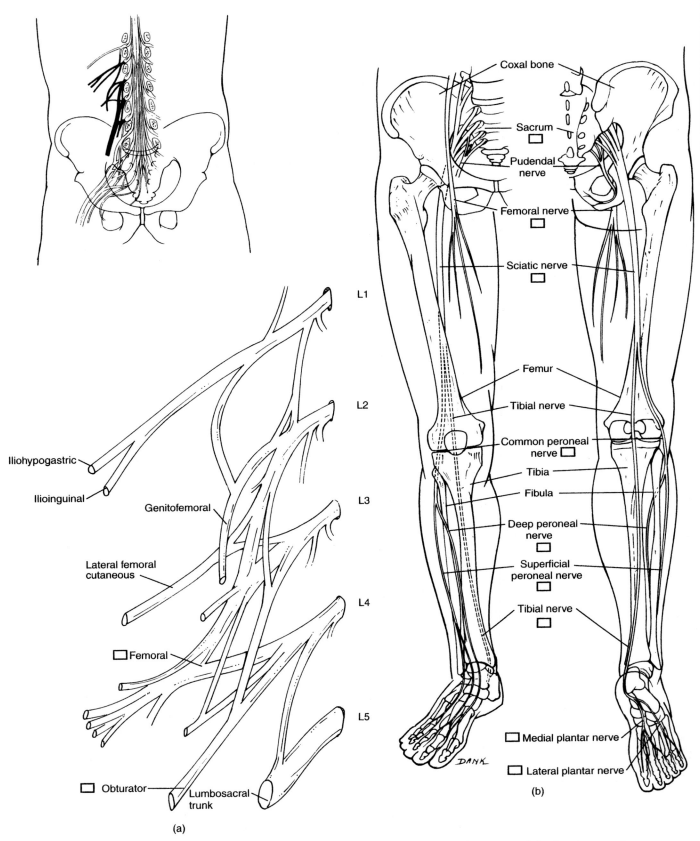

L1

L2

Iliohypogastric

Ilioinguinal

Genitofemoral

L3

Lateral femoral
cutaneous

L4

☐ Femoral

L5

☐ Obturator — Lumbosacral
trunk

(a)

Coxal bone

Sacrum ☐

Pudendal
nerve

Femoral nerve ☐

Sciatic nerve ☐

Femur

Tibial nerve

Common peroneal
nerve ☐

Tibia

Fibula

Deep peroneal
nerve ☐

Superficial
peroneal nerve ☐

Tibial nerve ☐

☐ Medial plantar nerve

DANK

☐ Lateral plantar nerve

(b)

Figure 13.16 Lumbar plexus. (a) Origin. (b) Distribution of nerves of the lumbar and sacral plexuses in (left) anterior (right) posterior views.

The **obturator nerve** leaves the pelvis through the obturator foramen and supplies the adductor (medial) muscles of the thigh and the skin on the medial surface of the thigh.

Other smaller nerves originating from the lumbar plexus are the **lateral femoral cutaneous, iliohypogastric, ilioinguinal,** and **genitofemoral nerves**.

The Sacral Plexus—Description [Figures 13.12, 13.16, and 13.17]

The sacral plexus is located inferior to the lumbar plexus and anterior to the sacrum. It originates from the ventral rami of spinal nerves L4, L5, and S1–S4.

The **sciatic nerve** is the largest nerve of the sacral plexus and is composed of two nerves, the **common peroneal** and **tibial nerves,** surrounded by a common sciatic sheath. Each sciatic nerve leaves the pelvis through the greater sciatic notch and descends deep to the gluteus maximus muscle. The nerve innervates the skin and hamstring muscles (thigh extensors and knee flexors) of the posterior thigh. In the lower third of the thigh near the popliteal fossa above the knee, the sciatic nerve separates into the **tibial** and **common peroneal nerves.**

The **tibial nerve** is larger than the common peroneal nerve and innervates most of the posterior thigh and leg muscles (flexors of the leg and foot). It also gives rise to the sural nerve, which supplies the skin on the posterior surface of the leg. The tibial nerve ends in the sole of the foot as the **medial** and **lateral plantar nerves,** which innervate the intrinsic muscles of the foot.

The **common peroneal nerve,** which divides into **superficial** and **deep branches,** passes on the lateral side of the popliteal fossa and supplies the muscles in the anterior and lateral muscles of the leg and foot (flexors of the foot and extensor of the toes). It also innervates the skin on the anterior surface of the leg and dorsum of the foot.

Other important nerves of the sacral plexus are the pudendal and inferior and superior gluteal nerves. The **pudendal nerve** enters the perineum through the lesser sciatic notch and innervates the muscles of the perineum and external genitalia. The **inferior gluteal nerve** innervates the gluteus maximus muscle of the buttock. The **superior gluteal nerve** innervates the gluteus medius and minimus and tensor fascia latae muscles.

The Lumbar and Sacral (Lumbosacral) Plexuses— Identification [Figures 13.12, 13.16, and 13.17]

Examine a model and the illustrations of the lumbar and sacral plexuses and color the nerves. If possible, examine the lumbar and sacral regions in a prepared cadaver and identify the following:

1. **femoral nerve**
2. **sciatic nerve**
3. **pudendal nerve**
4. **tibial nerve**
5. **common peroneal nerve**
6. **lateral plantar nerve**
7. **medial plantar nerve**
8. **obturator nerve**

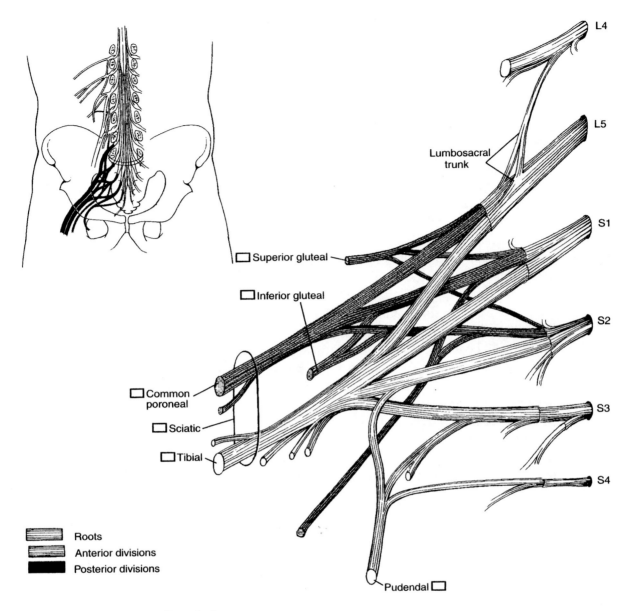

L4

L5

Lumbosacral
trunk

S1

☐ Superior gluteal

S2

☐ Inferior gluteal

☐ Common
poroneal

☐ Sciatic

☐ Tibial

S3

S4

▨ Roots
▨ Anterior divisions
■ Posterior divisions

Pudendal ☐

Figure 13.17 Sacral plexus.

CHAPTER SUMMARY AND CHECKLIST

I. **THE PERIPHERAL NERVOUS SYSTEM**

A. **Composition**
1. Nervous structures outside the central nervous system
2. Nerves arising from the brain are cranial nerves
3. Nerves arising from the spinal cord are spinal nerves
4. Most nerves are mixed afferent and efferent

B. **The Cranial Nerves**
1. Twelve pairs of cranial nerves originate from the brain
2. Most cranial nerves innervate the head and neck
3. Cranial nerves are designated by Roman numerals and names

II. THE CRANIAL NERVES

A. The Olfactory Nerves (I)
1. Originate from olfactory epithelium in the nose
2. Axons pass through cribriform plate to olfactory bulbs
3. Olfactory tracts enter the brain
4. Nerves are purely sensory for smell

B. The Optic Nerves (II)
1. Afferent nerves originate in retina of the eye
2. Pass through optic foramina into cranium
3. Axons from medial half of each retina cross over in optic chiasm
4. Nerves are purely sensory for vision

C. The Oculomotor Nerves (III)
1. Originate in midbrain, and fibers exit through superior orbital fissure
2. Primary function is motor for extrinsic eye muscle pairs
3. Fibers innervate iris and lens muscles

D. The Trochlear Nerves (IV)
1. The smallest cranial nerves, originate in midbrain
2. Pass through superior orbital fissure into orbit
3. Motor to one extrinsic eye muscle pair (superior oblique)

E. The Trigeminal Nerves (V)
1. The largest cranial nerves and major sensory nerves of the face
2. Originate on ventrolateral side of pons
3. Ophthalmic, maxillary, and mandibular subdivisions
4. Ophthalmic nerves are sensory to upper head and face
5. Maxillary nerves are sensory to maxillary region of face
6. Mandibular nerves are sensory to lower face
7. Motor fibers in mandibular branch innervate muscles of mastication

F. The Abducens Nerves (VI)
1. Originate near pons and medulla
2. Enter orbit through superior orbital fissure
3. Innervate one extrinsic eye muscle pair (lateral rectus)

G. The Facial Nerves (VII)
1. Originate at lower pons region
2. Exit cranium through stylomastoid foramina
3. Innervate the muscles of facial expression and muscles of scalp and platysma
4. Sensory fibers convey impulses from taste buds in anterior two-thirds of tongue

H. The Vestibulocochlear Nerves (VIII)
1. Purely sensory nerves
2. Fibers originate in auditory and equilibrium apparatus
3. Fibers leave inner ear through internal auditory meatus

I. **The Glossopharyngeal Nerves (IX)**
   1. Originate in lateral medulla
   2. Fibers exit cranium through jugular foramina
   3. Innervate muscles of tongue and pharynx
   4. Control swallowing reflex and saliva production
   5. Sensory to pharynx, soft palate, and taste buds in posterior third of tongue

J. **The Vagus Nerves (X)**
   1. Originate in lateral medulla
   2. Exit cranium through jugular foramina
   3. Enter thorax and abdomen to innervate visceral organs
   4. Innervate pharynx, larynx, lungs, heart, esophagus, and most of abdominal viscera
   5. Sensory fibers from innervated organs

K. **The Accessory Nerves (XI)**
   1. Dual origin in cranial and spinal nerves
   2. Cranial part arises from medulla
   3. Spinal part arises from segments of the cervical spinal cord
   4. Leave cranium through jugular foramina
   5. Cranial portion innervates pharynx, larynx, soft palate
   6. Spinal portion innervates sternocleidomastoid and trapezius muscles

L. **The Hypoglossal Nerves (XII)**
   1. Originate in medulla and pass through hypoglossal canals
   2. Innervate muscles of tongue
   3. Coordinate chewing, swallowing, and speech

III. **THE STRUCTURE OF THE SPINAL NERVES**

A. **Description**
   1. All spinal nerves are mixed sensory and motor and lie in intervertebral foramina
   2. Spinal nerves are named according to region of origin in the cord
   3. There are 31 pairs of spinal nerves from all regions
   4. Spinal nerves attached to cord by dorsal (afferent) and ventral (efferent) roots
   5. Sensory nerve cells located in dorsal root ganglion
   6. Dorsal and ventral roots merge to form spinal nerves
   7. Spinal nerves divide into ventral and dorsal rami

IV. **THE SPINAL NERVE PLEXUSES**

A. **The Cervical Plexus**
   1. Small plexus deep in the neck region
   2. Formed from ventral rami of C1–C5 nerves
   3. Nerves innervate most of the skin of the neck, head, and shoulder
   4. Most important branches are the phrenic nerves
      a. phrenic nerves innervate the diaphragm muscles

B. The Brachial Plexus
1. Provides major nerves to upper extremities
2. Located in the neck and axillary regions
3. Formed from ventral rami of C5–C8 and T1 nerves
4. Divided into roots, trunks, divisions, and cords
   a. branches from the roots form dorsal scapular and long thoracic nerves
   b. the suprascapular nerve branches from the trunk
5. Major nerves arise from the lateral, medial, and posterior cords
6. Posterior cord forms axillary and radial nerves
   a. smaller nerves are superior and inferior subscapular and thoracodorsal
7. Lateral cord forms musculocutaneous and lateral pectoral nerves
8. Medial cord forms the ulnar and medial pectoral nerves
9. Medial and lateral cords form the median nerve
10. Axillary nerves innervate shoulder muscles and skin
11. Musculocutaneous nerves innervate flexors of the arm and lateral skin of the forearm
12. Radial nerves pass around lateral epicondyles of the humerus
    a. Innervate extensor muscles of posterior upper arm, forearm, hand, and skin of posterolateral limb surface
13. Median nerves innervate most flexors of anterior forearm, lateral palm, two fingers, and skin of lateral hand
14. Ulnar nerves pass around medial condyles of humerus
    a. Innervate some flexors of anterior forearm, medial palm and hand, and skin on medial third of hand

C. The Lumbar Plexus
1. Originates from ventral rami of T12, L1–L4
2. Nerves innervate abdominal wall, external genitalia, and lower extremities
3. Major nerves are the obturator and femoral
   a. Femoral nerves supply anterior flexors of thigh and knee extensors
   b. Obturator nerves innervate adductor thigh muscles and skin on medial thigh

D. The Sacral Plexus
1. Nerves originate from ventral rami of L4–L5 and S1
2. Sciatic nerves are the largest and longest nerves of the body
   a. Composed of common peroneal and tibial nerves
   b. Innervate thigh extensors and knee flexors on posterior thigh
   c. Separate into tibial and common peroneal nerves
3. Tibial nerves innervate muscles of the posterior leg
4. In the sole of the foot, tibial nerves end as medial and lateral plantar nerves to the intrinsic muscles of the foot
5. Common peroneal nerves divide into superficial and deep
   a. Innervate muscles of anterior and lateral compartments of leg

Laboratory Exercises 13

NAME _____

LAB SECTION _____ DATE _____

LABORATORY EXERCISE 13.1
Cranial Nerves

Part I

List the names and numbers of the 12 cranial nerves.

1. _____   7. _____

2. _____   8. _____

3. _____   9. _____

4. _____   10. _____

5. _____   11. _____

6. _____   12. _____

Part II

Supply the cranial nerve(s) whose functions are described.

1. Movement (rolling) of the eyeballs

    a. _____

    b. _____

    c. _____

2. Smelling of odors _____

3. Provides sensory information for vision _____

4. The three subdivisions of the trigeminal nerve

    a. _____

    b. _____

    c. _____

5. Which subdivision of the trigeminal nerves performs the following:

    a. mastication of food _____

    b. innervation of upper head and face _____

    c. innervation of upper lips and upper teeth _____

6. Which nerve produces facial expressions? _____

7. Which nerve conveys information about hearing and balance?

   _____

8. Which nerve innervates most of the vital visceral organs?

   _____

9. Which nerve induces saliva secretion and is involved in swallowing?

   _____

10. Which nerve allows for vocalization and chewing?

    _____

11. Which nerve innervates the neck muscles? _____

12. Which are the three purely sensory cranial nerves?

    a. _____

    b. _____

    c. _____

Part III
Spinal Nerves and Spinal Cord Nerve Plexuses

1. The spinal nerves are located in the _____.

2. There are _____ pairs of spinal nerves in the human body.

3. Each spinal nerve is attached to the spinal cord by

   _____ and _____.

4. After passing through the vertebral canal, each spinal nerve divides into

   which two major branches?

   a. _____

   b. _____

5. List the major nerve plexuses in the human body.

   a. _____

b. _____

c. _____

d. _____

6. Which nerve plexuses are located in the following regions?

   a. deep in the neck _____

   b. neck and axilla of the arm _____

   c. psoas major muscle _____

   d. anterior to the sacrum _____

7. Which major nerves from the cervical and brachial plexuses innervate the following organs or structures?

   a. diaphragm _____

   b. all extensor muscles of posterior upper arm, forearm, and hand

   _____

   c. muscles and skin of the shoulder region _____

   d. flexors of the arm and skin on the lateral forearm

   _____

   e. flexors of anterior forearm, medial palm, and muscles of the hand

   _____

   f. flexors of anterior forearm, lateral palm, and first two fingers

   _____

8. Which major nerves from the lumbar and sacral plexuses innervate the following organs or structures?

   a. adductor muscles of the thigh and skin on the medial surface of the thigh

   _____

   b. thigh flexor and knee extensor muscles _____

c. posterior thigh and leg muscles (flexors of the leg and foot)

_____

d. muscles of the perineum and external genitalia _____

e. flexors of the foot and extensors of toes _____

9. Which is the largest nerve of the lumbar plexus? _____

10. Which is the largest nerve of the sacral plexus? _____

Part IV

Label and then color the three cranial nerves.

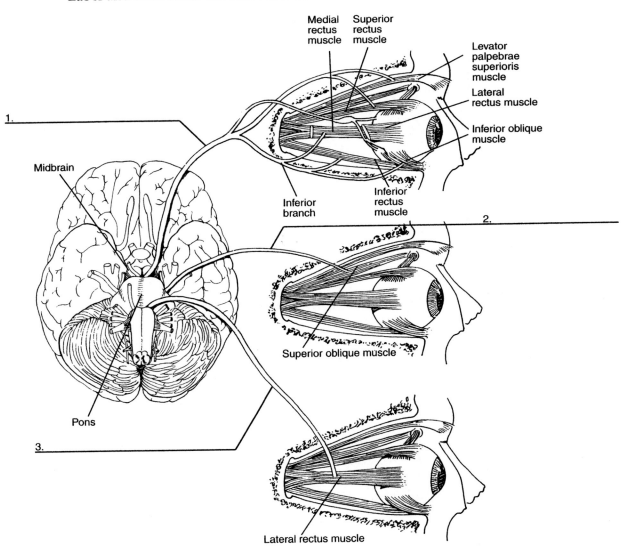

Figure 13.18 The cranial nerves innervate various skeletal muscles of the eyeball.

LABORATORY EXERCISE 13.2

Label the branches of the trigeminal nerve that innervate the designated structures and then color the parts.

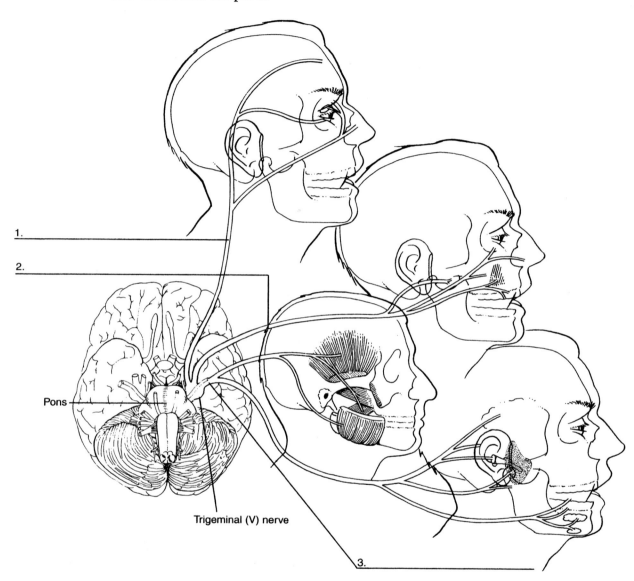

1. _____

2. _____

Pons

Trigeminal (V) nerve

3. _____

Figure 13.19 Distribution of the trigeminal nerve.

LABORATORY EXERCISE 13.3

Using the listed terms, label and then color the following structures.

Spinal cord
Ventral ramus
Anterior root
Dorsal ramus

Spinal nerve
Posterior root
Dorsal root ganglion

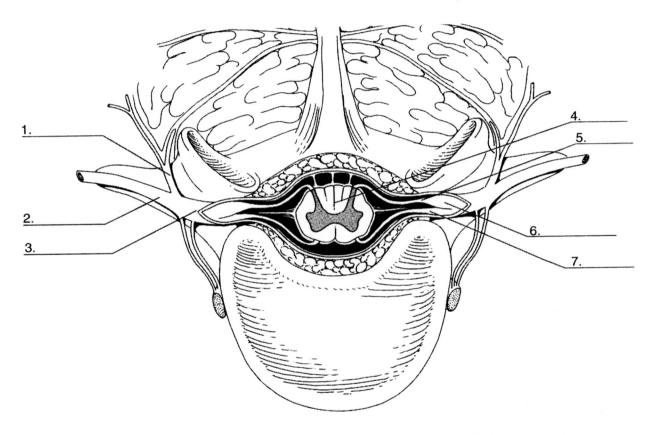

1.

2.

3.

4.

5.

6.

7.

Figure 13.20 Typical spinal nerve distribution through the vertebra (cross section).

LABORATORY EXERCISE 13.4

Using the listed terms, label and then color the cords and nerves of the brachial plexus.

Lateral, posterior, and medial cord   Radial nerve
Musculocutaneous nerve   Axillary nerve
Ulnar nerve   Median nerve

1. _____
2. _____
3. _____
4. _____
5. _____

6. _____
7. _____
8. _____

10. _____

11. _____

9. _____

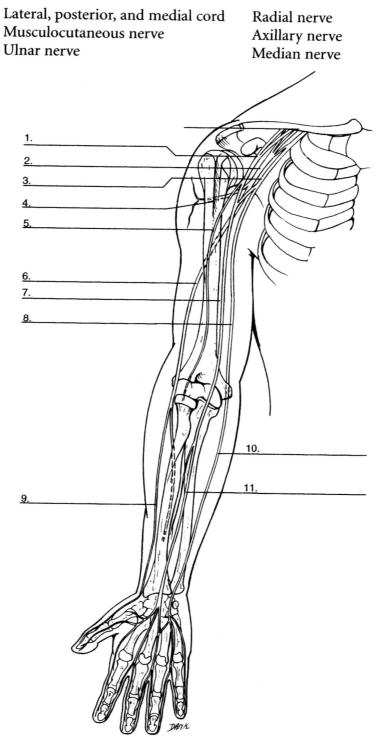

Figure 13.21 Distribution of the nerves of the brachial plexus in the arm.

LABORATORY EXERCISE 13.5

Using the listed terms, label and then color the nerves of the lumbar plexus.

Sciatic nerve
Tibial nerve
Tibial nerve
Medial plantar nerve
Pudendal nerve

Common peroneal nerve
Superficial peroneal nerve
Lateral plantar nerve
Femoral nerve
Deep peroneal nerve

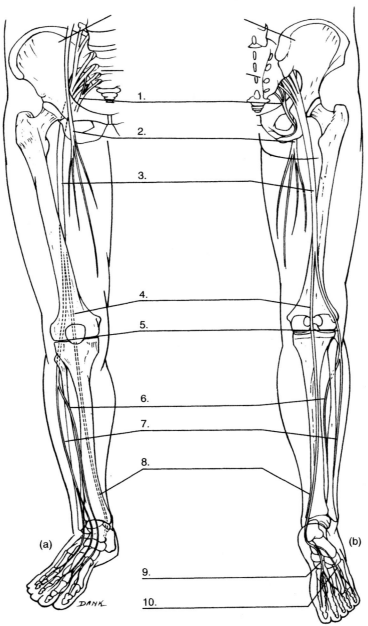

1.
2.
3.
4.
5.
6.
7.
8.
(a)
(b)
9.
10.

Figure 13.22 Anterior (a) and (b) posterior views of the distribution of the nerves of the lumbar plexus in the leg.

14

The Endocrine System

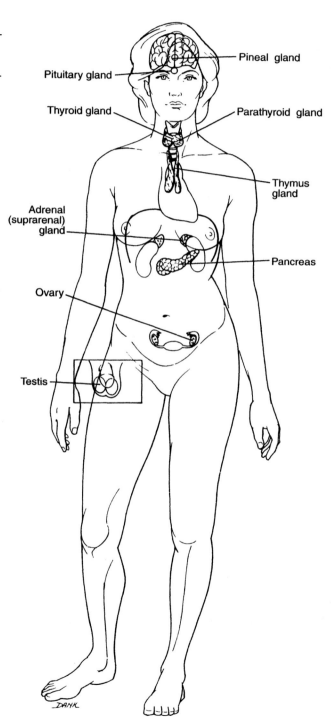

Objective

The objective of Chapter 14, "The Endocrine System," is to acquaint you with the location, distribution, and structure of the major endocrine organs in the human body.

Description

The endocrine system, in contrast to other body systems, consists of a few widely scattered glands that form distinct anatomical organs. The endocrine glands produce and secrete chemicals (hormones) directly into the bloodstream, in contrast to exocrine glands, which secrete their products into excretory ducts.

The major endocrine glands or organs are the single **pituitary (hypophysis) gland**, an unpaired **thyroid gland**, four **parathyroid glands**, and paired **adrenal (suprarenal) glands.** The pituitary gland is located in the head below the brain, the thyroid and parathyroid glands in the upper neck, and the adrenal glands in the upper abdominal cavity superior to the kidneys. The pituitary gland, by releasing different hormones, controls the function of the other endocrine organs (Figure 14.1). In addition to purely endocrine or exocrine organs are numerous organs of **mixed** type; that is, they perform both endocrine and exocrine functions. Such organs contain some isolated endocrine cells or tissues. Examples of mixed endocrine-exocrine organs are found in the digestive (**stomach, small intestine, pancreas**), reproductive (**ovaries, testes**), and urinary (**kidneys**) systems (Figure 14.1).

The Pituitary (Hypophysis) Gland—
Description [Figure 14.2]

The **pituitary gland,** or **hypophysis,** is a pea-sized organ located beneath the brain in a **sphenoid bone** depression called the **sella turcica.** The pituitary gland is connected to the hypothalamus of the diencephalon (brain) by a slender **hypophyseal stalk,** or **infundibulum.** The **dura mater** of the brain forms a diaphragm over the depression in which the pituitary gland is located; the stalk, or infundibulum, then passes through an opening in the dura mater to the gland. When the cadaver brain is removed from the cranial cavity, the stalk is usually torn.

The Pituitary (Hypophysis) Gland—
Identification [Figure 14.2]

Examine the interior and midsagittal section of a skull or model. If possible, examine the midsagittal section of a prepared human brain. Identify the following structures associated with the pituitary gland and then color them in the illustration:

1. **pituitary gland**
2. **infundibulum or hypophyseal stalk**
3. **hypothalamus**

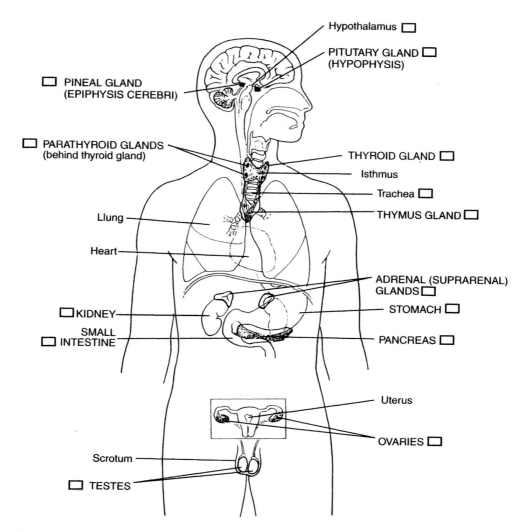

Figure 14.1 Endocrine organs and organs containing endocrine tissue.

4. **sella turcica**
5. **sphenoid bone**
6. **third ventricle**
7. **dura mater**

The Thyroid Gland—
Description *[Figures 14.3 and 14.4]*

The thyroid gland is a large, unpaired endocrine organ located anterior and lateral in the upper **trachea** below the **larynx.** Anteriorly, the gland has a H- or U-shape. The thyroid gland consists of **right** and **left lobes.** These are connected across the trachea by a thinner **isthmus** that lies on the second, third, and fourth **tracheal rings.** The sternohyoid and sternothyroid muscles lie in front of the thyroid gland; the **common carotid artery, internal jugular vein,** and **vagus nerve** lie posterolaterally; and the **trachea** lies posteriorly. Enclosing the thyroid gland is a connective tissue capsule (Figures 14.3 and 14.4).

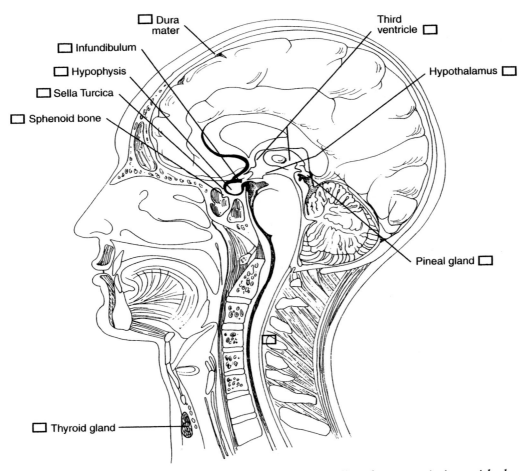

□ Dura mater

□ Infundibulum

□ Hypophysis

□ Sella Turcica

□ Sphenoid bone

Third ventricle □

Hypothalamus □

Pineal gland □

□ Thyroid gland

Figure 14.2 Location of the pituitary gland in the skull and its association with the brain structures. Note the location of the pineal and thyroid glands.

The Thyroid Gland—
Identification *[Figures 14.3 and 14.4]*

Examine a model and diagrams of the thyroid gland and color the associated parts. If possible, examine the prepared upper neck region of a cadaver. Identify the thyroid gland and its lobes, and the structures lying in close vicinity to it.

1. **right thyroid gland lobe**
2. **left thyroid gland lobe**
3. **thyroid gland isthmus**
4. **larynx**
5. **trachea**
6. **tracheal rings**
7. **vagus nerve**
8. **common carotid artery**
9. **internal jugular vein**

The Parathyroid Glands—Description *[Figure 14.4]*

In humans, the parathyroid glands consist of four separate structures closely associated with the thyroid gland. The parathyroid glands are embedded in the posteri-

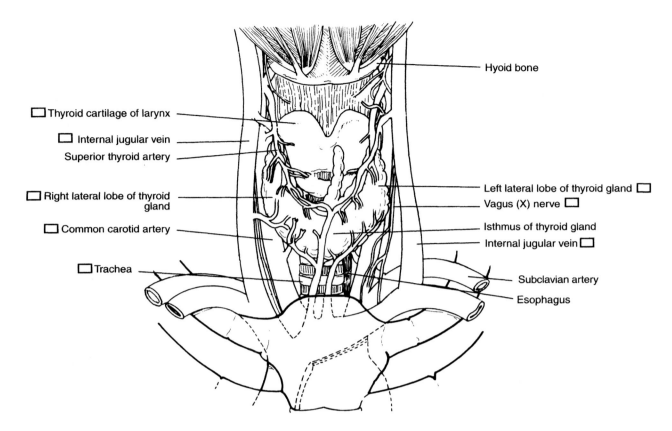

Thyroid cartilage of larynx ☐

Internal jugular vein ☐

Superior thyroid artery

Right lateral lobe of thyroid ☐
gland

Common carotid artery ☐

Trachea ☐

Hyoid bone

Left lateral lobe of thyroid gland ☐

Vagus (X) nerve ☐

Isthmus of thyroid gland

Internal jugular vein ☐

Subclavian artery

Esophagus

Figure 14.3 Anterior view of the thyroid gland and its associated structures and blood vessels.

or surface of the thyroid gland, with two superior and two inferior parathyroid glands in each thyroid lobe. The parathyroid glands are small and difficult to recognize in a cadaver because in prepared specimens their color is similar to that of the thyroid gland.

The Suprarenal (Adrenal) Glands—
Description *[Figure 14.5]*

Two **suprarenal glands** are situated retroperitoneally (behind the peritoneum). Each gland is located superiorly on each **kidney** and posteriorly, the glands rest against the **diaphragm**. The glands are surrounded by a connective tissue capsule and embedded in fat. Internally, the adrenal glands consist of an outer **cortex** and the inner **medulla**. The two portions have different structures, functions, and embryological origin; they are considered as two separate endocrine glands situated in one organ.

The Suprarenal (Adrenal) Glands—
Identification *[Figure 14.5]*

Examine a model and diagrams of the abdominal organs, kidneys, and adrenal glands and color them. If possible, examine the posterior abdominal wall cavity in a cadaver and identify the adrenal glands and these associated structures:

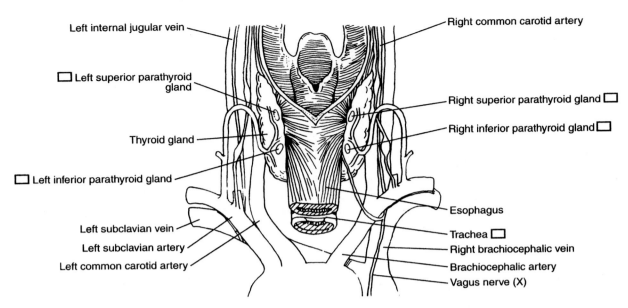

Figure 14.4 Location of the parathyroid glands in the thyroid gland, as seen in posterior view.

1. **right kidney**
2. **right adrenal gland**
3. **left kidney**
4. **left adrenal gland**
5. **diaphragm**

Other Endocrine Organs: The Pineal and Thymus Glands [Figure 14.1]

The **pineal gland** is a small structure attached to the roof of the **third ventricle** of the brain. This gland is believed to be endocrine, although its exact functions are obscure.

The **thymus gland** is a lymphoid organ, meaning its hormones play an important role in immunity from infection. The gland is important early in life because it causes certain lymphocytes to become immunocompetent and enables them to react against specific pathogens. The thymus is located in the superior mediastinum behind the sternum and between the lungs. It reaches maximum size by puberty, after which it decreases in size and is replaced by fat and connective tissue.

Functions of Endocrine Organs: A Brief Summary [Figure 14-6]

The pituitary gland consists of two parts: a glandular lobe called the **adenohypophysis** and a neural lobe called the **neurohypophysis**. The pituitary gland releases nine hormones into the bloodstream, seven from the adenohypophysis and two from the neurohypophysis. These hormones control the functions of numerous

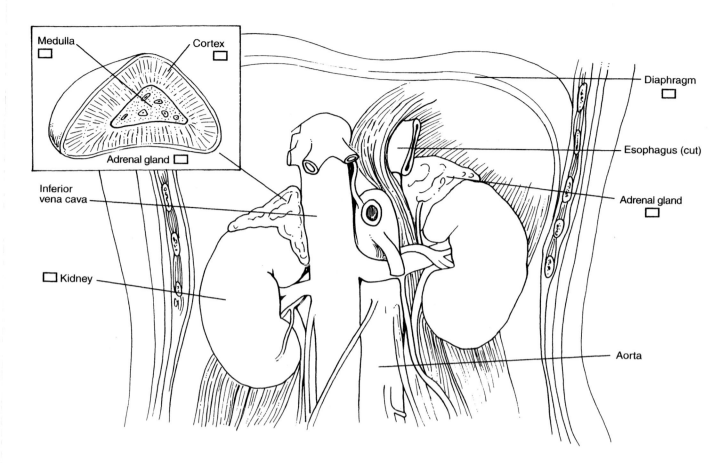

Figure 14.5 Location of the suprarenal (adrenal) glands on the posterior abdominal wall and inset of a gland.

organs. The hormones, their actions, and the affected organs, the target organs, are summarized in Figure 14.6.

The thyroid gland releases two hormones, **thyroxin** and **triiodothyronine**. These hormones stimulate cellular metabolism of carbohydrates, fats, and proteins and are essential for normal body development. The thyroid gland also releases the hormone **calcitonin**. Its function is to lower the blood calcium level by stimulating calcium uptake and deposition in the bone. Increased blood calcium levels trigger the release of calcitonin.

The parathyroid glands are essential to life. They produce **parathyroid hormone**, or **parathormone**. This hormone counteracts calcitonin and raises calcium levels by increasing calcium release from the bone and by absorbing calcium from the gastrointestinal tract. Parathyroid hormone is released from these glands in direct response to low blood calcium levels.

The cortex of the adrenal gland is essential to life; it produces **glucocorticoid hormones**, which affect the metabolism of proteins, fats, and glucose and **mineralcorticoid hormones**, which control the balance of sodium and potassium in the blood. The adrenal medulla secretes **epinephrine** and **norepinephrine**, which prepare the body for stress.

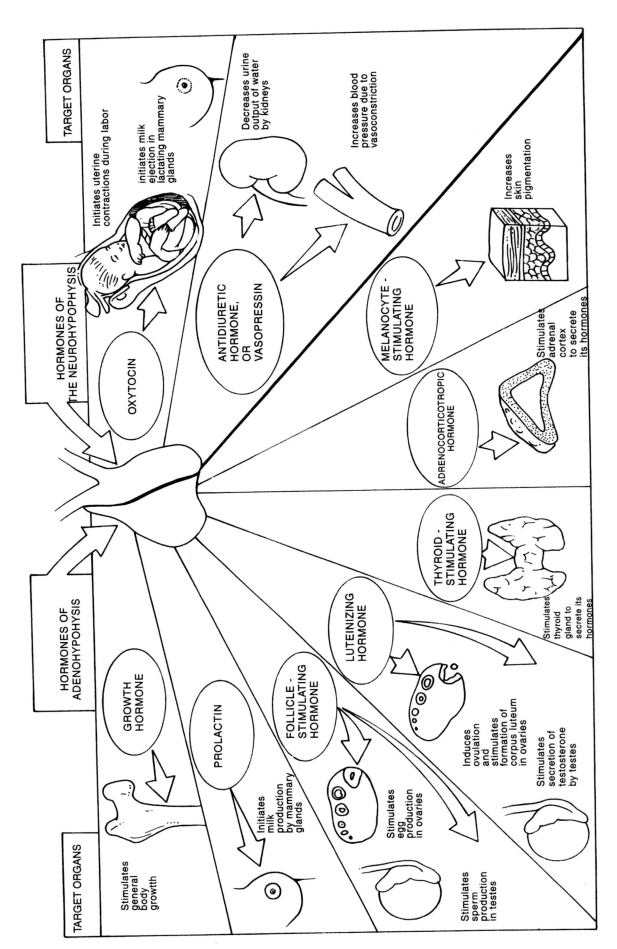

Figure 14.6 The pituitary (hypophysis) gland, its hormones, and their activities in the corresponding target organs.

CHAPTER SUMMARY AND CHECKLIST

I. **THE ENDOCRINE SYSTEM**

   A. **Four Purely Endocrine Glands or Organs**
1. Pituitary
2. Thyroid
3. Parathyroid
4. Adrenal (suprarenal)

   B. **Mixed Endocrine and Exocrine Organs**
1. Stomach
2. Small intestine
3. Pancreas
4. Ovaries
5. Testes
6. Kidneys

II. **THE ENDOCRINE GLANDS**

   A. **The Pituitary Gland (Hypophysis)**
1. Unpaired
2. Continuous with the brain
3. Adenohypophysis (glandular lobe) has epithelial origin
4. Neurohypophysis (neural lobe) has neural origin
5. Located in the sella turcica of the skull
6. Attached to base of the brain by the infundibulum

   B. **Thyroid Gland**
1. Unpaired
2. Located below larynx
3. Has two lateral lobes and an isthmus
4. Near sternohyoid and sternothyroid muscles

   C. **Parathyroid Glands**
1. Two pairs embedded in the posterior thyroid gland

   D. **Suprarenal (Adrenal) Glands**
1. Situated retroperitoneally and superior to kidneys
2. Contain cortex and medulla

III. **OTHER ENDOCRINE ORGANS OR GLANDS**

   A. **Pineal Gland**

   B. **Thymus Gland**
1. Immunologically important
2. Produces T lymphocytes
3. Atrophies in adults

IV. **FUNCTIONS OF ENDOCRINE ORGANS: SUMMARY**

   A. **Pituitary Gland**
1. Glandular pituitary adenohypophysis secretes seven hormones

2. Neural pituitary neurohypophysis secretes two hormones

B. Thyroid Gland
   1. Thyroxin and triiodothyronine stimulate cellular metabolism
   2. Calcitonin lowers blood calcium levels when they rise

C. Parathyroid Glands
   1. Essential to life
   2. Parathyroid hormone raises blood calcium levels when they fall

D. Adrenal Glands
   1. Cortex essential to life
   2. Glucocorticoids affect matabolism of proteins, fats, and glucose
   3. Mineralocorticoids control sodium and potassium balance
   4. Medulla secretes epinephrine and norepinephrine during stress

Laboratory Exercises 14

NAME _____

LAB SECTION _____ DATE _____

LABORATORY EXERCISE 14.1
The Endocrine Glands
Part I

1. What are the major endocrine glands in the body?

   a. _____

   b. _____

   c. _____

   d. _____

2. What are the mixed endocrine-exocrine organs?

   a. _____

   b. _____

   c. _____

   d. _____

   e. _____

   f. _____

3. Identify the endocrine or mixed glands described.

   a. situated above the kidney _____

   b. located in sella turcica _____

   c. bilobed gland located in the throat _____

   d. four glands posterior to thyroid _____

   e. paired organs in the scrotum _____

   f. paired organs near uterus _____

   g. attached to the roof of third ventricle _____

h. lymphoid organ that reaches maximum size during youth

_____

4. Identify the endocrine glands, hormones, or functions described:

   a. secretes oxytocin and antidiuretic hormone _____

   b. secretes seven hormones _____

   c. stimulates ovarian and/or testicular functions _____

   d. stimulates general body growth _____

   e. secretes calcitonin and thyroxin _____

   f. hormone that lowers blood calcium levels _____

   g. hormone that raises blood calcium levels _____

   h. endocrine glands that are essential to life _____

   i. gland and hormones that prepare body to resist stress _____

_____

Part II

Using the following endocrine and mixed glands and associated organs, label the illustration.

Pituitary gland
Thyroid gland
Parathyroid glands
Stomach
Testes
Pineal gland
Isthmus of thyroid gland

Adrenal glands
Small intestine
Hypothalamus
Thymus gland
Kidney
Ovaries
Pancreas

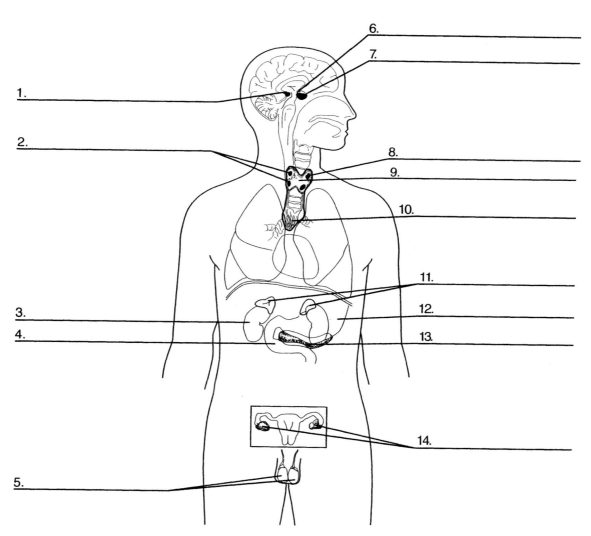

Figure 14.7 Endocrine and mixed glands.

Part Five

Regulation and Maintenance of the Human Body

15

The Cardiovascular System

THE HEART

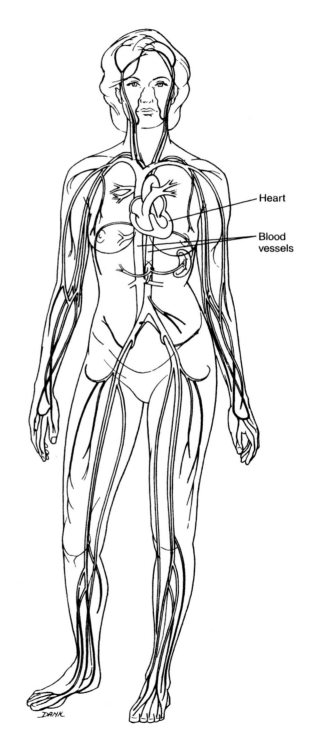

Heart

Blood
vessels

Objective

The objective of Chapter 15, "The Cardiovascular System: The Heart," is to introduce the:

1. **Location of the heart in the thorax and mediastinum**
2. **Structure of the pericardial sac and heart wall**
3. **Internal anatomy of the heart chambers**
4. **Major vessels that enter and leave the heart**

The Location of the Heart [Figure 15.1]

The heart is located in the thorax between the two lungs. The lungs are located in the pleural cavities and the heart in the **mediastinum**. The mediastinum is a median partition that separates the lungs from the heart. The region occupied by the heart and surrounding connective tissue, the **pericardium**, or the **pericardial sac**, is the **middle mediastinum**. (The thorax, its contents, and mediastinal borders are described in greater detail in Chapter 18, "The Respiratory System.")

In the thorax, the heart lies at an oblique angle in the mediastinum, with most of it left of the body's midline. The heart has a **base** and an **apex**. The **apex** is the pointed end next to the diaphragm and it projects anteriorly, inferiorly, and to the left. The **base** of the heart is the broad posterior surface that projects superiorly to the right. Large vessels enter (**superior** and **inferior vena cava**) and leave (**aorta** and **pulmonary trunk**) the base of the heart. The apex lies at about the fifth intercostal space, and the base is inferior to the second rib (Figure 15.1).

The Pericardium [Figures 15.1 and 15.2]

The **pericardium**, or **pericardial sac**, surrounds the heart and roots of the major vessels at the base. The pericardium protects the heart and consists of two membranous layers. The inner pericardial layer is the **serous pericardium** or **visceral pericardium**; it adheres to and covers the outer heart wall (Figure 15.2).

Great vessels enter and leave the heart at its base, where the visceral pericardium folds back and becomes continuous with the outer layer of the pericardium. This is the **parietal pericardium** and it consists of an outer **fibrous pericardium** and an inner **serous pericardium**. Between the parietal and visceral pericardium is the **pericardial cavity**, which contains a small amount of fluid secreted by the serous cells of the pericardial membranes. This fluid lubricates the visceral and parietal pericardium and facilitates rhythmic heart movements by minimizing friction (Figure 15.2).

The Heart Location and the Pericardium— Identification [Figures 15.1 and 15.2]

Examine a heart model and diagrams, and color the parts. Then examine a prepared thoracic cavity and cadaver heart, and identify the following:

1. **visceral pericardium**
2. **base of the heart**

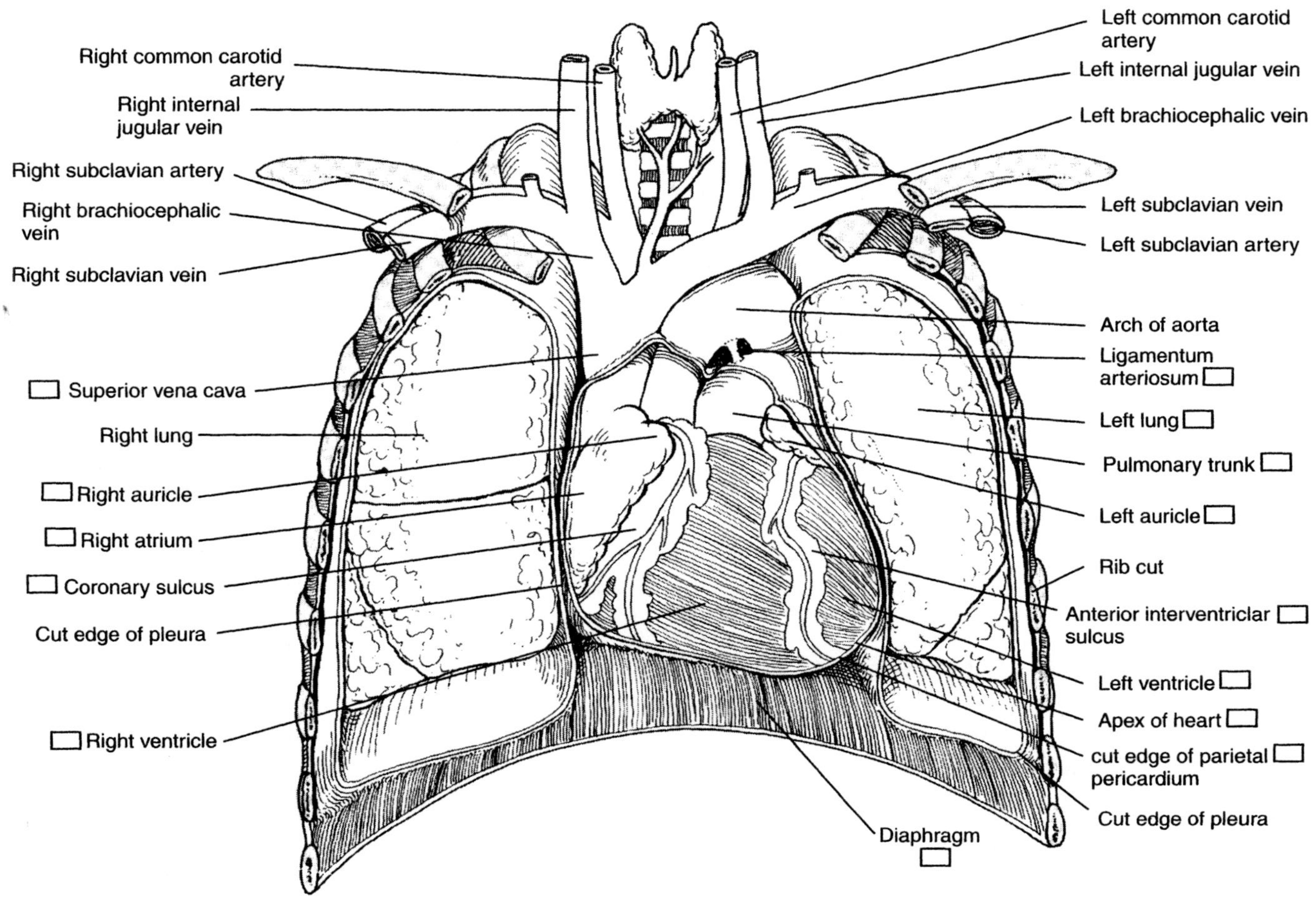
Right common carotid artery
Right internal jugular vein
Right subclavian artery
Right brachiocephalic vein
Right subclavian vein
Superior vena cava
Right lung
Right auricle
Right atrium
Coronary sulcus
Cut edge of pleura
Right ventricle
Left common carotid artery
Left internal jugular vein
Left brachiocephalic vein
Left subclavian vein
Left subclavian artery
Arch of aorta
Ligamentum arteriosum
Left lung
Pulmonary trunk
Left auricle
Rib cut
Anterior interventriclar sulcus
Left ventricle
Apex of heart
cut edge of parietal pericardium
Cut edge of pleura
Diaphragm

Figure 15.1 Location of the heart and associated blood vessels in the thoracic cavity. The pericardial sac has been cut to expose the heart.

3. **parietal pericardium**
4. **apex of the heart**

The Vascular Circuits—Description

The Pulmonary Circulation [Figures 15.1 and 15.3]

The deoxygenated venous blood brought to the right atrium by the superior and inferior venae cavae and by the coronary sinus flows into the right ventricle. Ventricular contraction ejects the venous blood into the **pulmonary trunk**, which emerges from the right ventricle and bifurcates into the **right** and **left pulmonary arteries**. The pulmonary arteries then carry the deoxygenated blood to the right and left lungs. After gaseous exchange in the alveoli of the lungs, the oxygenated blood then flows back to the left atrium of the heart through two **pulmonary veins**.

The Systemic Circulation

The left atrium delivers oxygenated blood to the left ventricle, from where it enters the systemic circulation. The vessels of the systemic circulation include the aorta,

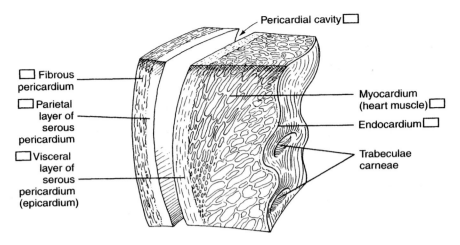

Figure 15.2 A section of the heart wall showing the pericardial and muscle layers.

all its major branches, the arterioles, capillary beds, venules, veins, venous sinuses, major veins, and the venae cavae, which return the blood to the heart.

The Heart Wall *[Figure 15.2]*

The heart wall consists of three layers. The **outer epicardium** was described above. The middle layer is called the **myocardium**. It is composed of cardiac muscle and is the thickest layer of the heart. The innermost layer of the heart is the **endocardium**, composed of a single layer of endothelial cells and supporting connective tissue. This layer is continuous with the endothelial lining of the blood vessels.

The Heart Chambers

The Atria [Figures 15.4, 15.5, and 15.6]

The mammalian heart has four separate chambers. The two superior chambers are the **right** and **left atria**; the two inferior chambers are the **right** and **left ventricles**. The atria are smaller and have thinner muscular walls than the ventricles. Each atrium has an anterolateral appendage or outpocketing called the **auricle**. The walls of the atria are smooth except anteriorly and in the auricles. In these areas, the walls exhibit a rough surface due to the irregular arrangement of cardiac muscles into ridges called the **musculi pectinati**, or **pectinate muscles**.

The atria are separated from the ventricles by grooves, or **sulci**. The **atrioventricular (coronary) sulcus**, visible on the heart, separates the atria from the ventricles. This sulcus corresponds to a ridge on the inner surface of the tria called the **crista terminalis**.

A thin muscular **interatrial septum** separates the two atria. This septum bears a prominent oval depression called the **fossa ovalis**, which is a remnant of the fetal **foramen ovale**. In the fetal heart, the foramen (opening) ovale directs blood from the right atrium to the left atrium. Normally, this foramen closes at birth. Also in the fetus, blood that enters the right ventricle is pumped into the pulmonary trunk. Most of the blood in this trunk is diverted from the nonfunctioning fetal lungs into the aorta by a short vessel called the **ductus arteriosus**. Although this vessel is not

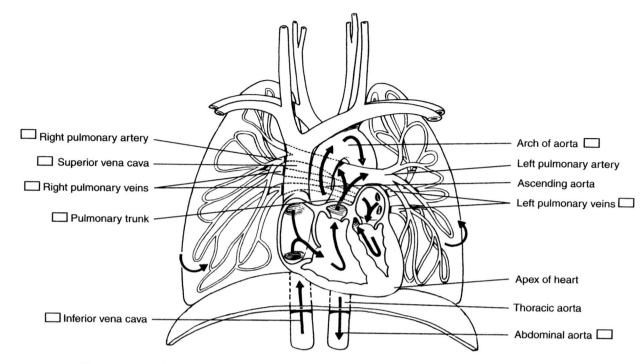

Right pulmonary artery ☐
Superior vena cava ☐
Right pulmonary veins ☐
Pulmonary trunk ☐

Arch of aorta ☐
Left pulmonary artery
Ascending aorta
Left pulmonary veins ☐

Apex of heart

Thoracic aorta

Abdominal aorta ☐

Inferior vena cava ☐

Figure 15.3 The pulmonary circulation.

part of the heart, it plays a vital role in fetal circulation by connecting the pulmonary trunk with the aorta. After birth, the blood is directed to the lungs, and the ductus arteriosus closes and atrophies. In the adult heart, this structure is recognized as the **ligamentum arteriosum** between the aorta and pulmonary trunk (Figure 15.1).

Three veins open into the right atrium: the **superior vena cava** (blood from the upper body regions), **inferior vena cava** (blood from the lower body regions), and **coronary sinus** (blood from the heart wall). Four **pulmonary veins** enter the left atrium: two right and two left pulmonary veins. They are best seen from the posterior side of the heart (Figure 15.5).

The Ventricles [Figures 15.4–15.6]

The walls of the ventricles are thicker than those of the atria. This thickness is necessary to force the blood into the systemic circulation. On the heart exterior, a shallow **anterior** and **posterior interventricular sulcus** separates right and left ventricles. Located within the atrioventricular and interventricular sulci are **coronary blood vessels**; these supply and drain the heart musculature.

Internally, the right and left ventricles are separated from each other by a thick partition called the **interventricular septum**, which has a **muscular** and a **membranous** portion. The larger muscular portion of this septum extends superiorly from the apex of the heart. The membranous portion of the septum is thin, short, and smooth and consists of connective tissue. It is located in the upper region of the interventricular septum, which separates the root of the aorta from the lower part of the right atrium and the lower part of the right ventricle. This portion of the septum is a common site for defects.

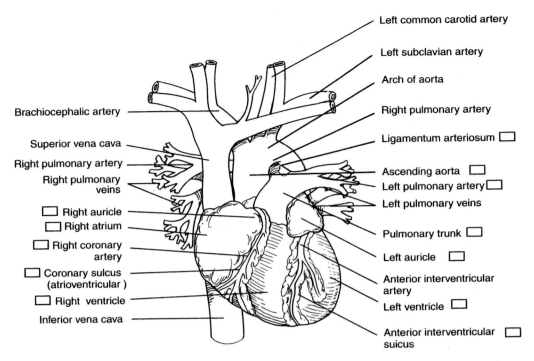

Left common carotid artery

Left subclavian artery

Arch of aorta

Right pulmonary artery

Ligamentum arteriosum ☐

Ascending aorta ☐

Left pulmonary artery ☐

Left pulmonary veins

Pulmonary trunk ☐

Left auricle ☐

Anterior interventricular
artery

Left ventricle ☐

Anterior interventricular ☐
suicus

Brachiocephalic artery

Superior vena cava

Right pulmonary artery

Right pulmonary
veins

☐ Right auricle

☐ Right atrium

☐ Right coronary
artery

☐ Coronary sulcus
(atrioventricular)

☐ Right ventricle

Inferior vena cava

Figure 15.4 Anterior view of the heart, its chambers, and the major blood vessels.

The inner surface of both ventricles is very irregular due to muscular ridges called **trabeculae carneae**. Some of these trabeculae project into the ventricles as the **papillary muscles**. The papillary muscles, in turn, attach to the fibrous bands of the heart valves, called the **chordae tendineae**. The papillary muscles and chordae tendineae prevent backflow of blood through the heart valves. The right ventricle has a muscular bundle called the **moderator band**, or **septomarginal trabecula**, which crosses the ventricle from the septum to the base of a large papillary muscle on the anterior wall.

The wall of the left ventricle is three to four times thicker than the right ventricle and can be identified in the cadaver by squeezing the inferior region of the heart. The left ventricle feels more solid and firm due to the thicker wall. The right ventricle compresses easily and feels softer because of its thinner wall.

The Heart Valves *[Figures 15.6 and 15.7]*

The heart valves prevent backflow of blood into the heart chambers and direct the flow of blood through these chambers in one direction.

The Atrioventricular Valves [Figures 15.6 and 15.7]

Two **atrioventricular valves (AV valves)** are located between the atria and ventricles. The right AV valve, situated between the right atrium and right ventricle, is called the **tricuspid valve** because it has three **cusps** or **valve flaps**: the **anterior, posterior,** and **septal** cusps. The free borders of each cusp, which project into the

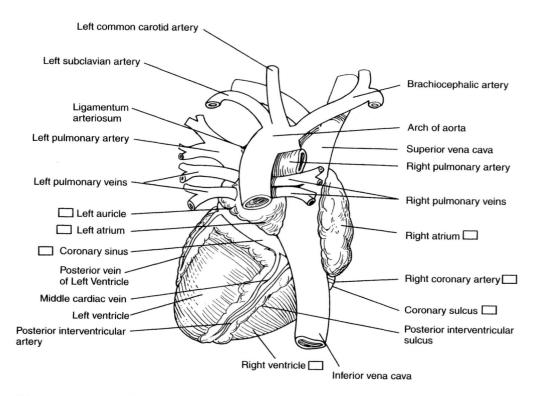

Left common carotid artery

Left subclavian artery

Ligamentum arteriosum

Left pulmonary artery

Left pulmonary veins

Left auricle

Left atrium

Coronary sinus

Posterior vein of Left Ventricle

Middle cardiac vein

Left ventricle

Posterior interventricular artery

Right ventricle

Inferior vena cava

Brachiocephalic artery

Arch of aorta

Superior vena cava

Right pulmonary artery

Right pulmonary veins

Right atrium

Right coronary artery

Coronary sulcus

Posterior interventricular sulcus

Figure 15.5 Base of the heart viewed from the posterior aspect illustrates the chambers and major blood vessels.

ventricle, are attached to fibrous strands, the **chordae tendineae**. The chordae tendineae anchor the valve cusps to the **papillary muscles** of the ventricle. Normally, three papillary muscles are found in the right ventricle.

The left AV valve, situated between the left atrium and left ventricle, is called the **bicuspid (mitral) valve** because it has two cusps: the **anterior** and **posterior** cusps. The two cusps are attached to the chordae tendineae and two papillary muscles of the left ventricle. When the ventricles contract, the papillary muscles also contract, hold the cusps firmly in place, and prevent blood from regurgitating back into the atria.

The Semilunar Valves [Figures 15.1, 15.6, and 15.7]

The great vessels that emerge from the heart, the pulmonary trunk and the aorta, also have valves at their bases, called the **pulmonary semilunar** and **aortic semilunar valves**, respectively. Their function is to prevent the backflow of blood from these vessels into the ventricles. Although the function of these valves is similar to that of the AV valves, their morphology is different.

The aortic and pulmonary semilunar valves have three crescent-shaped **half-moon cusps** attached directly to the vessel walls. In the two vessels, the valve cusps have similar names: the **right, left**, and **anterior** (pulmonary) or **posterior** (aortic) cusps.

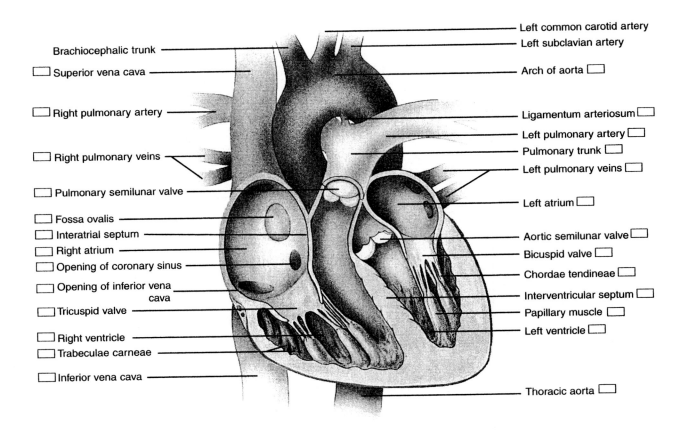

Brachiocephalic trunk

☐ Superior vena cava

☐ Right pulmonary artery

☐ Right pulmonary veins

☐ Pulmonary semilunar valve

☐ Fossa ovalis
☐ Interatrial septum
☐ Right atrium
☐ Opening of coronary sinus
☐ Opening of inferior vena cava
☐ Tricuspid valve

☐ Right ventricle
☐ Trabeculae carneae

☐ Inferior vena cava

Left common carotid artery
Left subclavian artery
Arch of aorta ☐

Ligamentum arteriosum ☐
Left pulmonary artery ☐
Pulmonary trunk ☐
Left pulmonary veins ☐

Left atrium ☐

Aortic semilunar valve ☐
Bicuspid valve ☐
Chordae tendineae ☐
Interventricular septum ☐
Papillary muscle ☐
Left ventricle ☐

Thoracic aorta ☐

Figure 15.6 A cross section of the heart shows the structure of the atria, ventricles, and various heart valves.

The Heart Wall, Chambers, Great Vessels, and Valves—Identification *[Figures 15.2–15.7]*

Examine a heart model and diagrams of the heart musculature and color the component parts. If possible, examine a prepared cadaver heart and its chambers, great vessels, valves, and musculature, and identify the following structures:

1. epicardium
2. myocardium
3. endocardium
4. right atrium
5. left atrium
6. right ventricle
7. left ventricle
8. bicuspid valve
9. tricuspid valve
10. AV cusps
11. pulmonary semilunar valve
12. aortic semilunar valve
13. interatrial septum
14. fossa ovalis
15. pectinate muscles

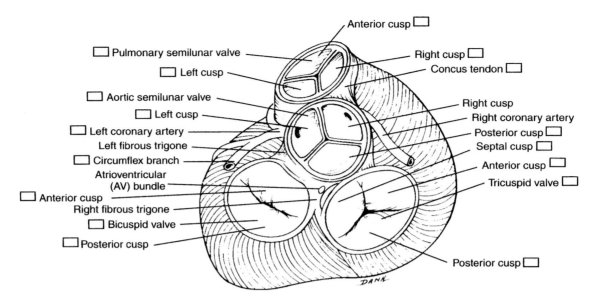

Figure 15.7 Atrioventricular and semilunar heart valves in superior view. The atria have been removed to expose the valves and the two coronary arteries.

16. **right and left auricles**
17. **atrioventricular sulcus**
18. **crista terminalis**
19. **trabeculae carneae**
20. **septomarginal trabeculae**
21. **chordae tendineae**
22. **papillary muscles**
23. **interventricular sulcus**
24. **interventricular septum**
25. **superior vena cava**
26. **inferior vena cava**
27. **pulmonary trunk**
28. **pulmonary arteries**
29. **pulmonary veins**
30. **ligamentum arteriosum**
31. **coronary sinus**

The Coronary (Cardiac) Vessels—Description

[Figures 15.7 and 15.8]

The Coronary Arteries

The heart carries its own blood supply to the myocardium via the **right** and **left coronary arteries** and their branches. These two major coronary arteries arise above the semilunar valves of the **ascending aorta** near its exit from the left ventricle.

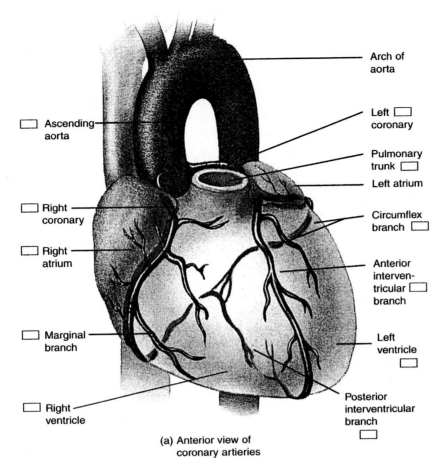

Figure 15.8(a) The coronary circulation showing coronary arteries.

Variation is observed in the distribution of the right and left coronary arteries. The most common pattern is that the **right coronary artery** arises behind the **right cusp** of the aortic semilunar valve and passes to the posterior surface of the heart in the **atrioventricular sulcus**. Here, the coronary artery descends toward the **apex** of the heart and terminates as the **posterior interventricular artery** in the **posterior interventricular sulcus**. A branch from the right coronary artery, the **marginal artery**, passes to the right ventricle, usually on the right margin of the heart.

The **left coronary artery** arises behind the **left cusp** of the aortic semilunar valve and enters the left **atrioventricular sulcus**. After passing a short distance to the left, the artery divides into the **anterior interventricular artery** and the **circumflex artery**. The anterior interventricular artery passes inferiorly in the anterior interventricular sulcus toward the apex of the heart. The circumflex artery remains in the coronary sulcus and passes to the left margin of the heart. It runs in the posterior atrioventricular sulcus and supplies branches to the posterior part of the left ventricle (Figure 15.8a).

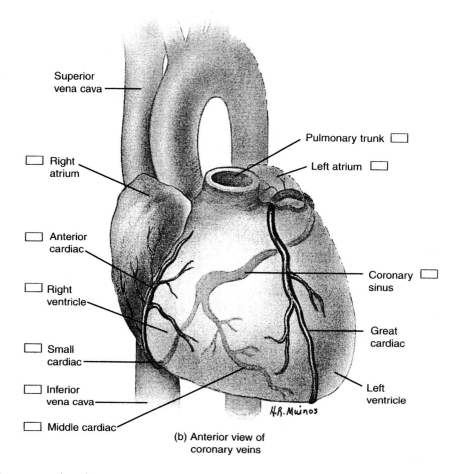

Superior
vena cava

Pulmonary trunk ☐

Right
atrium ☐

Left atrium ☐

Anterior
cardiac ☐

Coronary ☐
sinus

Right
ventricle ☐

Great
cardiac

Small
cardiac ☐

Inferior
vena cava ☐

Left
ventricle

H.R. Muinos

Middle cardiac ☐

(b) Anterior view of
coronary veins

Figure 15.8(b) The coronary circulation showing cardiac veins.

The Cardiac Veins [Figures 15.5 and 15.8]

The venous blood from the heart returns to the right atrium via the cardiac veins, which largely parallel the arteries in the sulci. The anterior surface of the heart drains through the **great cardiac vein**, which begins at the apex and ascends in the anterior interventricular sulcus alongside the anterior interventricular artery. The great cardiac vein ends by entering and becoming continuous with the **coronary sinus**, which runs in the posterior atrioventricular sulcus.

The posterior surface of the heart is drained by the **posterior vein** of the **left ventricle**, located in the vicinity of the **circumflex artery**, and the **middle cardiac vein**, located in the **posterior interventricular sulcus** alongside the **posterior interventricular artery**. A **small cardiac vein**, located in the right margin of the heart near the **marginal artery**, enters the coronary sinus on the posterior side of the heart. These veins, similar to the great cardiac vein, drain into the coronary sinus. The **coronary sinus** is a thin-walled venous channel located in the posterior atrioventricular sulcus. It drains the venous blood from the heart wall into the right atrium (Figure 15.6).

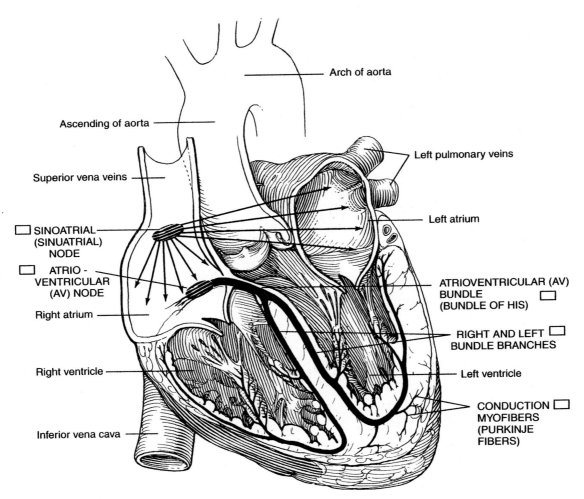

Figure 15.9 Conduction system of the heart: nodes and bundles.

Conducting System of the Heart *[Figure 15.9]*

The components of the conducting system are the **sinoatrial (SA) node**, the **atrioventricular (AV) node**, the **atrioventricular bundle**, the **bundle branches**, and the **Purkinje fibers**. The SA node is located in the wall of the right atrium near the entrance of the superior vena cava. The AV node is located in the inferior portion of the interatrial septum near the entrance of the coronary sinus. The atrioventricular bundle starts at the AV node and divides into right and left branches that descend toward the apex of the heart in the interventricular septum. The branches terminate by spreading through the ventricles as the Purkinje fibers. The conducting system is not visible in a dissected heart.

The Coronary (Cardiac) Vessels—Identification
[Figures 15.4, 15.5, 15.7, and 15.8]

Examine the coronary vessels on a heart model and the diagrams. If possible, examine a prepared cadaver heart and identify the following arteries and veins:

   1. **right coronary artery**

2. marginal artery
3. posterior interventricular artery
4. left coronary artery
5. anterior interventricular artery
6. circumflex artery
7. great cardiac vein
8. middle cardiac vein
9. small cardiac vein
10. coronary sinus

CHAPTER SUMMARY AND CHECKLIST

I. THE HEART

A. **The Location of the Heart**
   1. Located in the middle mediastinum of the thorax
   2. The apex points left and lies next to the diaphragm
   3. Base is the broad, posterior end with large vessels

B. **Pericardium**
   1. A protective connective tissue sac around the heart
   2. Inner layer is serous and forms the visceral pericardium
   3. Outer layer is the parietal pericardium
   4. Between the layers lies the pericardial cavity
   5. Fluid in the cavity lubricates membranes

II. THE VASCULAR CIRCUITS

A. **Pulmonary Circulation**
   1. From right side of heart to lungs via pulmonary trunk
   2. From lungs to left side of heart via pulmonary veins

B. **Systemic Circulation**
   1. From left ventricle via aorta to all systems of body
   2. Return to heart via venous tributaries and venae cavae

III. THE HEART WALL, CHAMBERS, AND VALVES

A. **The Heart Wall**
   1. Epicardium is the thin outer layer
   2. Myocardium is the thick, muscular middle layer
   3. Endocardium is the innermost, endothelial layer
   4. Two superior chambers are the right and left atria
   5. Two inferior chambers are the right and left ventricles

B. **The Atria**
   1. Atrioventricular sulcus separates atria from ventricles
   2. Interatrial septum separates atria and exhibits fossa ovalis
   3. Three veins open into the right atrium (superior and inferior venae cavae, coronary sinus)
   4. Four pulmonary veins enter left atrium

C. **The Ventricles**
   1. Separated from each other by interventricular septum
   2. Inner surface bears projections called trabeculae carneae and septomarginal trabecula
   3. Left ventricle three to four times thicker
   4. Interventricular sulcus between ventricles

D. **The Heart Valves**
   1. Right atrioventricular valve is tricuspid (three cusps)
   2. Left atrioventricular valve is bicuspid (two cusps)
   3. Fibrous chordae tendineae attach to cusps
   4. Cardiac papillary muscles attach to chordae tendineae

E. **The Semilunar Valves**
   1. Located at proximal ends of the aorta and pulmonary trunk
   2. Contain three half-moon cusps per valve
   3. Open and close following ventricular contraction or relaxation

IV. **THE CORONARY (CARDIAC) VESSELS**

A. **The Coronary Arteries**
   1. Right and left coronary arteries arise from the aorta
   2. Right coronary artery passes to posterior heart
      a. terminates as the posterior interventricular artery
   3. Left coronary artery divides short distance from origin
      a. branches are anterior interventricular and circumflex

B. **The Cardiac Veins**
   1. Great cardiac vein drains anterior surface of heart
      a. terminates by entering the coronary sinus
   2. Veins drain the posterior surface of the heart:
      a. posterior vein of left ventricle
      b. middle cardiac vein
   3. Small cardiac vein drains right margin of the heart
   4. Coronary sinus drains venous blood from the heart
      a. empties into right atrium

Laboratory Exercises 15

NAME _____

LAB SECTION _____ DATE _____

LABORATORY EXERCISE 15.1
The Heart

Part I

Match the description with the answer.

1. Location of heart to body's midline_____   A. myocardium
2. Location of heart in the thorax_____   B. apex
3. Pointed end of heart next to diaphragm_____   C. pulmonary veins
4. Protective cover of the heart_____   D. pulmonary arteries
5. Outer membrane of the heart_____   E. pericardium
6. Space between heart membranes_____   F. mediastinum
7. Carry deoxygenated blood to lungs_____   G. epicardium
8. Carry oxygenated blood to the heart_____   H. pericardial cavity
9. Muscle layer of heart_____   I. left side
10. Innermost heart layer_____   J. endocardium

Part II

1. The upper two heart chambers are called_____

2. The thicker lower heart chamber is the_____

3. The anterolateral appendage of the heart is the_____

4. What separates atria from ventricles?_____

5. What separates the two atria?_____

6. The rough atrial surface is due to_____

7. The oval depression in the atrial septum is the_____

8. Which fetal remnant is found between the pulmonary trunk and aorta?

   _____

9. Which vessels drain into the right atrium?

   a. _____

   b. _____

   c. _____

10. What separates the two ventricles?_____

11. What two structures attach to the heart valves to prevent the backflow of blood?

    a. _____

    b. _____

12. The right atrioventricular valve is called _____

    because it has _____ (number) of cusps.

13. What structures associated with the heart contain semilunar valves?

    a. _____

    b. _____

14. Which two major vessels supply the heart muscles with blood?

    a. _____

    b. _____

15. What is the major source of the blood vessels that supply the heart?

    _____

16. The anterior and posterior surfaces of the heart are drained by which three major veins?

    a. _____

    b. _____

    c. _____

17. The major components of the conducting system of the heart are the

    a. _____

    b. _____

    c. _____

    d. _____

    e. _____

Part III

Using the listed terms, label the major vessels and structures of the heart and then color them.

Superior vena cava
Right auricle
Left auricle
Right ventricle
Inferior vena cava
Right pulmonary artery
Left pulmonary artery
Arch of aorta
Right atrium

Anterior interventricular sulcus
Coronary sulcus
Anterior interventricular artery
Right pulmonary veins
Left pulmonary veins
Pulmonary trunk
Ligamentum arteriosum
Left ventricle
Right coronary artery

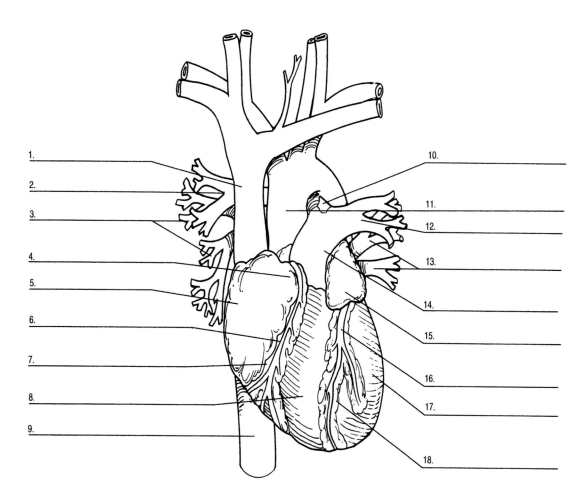

1. _____
2. _____
3. _____
4. _____
5. _____
6. _____
7. _____
8. _____
9. _____

10. _____
11. _____
12. _____
13. _____
14. _____
15. _____
16. _____
17. _____
18. _____

Figure 15.10 Anterior view of the heart illustrates the major vessels and chambers.

LABORATORY EXERCISE 15.2

Using the listed terms, label the major vessels and structures of the heart and then color them.

Left pulmonary artery
Arch of aorta
Right pulmonary artery
Middle cardiac vein
Posterior interventricular artery
Inferior vena cava
Left pulmonary veins
Superior vena cava
Left atrium

Left ventricle
Posterior interventricular sulcus
Right pulmonary veins
Left auricle
Right atrium
Coronary sinus
Right ventricle
Right coronary artery

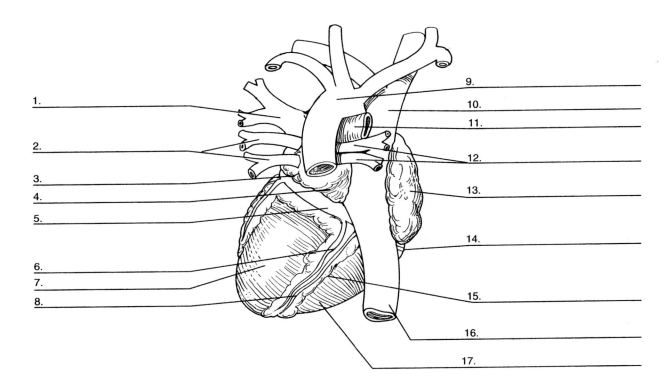

1. _____
2. _____
3. _____
4. _____
5. _____
6. _____
7. _____
8. _____

9. _____
10. _____
11. _____
12. _____
13. _____
14. _____
15. _____
16. _____
17. _____

Figure 15.11 Posterior view of the heart illustrates the major vessels and chambers.

LABORATORY EXERCISE 15.3

Using the listed terms, label the structures in the interior of the heart and then color them.

Fossa ovalis
Bicuspid valve
Papillary muscle
Right atrium
Opening of coronary sinus
Interventricular septum
Pulmonary semilunar valve
Aortic semilunar valve

Left atrium
Left ventricle
Trabeculae carneae
Interatrial septum
Opening of inferior vena cava
Tricuspid valve
Right ventricle
Chorda tendineae

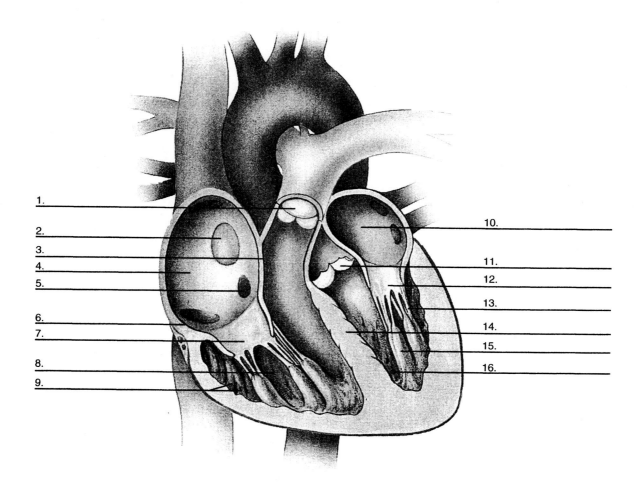

1.
2.
3.
4.
5.
6.
7.
8.
9.
10.
11.
12.
13.
14.
15.
16.

Figure 15.12 A cross section of the heart illustrates its internal anatomy.

LABORATORY EXERCISE 15.4

Using the listed terms, label the arteries, veins, and associated vessels and then color them.

Arteries

Right coronary
artery
Circumflex artery
Posterior interven-
tricular artery
Left coronary artery
Anterior interven-
tricular artery
Marginal artery

Veins

Anterior cardiac
vein
Great cardiac vein
Small cardiac vein
Middle cardiac vein
Coronary sinus

Associated Vessels

Arch of aorta
Inferior vena cava
Pulmonary trunk
Superior vena cava

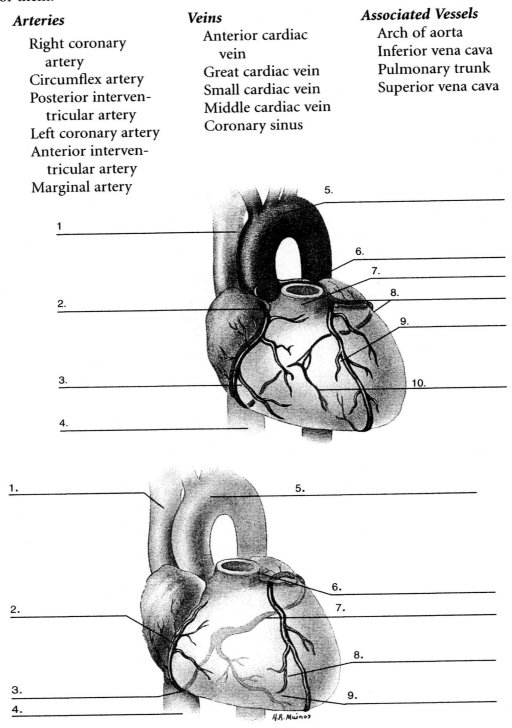

Figure 15.13 Circulation of the heart, showing (top) coronary arteries and (bottom) coronary veins.

16

The Cardiovascular System

THE MAJOR ARTERIES

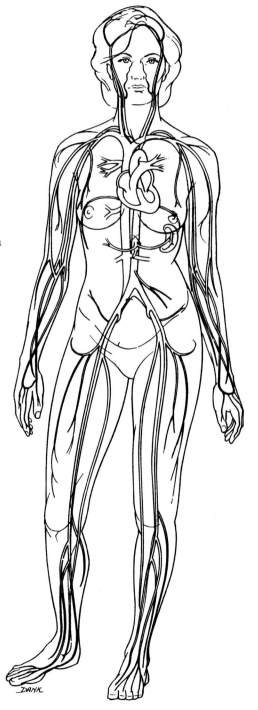

Objective

The objective of Chapter 16, "The Cardiovascular System: The Major Arteries," is for you to be able to identify the major arteries that supply the following regions:

1. Head, neck, and upper extremities
2. Thoracic and abdominal cavities
3. Pelvis and lower extremities

The Major Systemic Arteries *[Figure 16.1]*

The arteries carry blood away from the heart, including oxygenated blood in the aorta and deoxygenated blood in the pulmonary trunk and pulmonary arteries. Arteries have thicker walls and smaller lumina (openings) than veins. The major blood vessel that exists the left ventricle is the aorta; all systemic arteries arise from the aorta. For descriptive purposes, the aorta is subdivided into different regions. The **ascending aorta** leaves the heart and ascends toward the head. The **aortic arch** is the archlike portion of the aorta that curves inferiorly. The **descending aorta** continues caudally. In the thorax, the aorta is called the **thoracic aorta**; below the diaphragm it is called the **abdominal aorta**.

The Ascending Aorta, Aortic Arch, and Arteries of the Head and Neck—Description

[Figures 16.1 and 16.2]

The **ascending aorta** is short; it emerges from the **left ventricle** and passes superiorly deep to the **pulmonary trunk**. The ascending aorta gives rise to the **right** and **left coronary arteries** that supply the heart.

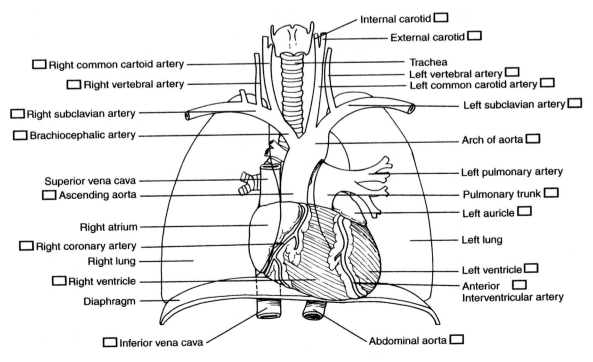

Figure 16.1 The heart, coronary vessels, and ascending aorta with its branches in anterior view.

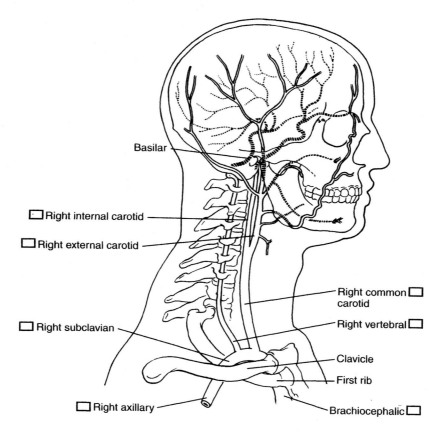

Figure 16.2 Arteries of the aortic arch and the head and neck.

As the ascending aorta arches dorsally to the left, it forms the **aortic arch**. Three major vessels arise from the aortic arch: the **brachiocephalic artery** (innominate artery), the **left common carotid artery**, and the **left subclavian artery**. These vessels deliver arterial blood to the head, neck, and upper limbs.

The brachiocephalic artery is the first branch off the aortic arch; it further subdivides into the **right subclavian artery** and **right common carotid artery**. The **left common carotid artery** is the second branch, and the **left subclavian artery** is the third branch off the aortic arch. Before passing into the axillary region of the upper arms, the right and left subclavian arteries give rise to **vertebral arteries**.

The right and left common carotid arteries ascend in the neck on either side of the trachea and give off numerous smaller branches to the neck. In the upper neck, the right and left common carotid arteries each divide into **internal** and **external carotid arteries**. The external carotid arteries ascend to supply the upper head, face, and neck regions, and the internal carotid arteries enter the cranial cavity to supply the brain.

The Ascending Aorta, Aortic Arch, and Arteries of the Head and Neck—Identification
[Figures 16.1 and 16.2]

Examine a model and the diagrams of the aorta and its major branches that supply the head and neck. Color these vessels. If possible, examine the vessels in a prepared cadaver and identify the aorta and its segments and branches in the head and neck.

1. ascending aorta
2. aortic arch
3. descending aorta
4. thoracic aorta
5. brachiocephalic artery
6. left common carotid artery
7. left subclavian artery
8. right common carotid artery
9. right subclavian artery
10. internal carotid artery
11. external carotid arteries
12. vertebral arteries

Arteries of the Brain—Description [*Figures 16.2 and 16.3*]

The brain is supplied with arterial blood by the **right** and **left internal carotid arteries** and the **right** and **left vertebral arteries**. These arteries enter the cranial cavity and join on the inferior surface of the brain to form a circular anastomosis (**circle of Willis**) around the pituitary gland. Numerous branches from these vessels supply all the superficial cortical and deep regions of the brain.

The **vertebral arteries**, which arise from the **subclavian arteries**, ascend through the **transverse foramina** of the cervical vertebrae and enter the cranial cavity through the **foramen magnum** of the skull. The right and left vertebral arteries join on the inferior surface near the **pons** of the brain and form the **basilar artery**. The basilar artery continues forward over the pons, branching into smaller arteries to supply the inferior surfaces of the brain. The basilar artery terminates by dividing into the **right** and **left posterior cerebral arteries**. These arteries supply the posterior regions of the cerebrum (brain). **Posterior communicating arteries** arise from the posterior cerebral arteries and connect these vessels with the nearby **internal carotid arteries**.

The **internal carotid arteries** enter the inferior cranial cavity through the **carotid canal** of the temporal bone and give terminal branches to the brain. The

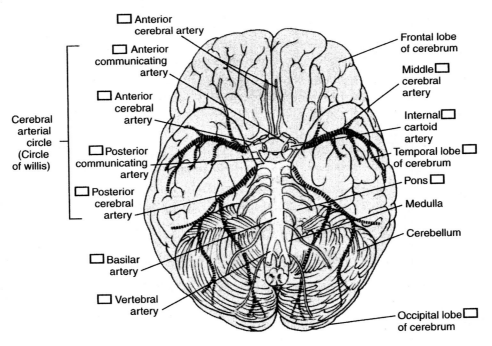

Figure 16.3 Arteries of the brain seen from the inferior view.

two major branches of the internal carotid arteries are the **anterior cerebral** and **middle cerebral arteries**. The anterior cerebral arteries supply the medial surfaces of the anterior cerebral hemispheres. A short **anterior communicating artery** connects the right and left anterior cerebral arteries. The middle cerebral arteries continue laterally over the brain surface to supply the lateral regions of the temporal and parietal lobes. Thus, the small **anterior** and **posterior communicating arteries** connect the **anterior** and **posterior cerebral vessels** around the pituitary gland. This junction forms the circular anastomosis called the **cerebral arterial circle (circle of Willis)**.

Arteries of the Brain—Identification *[Figure 16.3]*

Examine the arteries on the inferior surface of a brain model and the diagrams, and then color them. If possible, examine the inferior surface of a human brain and identify the following major arteries:

1. **cerebral arterial circle (circle of Willis)**
2. **vertebral arteries**
3. **basilar artery**
4. **posterior cerebral arteries**
5. **posterior communicating arteries**
6. **internal carotid arteries**
7. **middle cerebral arteries**
8. **anterior cerebral arteries**
9. **anterior communicating artery**

Arteries of the Upper Extremities— Description *[Figures 16.4 and 16.5]*

The upper extremities are supplied by vessels from the **subclavian arteries**. The right **subclavian artery** is a branch of the **brachiocephalic artery**, and the left subclavian artery is a direct branch from the **aortic arch**. Arising also from the subclavian arteries and descending anteriorly into the thoracic cavity on each side of the sternum are the **internal thoracic arteries** (described below).

The subclavian artery is given different names as it passes through different regions of the upper limb. The **subclavian artery** exits the thoracic cavity and passes on the superior surface of the first rib, forming a **subclavian groove**. After crossing the first rib and passing inferior to the clavicle, the subclavian artery enters the axilla (armpit) and becomes the **axillary artery**. The axillary artery supplies the upper thoracic wall and shoulder. On exiting the axilla, the axillary artery continues into the upper arm medial to the humerus as the **brachial artery**, giving off smaller branches to the arm (deep brachial, inferior, and superior collateral).

The brachial artery crosses the elbow on the anterior surface and divides into the lateral **radial** and medial **ulnar artery**. These two vessels descend toward the palm, where they anastomose to form two **palmar arches**, the **deep palmar arch** and the **superficial palmar arch**. Arising from these palmar arches are the **digital arteries**, which supply blood to the fingers.

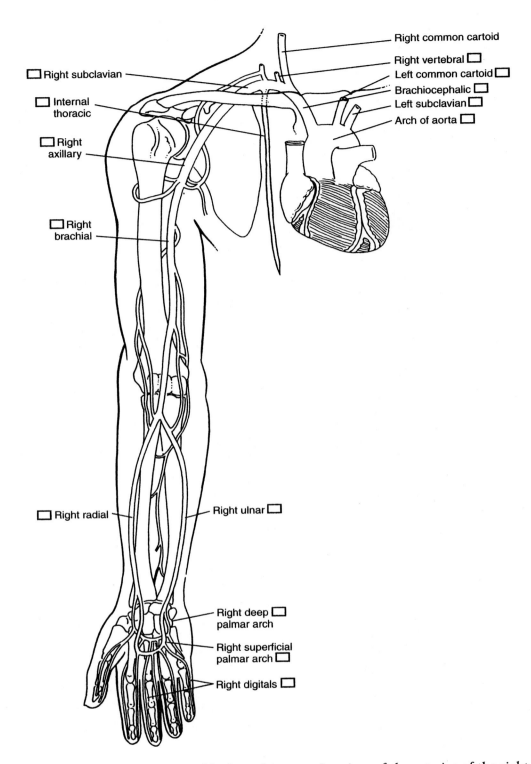

Right common cartoid

Right vertebral ☐

Left common cartoid ☐

Brachiocephalic ☐

Left subclavian ☐

Arch of aorta ☐

☐ Right subclavian

☐ Internal thoracic

☐ Right axillary

☐ Right brachial

☐ Right radial

Right ulnar ☐

Right deep ☐ palmar arch

Right superficial palmar arch ☐

Right digitals ☐

Figure 16.4 Arch of the aorta and its branches: anterior view of the arteries of the right upper extremity.

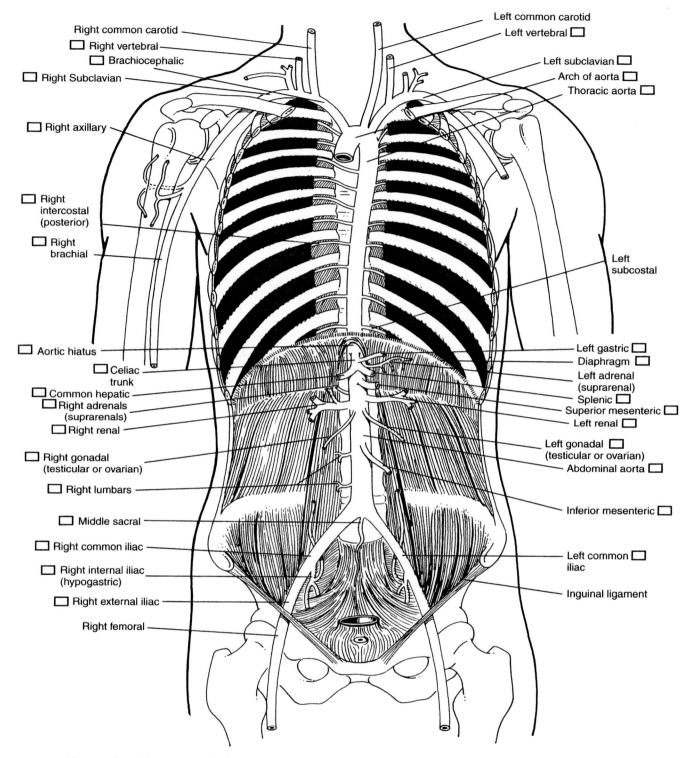

Right common carotid
☐ Right vertebral
☐ Brachiocephalic
☐ Right Subclavian

☐ Right axillary

☐ Right intercostal (posterior)

☐ Right brachial

☐ Aortic hiatus
☐ Celiac trunk
☐ Common hepatic
☐ Right adrenals (suprarenals)
☐ Right renal

☐ Right gonadal (testicular or ovarian)

☐ Right lumbars

☐ Middle sacral

☐ Right common iliac

☐ Right internal iliac (hypogastric)

☐ Right external iliac

Right femoral

Left common carotid
Left vertebral ☐
Left subclavian ☐
Arch of aorta ☐
Thoracic aorta ☐

Left subcostal

Left gastric ☐
Diaphragm ☐
Left adrenal (suprarenal)
Splenic ☐
Superior mesenteric ☐
Left renal ☐

Left gonadal ☐ (testicular or ovarian)
Abdominal aorta ☐

Inferior mesenteric ☐

Left common ☐ iliac

Inguinal ligament

Figure 16.5 Thoracic and abdominal aorta and their principal branches in anterior view.

Arteries of the Upper Extremities— Identification *[Figures 16.4 and 16.5]*

Examine the arteries in the upper extremities of a model and the diagrams, and then color them. If possible, examine the upper limbs in a prepared cadaver and identify the following major arteries:

1. **right brachiocephalic artery**
2. **subclavian arteries**
3. **axillary arteries**
4. **brachial arteries**
5. **radial arteries**
6. **ulnar arteries**
7. **palmar arches**

Arteries of the Thoracic Aorta and Thoracic Cavity—Description *[Figure 16.5]*

The **thoracic aorta**, a continuation of the aortic arch, descends into the thorax anterior to the twelve thoracic vertebrae on the left side of the vertebral column. Nine pairs of **posterior intercostal arteries** arise posteriorly from the thoracic aorta and run anteriorly between the intercostal muscles of the ribs until they anastomose with the **anterior intercostal arteries** from the **internal thoracic arteries** (Figure 16.4) The internal thoracic arteries arise from the subclavian arteries and descend toward the diaphragm on each side of the sternum inside the thoracic cavity. They supply the thoracic and intercostal muscles, the mediastinum, and the diaphragm. One of the terminal arteries of the internal thoracic arteries, the **superior epigastric artery**, supplies the muscles of the anterior abdominal wall (rectus abdominis muscles).

Arteries of the Thoracic Aorta and the Thoracic Cavity—Identification *[Figure 16.5]*

Examine the arteries in the interior thoracic cavity on a model and on the diagrams, and then color them. If possible, examine the thorax in a prepared cadaver and identify the following arteries:

1. **thoracic aorta**
2. **anterior intercostal arteries**
3. **posterior intercostal arteries**
4. **internal thoracic arteries**

Arteries of the Abdominal Aorta— Description *[Figures 16.5 and 16.6]*

The thoracic aorta leaves the thorax through the **aortic hiatus** (opening) of the diaphragm and enters the abdominopelvic cavity. The aorta descends in the abdomen on the anterior surface of the lumbar vertebrae to the fourth lumbar

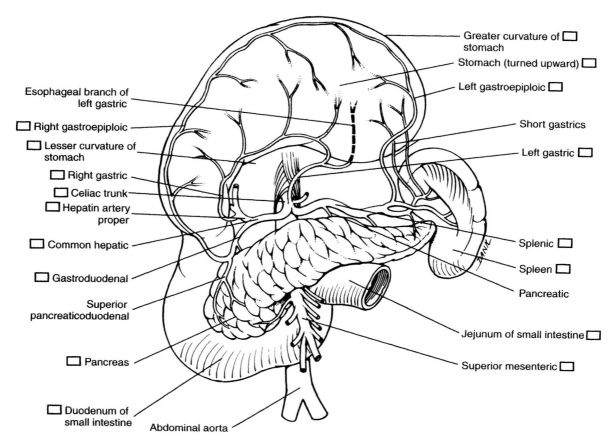

Esophageal branch of left gastric

☐ Right gastroepiploic

☐ Lesser curvature of stomach

☐ Right gastric

☐ Celiac trunk

☐ Hepatin artery proper

☐ Common hepatic

☐ Gastroduodenal

Superior pancreaticoduodenal

☐ Pancreas

☐ Duodenum of small intestine

Abdominal aorta

Greater curvature of ☐ stomach

Stomach (turned upward) ☐

Left gastroepiploic ☐

Short gastrics

Left gastric ☐

Splenic ☐

Spleen ☐

Pancreatic

Jejunum of small intestine ☐

Superior mesenteric ☐

Figure 16.6 (a) Main arterial branches to viscera from the celiac trunk. The stomach has been turned upward to expose celiac trunk and its branches.

vertebra. Here, the abdominal aorta divides into **right** and **left common iliac arteries** and a small, single **middle sacral artery**, which descends to supply the sacrum and coccyx.

The first vessels from the anterior abdominal aorta are a pair of small **inferior phrenic arteries**; these supply the inferior surface of the diaphragm. On its posterior surface, **lumbar arteries** arise from each side of the abdominal aorta to supply the muscles and spinal cord of the lumbar region.

Below the inferior phrenic arteries is the single (unpaired), short **celiac trunk (artery)**. It divides immediately into the **left gastric, splenic,** and **common hepatic arteries**. The **left gastric artery** reaches the esophagus-stomach junction on the left and then descends along the lesser curvature of the stomach; this artery supplies the lower third of the esophagus and the upper left portion of the stomach. The splenic artery runs left along the superior margin of the pancreas and supplies the pancreas and stomach and then the spleen. A large branch from the splenic artery, the **left gastroepiploic artery**, also supplies the greater curvature of the stomach.

The **common hepatic artery** divides into the **hepatic artery proper, gastroduodenal artery,** and **right gastric artery**. The right gastric artery supplies the lesser curvature of the stomach and anastomoses (unites) with the left gastric artery. The gastroduodenal artery supplies the pancreas and gives rise to the **right gastroepiploic artery**. This artery supplies the greater curvature of the stomach where it anastomoses with the left gastroepiploic branch of the splenic artery. After giving

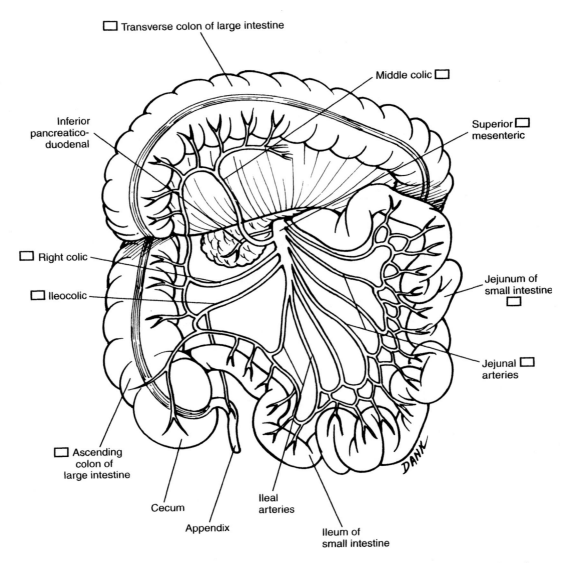

□ Transverse colon of large intestine

Middle colic □

Inferior
pancreatico-
duodenal

Superior □
mesenteric

□ Right colic

□ Ileocolic

Jejunum of
small intestine
□

Jejunal □
arteries

□ Ascending
colon of
large intestine

Cecum

Ileal
arteries

Appendix

Ileum of
small intestine

Figure 16.6 (b) Principal branches of the superior mesenteric artery to the small and large intestine.

off these branches, the common hepatic artery branches into the **right** and **left hepatic arteries**. These vessels supply both lobes of the liver and the gallbladder via the cystic artery.

The second unpaired vessel immediately below the celiac artery is the **superior mesenteric artery**. It branches into several principal vessels. The **jejunal** and **ilial** arteries supply the small intestine. The **ileocecal artery** supplies the terminal portion of the ileum and the cecum. The **right colic artery** supplies the ascending colon, and the **middle colic artery** supplies the transverse colon (Figure 16.6b).

Inferior to the superior mesenteric artery are the **renal arteries**, which arise on the lateral side of the aorta and enter the kidneys. Below the renal arteries are the

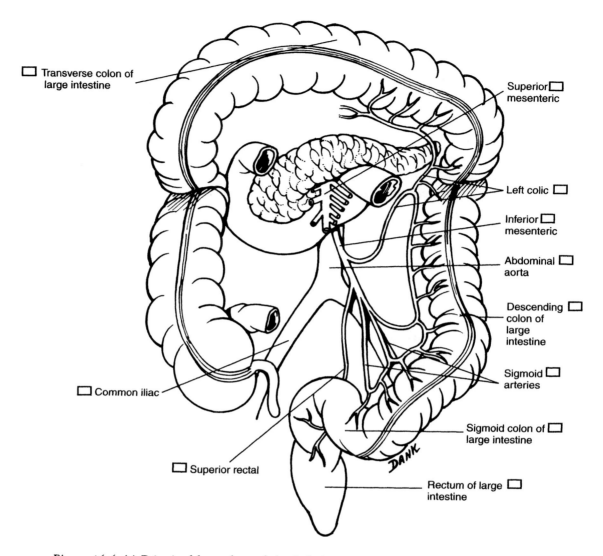

Transverse colon of large intestine

Superior ☐ mesenteric

Left colic ☐

Inferior ☐ mesenteric

Abdominal ☐ aorta

Descending ☐ colon of large intestine

Sigmoid ☐ arteries

Sigmoid colon of ☐ large intestine

Rectum of large ☐ intestine

☐ Common iliac

☐ Superior rectal

Figure 16.6 (c) Principal branches of the inferior mesenteric artery to the descending colon and superior region of the rectum.

gonadal vessels; they supply the reproductive organs. In the female, the **ovarian arteries** descend into the pelvic cavity and supply the ovaries. In the male, the **testicular arteries** descend into the pelvic cavity, pass through the inguinal canal, and enter the scrotum to supply the testes (Figure 16.5)

The last major branch of the abdominal aorta before the aorta divides into the right and left common iliac arteries is the unpaired **inferior mesenteric artery**. This vessel gives rise to the **left colic, sigmoid**, and **superior rectal arteries**. The left colic artery supplies the transverse and descending colon. The sigmoid arteries supply the sigmoid colon, and the superior rectal artery supplies the upper region of the rectum (Figure 16.6c).

Arteries of the Abdominal Aorta—
Identification *[Figures 16.5 and 16.6]*

Examine a model and the diagrams of the abdominal aorta and its arteries, and then color them. If possible, examine the abdominal cavity in a prepared cadaver and identify the following arteries:

1. abdominal aorta
2. aortic hiatus
3. celiac trunk
4. splenic artery
5. left gastroepiploic artery
6. left gastric artery
7. common hepatic artery
8. left hepatic artery
9. right hepatic artery
10. right gastric artery
11. right gastroepiploic artery
12. superior mesenteric artery
    a. ileocolic artery
    b. right colic artery
    c. middle colic artery
13. renal arteries
14. ovarian (gonadal) arteries
15. testicular (gonadal) arteries
16. inferior mesenteric artery
    a. left colic artery
    b. sigmoid artery
    c. superior rectal artery
17. lumbar arteries

Arteries of the Pelvis and Lower Extremities—
Description *[Figure 16.7]*

The **right** and **left common iliac arteries** are short; they terminate in the pelvis by branching into larger **external** and smaller **internal iliac arteries**. The **internal iliac arteries** branch extensively and supply the pelvic viscera, reproductive and urinary organs, external genitalia, and gluteal, lumbar, and medial thigh muscles.

The **external iliac arteries** are the major source of blood to the lower limbs. Each external iliac artery leaves the pelvis and enters the thigh beneath the inguinal ligament. In the thigh, the external iliac artery becomes the **femoral artery**, which passes inferomedially across the thigh. In the thigh, the femoral artery gives off several branches to the surrounding skin and muscles. In the lower thigh, the femoral artery descends across the posterior knee, becomes the **popliteal artery**, and supplies the knee joint and leg muscles.

Inferior to the knee joint, the popliteal artery branches into an **anterior tibial artery** and **posterior tibial artery**. The anterior tibial artery descends anteriorly between the tibia and fibula bones and supplies the anterior leg muscles. The artery terminates on the dorsum of the foot as the **dorsalis pedis artery**.

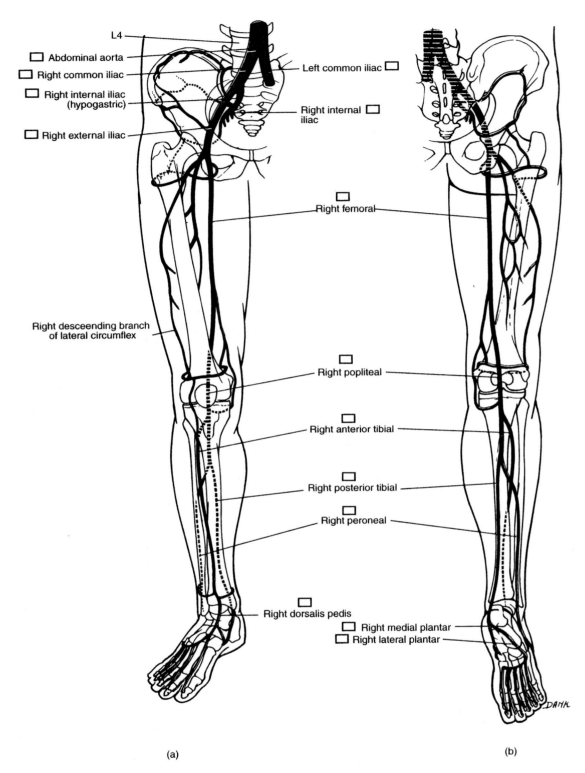

L4

☐ Abdominal aorta
☐ Right common iliac
☐ Right internal iliac (hypogastric)
☐ Right external iliac

Left common iliac ☐

Right internal iliac ☐

☐ Right femoral

Right desceending branch of lateral circumflex

☐ Right popliteal

☐ Right anterior tibial

☐ Right posterior tibial

☐ Right peroneal

☐ Right dorsalis pedis

☐ Right medial plantar
☐ Right lateral plantar

DANK

(a)   (b)

Figure 16.7 (a) Anterior and (b) posterior views of the arteries of the pelvis and right lower extremity.

A short distance from its origin, the **posterior tibial artery** gives off the **peroneal artery**, which supplies the peroneal muscles in the lateral region of the leg. The posterior tibial artery supplies the flexor muscles of the posterior leg. The posterior tibial artery divides behind the medial ankle into the **lateral** and **medial plantar arteries** to supply the muscles and sole of the foot. The lateral plantar artery anastomoses with the **medial plantar artery** to form the **plantar arch**. Digital arteries arising from the plantar arch supply blood to the toes.

Arteries of the Pelvis and Lower Extremities— Identification [Figure 16.7]

Examine a model and the diagrams of the arteries in the lower extremities, and then color them. If possible, examine the lower limb of a cadaver and identify the following arteries:

1. **common iliac arteries**
2. **external iliac arteries**
3. **internal iliac arteries**
4. **femoral arteries**
5. **popliteal arteries**
6. **anterior tibial arteries**
7. **posterior tibial arteries**
8. **peroneal arteries**
9. **dorsalis pedis arteries**
10. **plantar arteries**
11. **plantar arch**

CHAPTER SUMMARY AND CHECKLIST

I. **MAJOR SYSTEMIC ARTERIES**

A. **Ascending Aorta, Aortic Arch, and Arteries of Head and Neck**
1. Left and right coronary arteries from the ascending aorta
2. Major branches of the aortic arch:
    a. brachiocephalic: subdivides into right subclavian and right common carotid arteries
    b. left common carotid artery: second branch
    c. left subclavian artery: third branch
3. Right and left subclavian arteries give rise to vertebral arteries
4. Common carotids divide into internal and external carotid arteries
    a. external carotid arteries supply head, face, and neck
    b. internal carotid arteries supply the brain

II. **ARTERIES OF THE BRAIN**

A. **Internal Carotid Arteries**
1. Enter cranial cavity through carotid canal
2. Branch into anterior and middle cerebral arteries
    a. anterior communicating artery connects anterior cerebral arteries

3. Anterior cerebral arteries supply medial cerebral hemispheres
4. Middle cerebral arteries supply lateral hemispheres of temporal and parietal lobes

B. Vertebral Arteries
1. Enter cranial cavity through foramen magnum
2. Unite on inferior brain surface to form basilar artery
3. Basilar divides into posterior cerebral arteries
4. Posterior communicating artery connects basilar artery with arteries from internal carotid
   a. form cerebral arterial circle (of Willis)

III. ARTERIES OF UPPER EXTREMITIES

A. Major Branches
1. Subclavian artery exits the thorax
2. In armpit, subclavian becomes axillary artery
3. Axillary continues into upper arm as brachial artery
4. Brachial divides into ulnar and radial arteries in cubitus
5. In the palm, ulnar and radial arteries anastomose to form palmar arches (deep and superficial)

IV. ARTERIES OF THE THORACIC AORTA AND THORACIC CAVITY

A. Thoracic Cavity
1. Posterior intercostal arteries arise from aorta
2. Anterior intercostal arteries arise from internal thoracic arteries
3. Internal thoracic arteries arise from subclavian arteries

V. ARTERIES OF THE ABDOMINAL AORTA

A. Major Arteries
1. Celiac trunk (artery) divides into three arteries:
   a. left gastric to stomach
   b. splenic artery to spleen, pancreas, and stomach
   c. common hepatic to liver, gallbladder, duodenum, and stomach
2. Superior mesenteric artery branches into:
   a. jejunal and ileal arteries to small intestine
   b. ileocecal artery to ileum and cecum
   c. right colic artery to ascending colon
   d. middle colic artery to transverse colon
3. Renal arteries to kidneys
4. Gonadal arteries (testicular and ovarian) to gonads
5. Inferior mesenteric artery branches into:
   a. left colic artery to transverse and descending colon
   b. sigmoid artery to sigmoid colon
   c. superior rectal artery to upper rectum

VI. ARTERIES OF THE PELVIS AND LOWER EXTREMITIES

A. Common Iliac Arteries
1. Aorta terminates by branching into common iliac arteries

2. Common iliac arteries branch into external and internal iliac arteries
3. Internal iliac arteries supply pelvic viscera
4. External iliac arteries supply lower limb
   a. femoral artery to thigh region
   b. popliteal artery across posterior surface of knee
   c. popliteal artery branches into anterior and posterior tibial arteries
   d. anterior tibial artery terminates as dorsalis pedis artery on dorsum of foot
   e. posterior tibial artery branches to lateral and medial plantar arteries to supply muscle and sole of foot

Laboratory Exercises 16

NAME _____

LAB SECTION _____ DATE _____

LABORATORY EXERCISE 16.1
The Major Arteries
Part I

1. What major artery emerges from the left ventricle?

   _____

2. What is the aorta called below the diaphragm?

   _____

3. Which three major arteries arise from the aortic arch?

   a. _____

   b. _____

   c. _____

4. From what artery do the right subclavian and right common carotid arteries

   branch? _____

5. Into which two major arteries do the common carotid arteries branch?

   a. _____

   b. _____

6. Which major arteries ascend toward the head in the transverse foramina of

   the cervical vertebra? _____

7. Which four major arteries enter the cranium to supply the brain?

   a. _____

   b. _____

   c. _____

   d. _____

8. What vessel is formed by vertebral arteries that join in the brain?

   _____

9. Which arteries anastomose on the inferior side of the brain to form the anterior arterial circle (of Willis)?

a. _____

b. _____

c. _____

d. _____

10. The middle cerebral artery is a branch of the _____ and the posterior cerebral artery is a branch of the _____.

Part II

Match the description of the vessel in the left column with the name of the vessel in the right column.

1. Supplies vessels to upper extremities _____

2. Major artery in the armpit_____

3. Major artery in the upper arm _____

4. Two arteries that descend in the forearm

   _____, _____

5. Gives rise to posterior intercostal arteries _____

6. Gives rise to anterior intercostal arteries _____

7. Aorta terminates by dividing into _____

8. Unpaired vessels that gives rise to arteries of liver, stomach, and spleen _____

9. Supplies the greater curvature of stomach _____

10. Supplies the lesser curvature of stomach _____

11. Brings blood to liver and gallbladder _____

12. Major unpaired vessel that gives rise to vessels that supply the small and large intestines _____

13. Artery that supplies ascending colon _____

14. Transverse colon received blood for this artery _____

15. Kidneys are supplied by this artery _____

16. Last unpaired vessel of aorta to intestines _____

17. Sigmoid colon supplied by _____

18. Major source of blood to lower limbs _____

19. External iliac artery becomes this vessel _____

20. Muscles of anterior side of leg are supplied by _____

21. Major artery of the thigh _____

A. Common iliac
B. Celiac trunk
C. Gastroepiploic
D. Brachial
E. Axillary
F. Subclavian
G. Ulnar
H. Radial
I. Aorta
J. Internal thoracic
K. Gastric
L. Anterior tibial
M. Femoral
N. External iliac
O. Sigmoid
P. Inferior mesenteric
Q. Renal
R. Ileocolic
S. Middle colic
T. Hepatic
U. Superior mesenteric

Using the listed terms, label the following arteries of the ascending aorta and then color them.

Ascending aorta
Internal carotid
Right subclavian
Left subclavian
Arch of aorta
Right common carotid

External carotid
Left common carotid
Brachiocephalic artery
Right vertebral
Left vertebral
Abdominal aorta

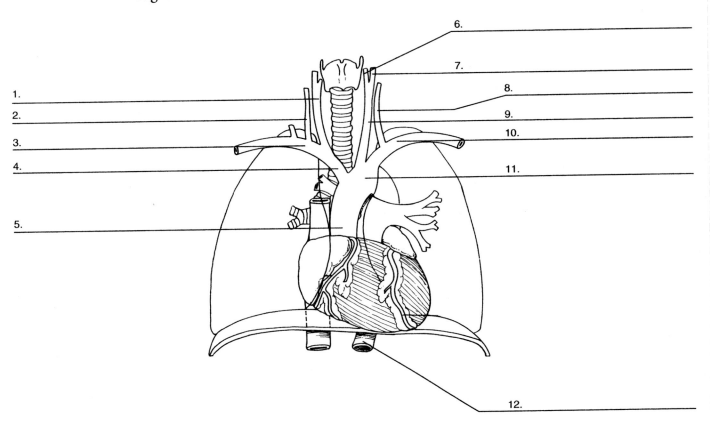

Figure 16.8 The ascending aorta and major arteries to the head and neck.

LABORATORY EXERCISE 16.2

Using the listed terms, label the following arteries of the aortic arch on the right side and then color them.

Subclavian   Brachiocephalic
Vertebral   Basilar
External carotid   Internal carotid
Common carotid

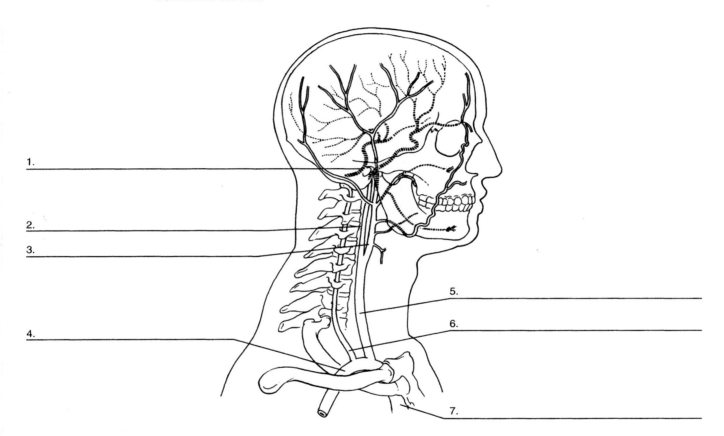

1. _____

2. _____

3. _____

4. _____

5. _____

6. _____

7. _____

Figure 16.9 Arteries of the aortic arch that supply the head and neck on the right side of the body.

LABORATORY EXERCISE 16.3

Using the listed terms, label the following arteries of the brain and then color them.

Anterior cerebral
Anterior communicating
Basilar
Posterior cerebral
Middle cerebral

Vertebral
Anterior cerebral
Internal carotid
Posterior communicating

1.

2.

3.

4.

5.

6.

7.

8.

9.

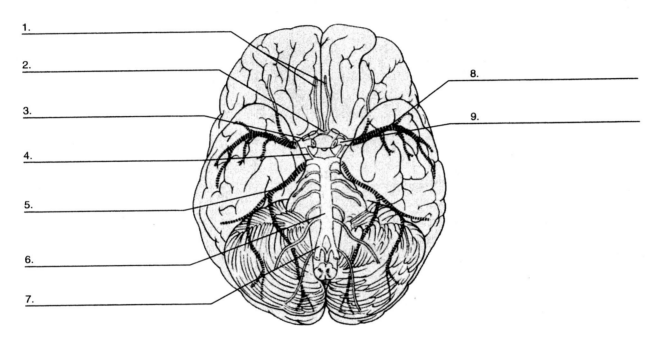

Figure 16.10 Major arteries of the brain, seen from the inferior side.

LABORATORY EXERCISE 16.4

Using the listed terms, label the following arteries of the aorta and then color them.

Brachiocephalic
Abdominal aorta
Right axillary
Common hepatic
Superior mesenteric
Right gonadal
Middle sacral
Right internal iliac
Left subclavian
Thoracic aorta
Right brachial

Splenic
Inferior mesenteric
Left common iliac
Right femoral
Arch of aorta
Right intercostal
Celiac trunk
Left gastric
Left renal
Right lumbar
Right external iliac

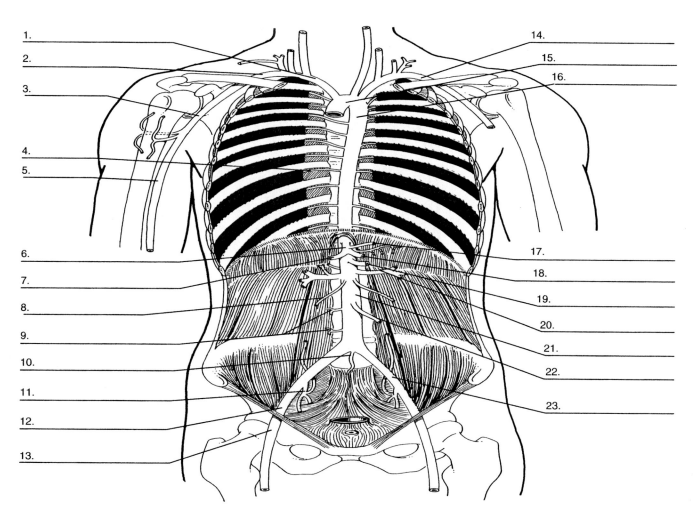

Figure 16.11 The aorta and its main branches to the upper extremity, thorax, abdomen, and lower extremity.

LABORATORY EXERCISE 16.5

Using the listed terms, label the following arteries of the celiac trunk and then color them.

Right gastroepiploic
Right gastric
Common hepatic
Left gastroepiploic
Left gastric

Splenic
Celiac trunk
Hepatic artery proper
Gastroduodenal

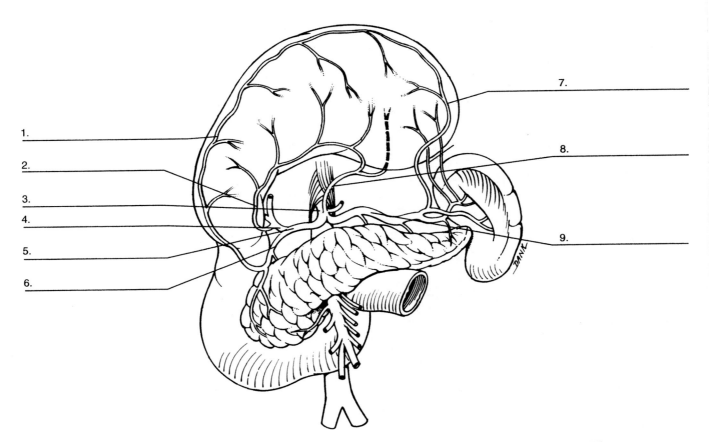

1.

2.

3.

4.

5.

6.

7.

8.

9.

Figure 16.12 Main arteries of the celiac trunk that supply the abdominal organs. The stomach has been turned upward to expose the celiac trunk and its branches.

LABORATORY EXERCISE 16.6

Using the listed terms, label the following arteries and then color them.

Superior mesenteric
Right colic
Superior rectal
Ileocolic
Jejunal

Middle colic
Sigmoid
Ilial
Left colic
Inferior mesenteric

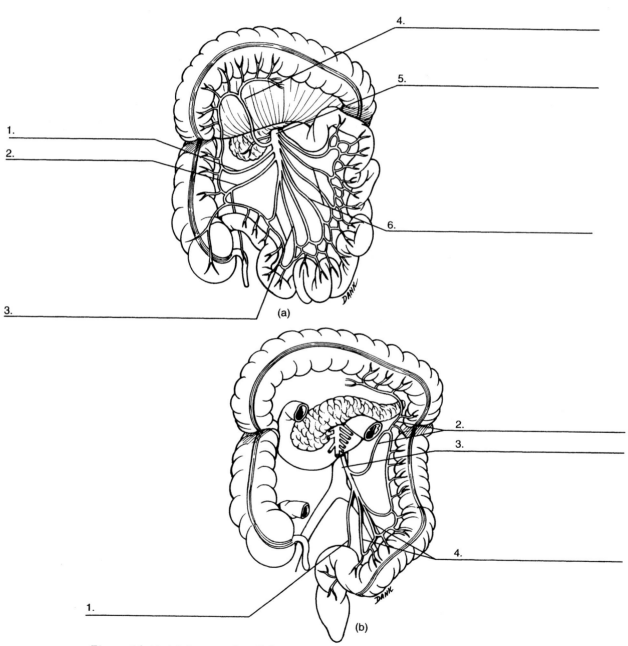

Figure 16.13 Main vessels of the superior (top) and inferior mesenteric (bottom) arteries to small and large intestines.

17

The Cardiovascular System

THE MAJOR VEINS

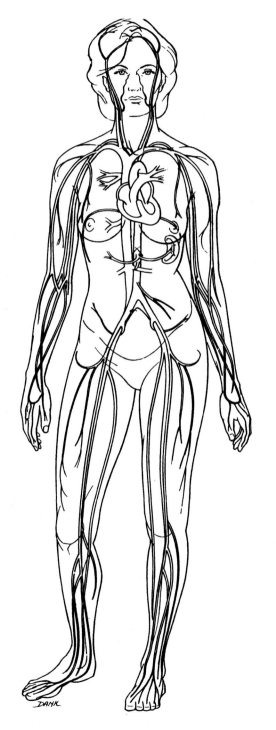

Objective

The objective of Chapter 17, "The Cardiovascular System: The Major Veins," is for you to be able to identify the major veins of the human body that drain the following areas:

1. **Head, neck, and upper extremities**
2. **Thoracic and abdominal cavities**
3. **Pelvis and lower extremities**

The Veins of the Heart *[Figures 17.1]*

Veins carry blood to the heart, including deoxygenated blood in the two venae cavae and oxygenated blood in the pulmonary veins. Veins join or unite as **tributaries**, in contrast to arteries, which **branch**. Veins are more numerous than arteries, thinner, and larger in lumen than arteries. To ensure that blood flows toward the heart, veins of the extremities have valves, which prevent backflow of blood. Large veins and veins in the head have no valves. Major veins usually run alongside arteries and have similar or same names.

Three major vessels deliver venous blood to the right atrium: the **superior vena cava**, **inferior vena cava**, and **coronary sinus**. The superior vena drains the **upper body** above the **diaphragm**, except the lungs. The **inferior vena cava** drains the **lower body** below the **diaphragm**. The inferior vena cava pierces the tendon of the diaphragm and enters the right atrium inferiorly. The **coronary sinus** drains the heart musculature via the cardiac veins. Veins that enter the left atrium are the

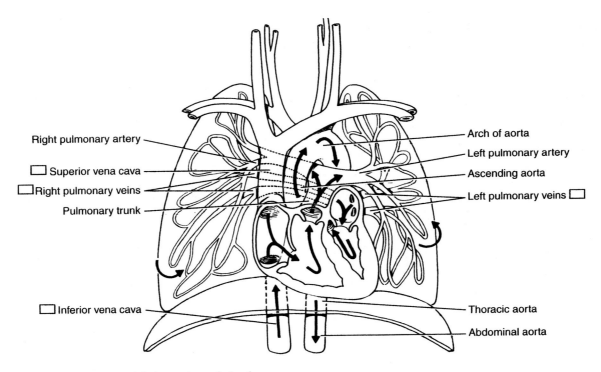

Figure 17.1 Major veins of the heart.

pulmonary veins, two from each lung. These are the only vessels that carry oxygenated blood to the heart.

Veins of the Head and Neck—
Description [Figure 17.2]

Blood from the scalp, portions of the face, and superficial neck region returns to the heart via the **external jugular veins**. These veins descend laterally on each side of the neck and empty into the **right** and **left subclavian veins**.

The **internal jugular veins** drain blood from the brain, meninges, and deep regions of the face and neck. These veins are larger than the external jugular veins and located deeper in the neck. The internal jugular veins start at the base of the skull in the **jugular foramen** as a continuation of the **venous sigmoid sinuses**.

The sinuses of the skull and brain are specialized venous channels located between the layers of the dura mater. In contrast to other veins in the body, the sinuses cannot collapse and remain open due to the rigid connective tissue walls of the dura mater. Other sinuses that drain into the sigmoid sinuses and then the

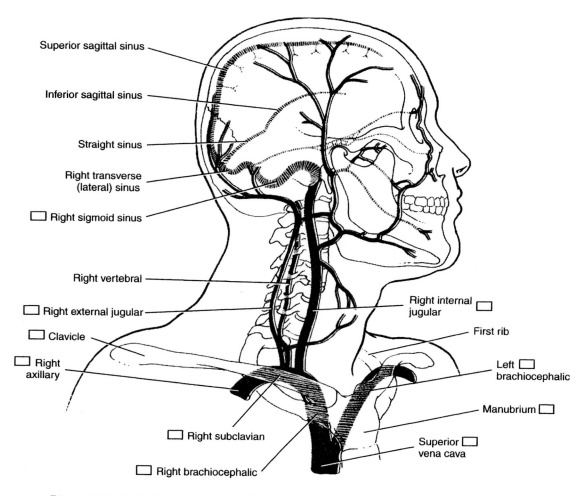

Figure 17.2 Right lateral view of the veins of the head and neck.

internal jugular veins are the superior sagittal, inferior sagittal, straight, and transverse sinuses.

The internal jugular veins descend on each side of the neck adjacent to the **common carotid artery** and the **vagus nerve**. Posterior to the clavicle, the **internal jugular veins** join the **right** and **left subclavian veins**, forming **right** and **left brachiocephalic veins**. Brachiocephalic veins merge to form the **superior vena cava**.

Veins of the Upper Extremities—
Description *[Figure 17.3]*

The upper limbs have both superficial and deep veins. Superficial veins are under the skin and highly variable. Deep veins usually run alongside arteries of the same region and bear the same names.

The superficial veins of the upper extremity are the **basilic** and **cephalic veins;** they originate from veins that drain the dorsal and palmar surfaces of the hand and run independently of arteries. The **cephalic vein** ascends on lateral side of the hand, forearm, and arm and empties into the **axillary vein** in the shoulder below the **clavicle**.

The **basilic vein** drains the medial side of the hand and forearm, and ascends medially to join the **axillary vein**. In the **cubital fossa** (elbow), a superficial **median cubital vein** ascends from the **cephalic vein** and connects with the **basilic vein**. The basilic vein is commonly used for drawing blood and injecting fluids in clinical settings.

The deep lateral **radial vein** of the forearm and the medial **ulnar vein** drain blood from the **superficial** and **deep palmar venous arches** of the hand. The **radial** and **ulnar veins** unite in the cubital fossa to form the **brachial vein**, which drains into the axillary vein. The axillary vein crosses the first rib and becomes the **subclavian vein**, which unites with the **external jugular vein** of that side to form the **brachiocephalic vein**.

Veins of the Head, Neck, and Upper
Extremities—Identification *[Figures 17.2 and 17.3]*

Examine models and the diagrams of veins that drain the head, neck, and upper extremities, and then color them. If possible, examine the thoracic cavity, head, neck, and one of the upper limbs in a prepared cadaver. Identify the following major veins:

1. **superior vena cava**
2. **brachiocephalic veins**
3. **internal jugular veins**
4. **external jugular veins**
5. **subclavian veins**
6. **cephalic vein**
7. **axillary vein**
8. **brachial vein**
9. **ulnar vein**
10. **radial vein**

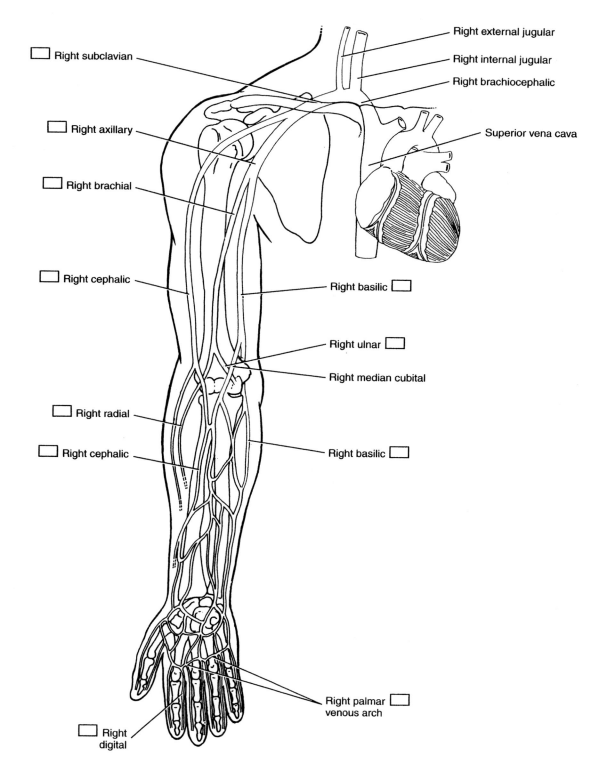

Right subclavian ☐

Right axillary ☐

Right brachial ☐

Right cephalic ☐

Right radial ☐

Right cephalic ☐

Right digital ☐

Right external jugular

Right internal jugular

Right brachiocephalic

Superior vena cava

Right basilic ☐

Right ulnar ☐

Right median cubital

Right basilic ☐

Right palmar ☐
venous arch

Figure 17.3 Anterior view of the principal veins of the right upper extremity.

11. basilic vein
12. median cubital vein
13. palmar venous arch

Veins of the Thorax—Description *[Figure 17.4]*

The veins that drain most of the thoracic wall are part of the **azygos system**. It consists of the **azygos, hemiazygos,** and **accessory hemiazygos veins**.

The **azygos vein** is unpaired and originates in the upper right posterior abdominal wall. It enters the thorax through the diaphragm with the aorta. The azygos vein ascends along the posterior abdominal and thoracic walls on the right side of the vertebral column. The vein passes over the root of the right lung and empties into the **superior vena cava** at the level of the fourth thoracic vertebra.

In the abdomen, the azygos vein receives most of the blood from the **ascending lumbar veins**. These drain most of the right abdominal cavity and connect the azygos vein with the common iliac vein and inferior vena cava. In the thorax, blood from the **right posterior intercostals, accessory hemiazygos,** and **hemiazygos veins** drain into the **azygos vein**.

The **hemiazygos vein** drains the left posterior regions of the upper abdominal wall via the **ascending lumbar vein** and the posterior thorax via the lower four **intercostal veins**. The **hemiazygos vein** ascends on the left side of the vertebral column, crosses the thoracic vertebra to the right, and empties into the azygos vein in the middle of the thorax.

The **accessory hemiazygos** vein drains the upper left region of the thorax, via the **intercostal veins**, and then crosses the vertebral column to empty into the azygos vein. Blood from the upper intercostal muscles drains into either the **left brachiocephalic vein** or the **accessory hemiazygos vein**.

Veins of the Thorax—Identification *[Figure 17.4]*

Examine a model and the diagrams of the thorax and veins of the posterior body wall, and then color them. If possible, examine the thorax in a prepared cadaver and identify the following veins:

1. posterior intercostal veins
2. azygos vein
3. hemiazygos vein
4. accessory hemiazygos vein

The Hepatic Portal System— Description *[Figure 17.5]*

Venous blood from the capillaries of the stomach, small intestine, large intestine, pancreas, gallbladder, and spleen first enters the liver via the **hepatic portal vein**. In the liver, the venous blood is distributed into the liver sinusoids and a second capillary network. It is then collected into the **hepatic veins** (right and left) before it enters the **inferior vena cava**. The blood flow from one capillary network via veins

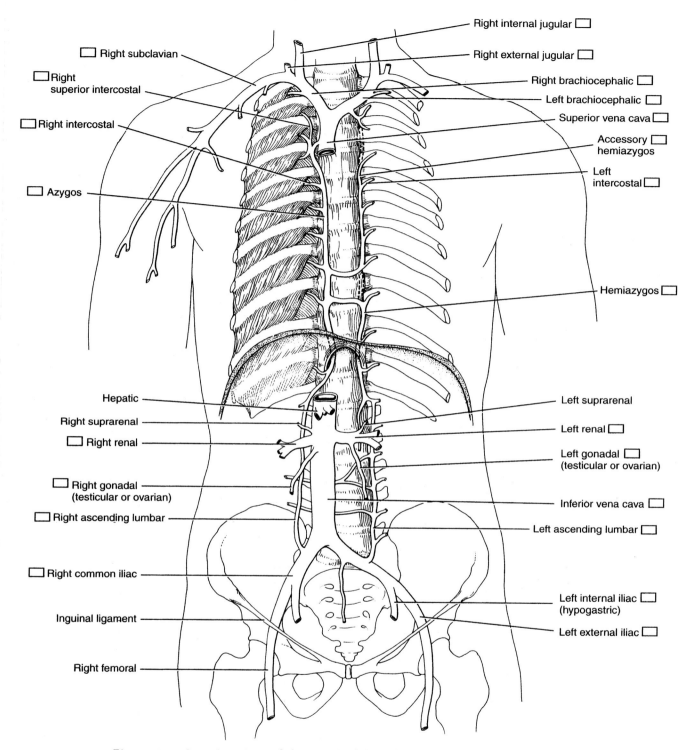

Right subclavian ☐

Right ☐
superior intercostal

☐ Right intercostal

☐ Azygos

Right internal jugular ☐

Right external jugular ☐

Right brachiocephalic ☐

Left brachiocephalic ☐

Superior vena cava ☐

Accessory ☐
hemiazygos

Left ☐
intercostal

Hemiazygos ☐

Hepatic

Right suprarenal

☐ Right renal

☐ Right gonadal
(testicular or ovarian)

☐ Right ascending lumbar

☐ Right common iliac

Inguinal ligament

Right femoral

Left suprarenal

Left renal ☐

Left gonadal ☐
(testicular or ovarian)

Inferior vena cava ☐

Left ascending lumbar ☐

Left internal iliac ☐
(hypogastric)

Left external iliac ☐

Figure 17.4 Anterior view of the veins of the thorax, abdomen, and pelvis.

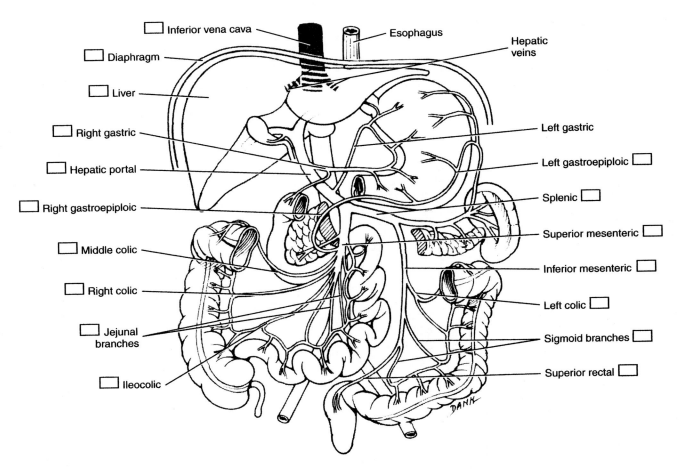

Inferior vena cava

Diaphragm

Liver

Right gastric

Hepatic portal

Right gastroepiploic

Middle colic

Right colic

Jejunal branches

Ileocolic

Esophagus

Hepatic veins

Left gastric

Left gastroepiploic

Splenic

Superior mesenteric

Inferior mesenteric

Left colic

Sigmoid branches

Superior rectal

Figure 17.5 The hepatic portal system and the major gastrointestinal veins.

into another constitutes a **portal system**. The liver has a dual blood supply: oxygenated blood from the **hepatic artery** and nutrient-rich blood from the **portal vein**.

The Hepatic Portal Veins—
Description *[Figure 17.5]*

The **hepatic portal vein** is normally formed by a joining of the **superior mesenteric** and **splenic veins**. The **superior mesenteric vein** drains the stomach, small intestine, cecum, and ascending and transverse colon. This is accomplished via the left gastroepiploic, jejunal, ileocolic, right colic, and middle colic veins.

The **splenic vein** receives blood from the spleen, pancreas, part of the stomach, and descending and sigmoid colon via the inferior mesenteric vein. The veins that drain the stomach are the **right** and **left gastric** and **right** and **left gastroepiploic**.

The **inferior mesenteric vein** drains the rectum, sigmoid colon, and descending colon. This is accomplished via the **left colic, sigmoid,** and **superior rectal veins**. The inferior mesenteric vein then joins the splenic vein just before it unites with the **superior mesenteric vein**.

The Hepatic Portal Veins— Identification [Figure 17.5]

Examine a model and the diagrams of the hepatic portal veins and then color them. If possible, examine the abdominal cavity of a prepared cadaver and identify the following veins:

1. hepatic veins
2. hepatic portal vein
3. left gastric vein
4. right gastric vein
5. splenic vein
6. superior mesenteric vein
   a. left gastroepiploic vein
   b. jejunal branches
   c. ileocolic vein
   d. right colic vein
   e. middle colic vein
7. inferior mesenteric vein
   a. left colic vein
   b. sigmoid veins
   c. superior rectal veins

Veins of the Abdominal and Pelvic Regions— Description [Figure 17.4]

Venous blood from the abdominopelvic viscera and abdominal walls returns to the heart via the **inferior vena cava**. It is formed by a confluence of the **right** and **left common iliac veins** at the fifth lumbar vertebra. Near the sacroiliac joint, the **external iliac vein** unites with the **internal iliac vein** to form the common iliac vein. The internal iliac veins drain the deep pelvic viscera, genital organs, and pelvic muscles. As the inferior vena cava ascends through the abdomen, it receives tributaries from veins that correspond to most of the arteries that arise from the descending aorta. The inferior vena cava then passes posterior to the liver in a groove, penetrates the diaphragm, and immediately enters the inferior border of the right atrium (Figures 17.1 and 17.4).

The veins that drain into the inferior vena cava are the right and left hepatic veins from the liver; the four **lumbar veins** from the posterior abdominal wall; the **right gonadal veins** (in males, the **testicular**, and in females, the **ovarian veins**) from the gonads (the left gonadal veins drain into the left renal vein); the renal veins from the kidneys; the right suprarenal vein from the right adrenal gland; and the right inferior phrenic vein from the inferior diaphragm.

Veins of the Abdominal and Pelvic Regions— Identification [Figures 17.4 and 17.5]

Examine a model and the diagrams of the posterior abdominal and pelvic veins, and then color them. If possible, identify the following veins in a prepared cadaver:

1. internal iliac veins
2. external iliac veins
3. common iliac veins
4. inferior vena cava
5. lumbar veins
6. gonadal veins (ovarian or testicular)
7. renal veins

Veins of the Lower Extremities—
Description *[Figure 17.6]*

Like the upper extremities, the lower extremities contain both deep and superficial veins. The deep veins travel alongside the arteries and bear the same names.

Two large superficial veins that drain the lower limbs are the **small** and **great saphenous veins**. The great saphenous vein is the longest vein in the body. It originates from the medial side of **dorsal venous arch** in the foot and ascends along the leg and thigh. It drains into the **femoral vein** below the inguinal ligament. The **small saphenous vein** originates in the lateral **dorsal venous arch**, ascends posteriorly along the calf of leg, and empties into the **popliteal vein** posterior to the knee.

The **posterior** and **anterior tibial veins** originate from the veins in the foot and ascend deep in the posterior muscle of the leg, posteriorly and anteriorly to the tibia, respectively. Posterior to the knee, the **anterior** and **posterior tibial veins** form the **popliteal vein**. Superior to the knee, the popliteal vein becomes the **femoral vein**.

Near the groin, the femoral vein receives blood from the great saphenous vein, after which it becomes the **external iliac vein** as it passes under the inguinal ligament and enters the abdominopelvic cavity. At the fifth lumbar vertebra, the left and right common iliac veins join to form the large inferior vena cava, which ascends toward the heart.

Veins of the Lower Extremities—
Identification *[Figure 17.6]*

Examine a model and the diagrams of the veins in the lower extremities and pelvis, and then color them. If possible, examine the lower abdominal and pelvic cavities and the lower limb in a prepared cadaver and identify the following veins:

1. external iliac vein
2. great saphenous vein
3. femoral vein
4. popliteal vein
5. small saphenous vein
6. anterior tibial vein
7. posterior tibial vein
8. dorsal venous arch

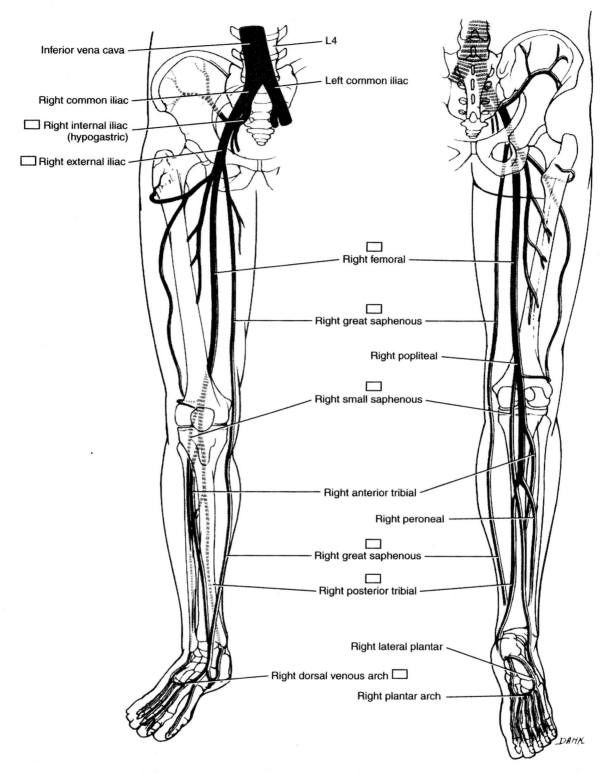

Figure 17.6 (Left) Anterior and (right) posterior views of the veins of the pelvis and right lower extremity.

Inferior vena cava

L4

Right common iliac

Left common iliac

☐ Right internal iliac (hypogastric)

☐ Right external iliac

☐ Right femoral

☐ Right great saphenous

Right popliteal

☐ Right small saphenous

Right anterior tribial

Right peroneal

☐ Right great saphenous

☐ Right posterior tribial

Right lateral plantar

Right dorsal venous arch ☐

Right plantar arch

DANK

CHAPTER SUMMARY AND CHECKLIST

I. **THE MAJOR VEINS**

  A. **Veins of the Heart**
1. Superior vena cava delivers blood from upper body region
2. Inferior vena cava brings blood from lower body region
3. The coronary sinus drains heart musculature
4. Pulmonary veins drain the lungs

II. **VEINS OF THE HEAD AND NECK**

  A. **External Jugular Veins**
1. Drain scalp, superficial face, and neck
2. Empty into right and left subclavian veins

  B. **Internal Jugular Veins**
1. Drain brain, meninges, and deep neck and face
2. Empty into right and left subclavian veins

  C. **Internal Jugular and Subclavian Veins Form Brachiocephalic Veins**

  D. **Superior Vena Cava**
1. Formed by union of right and left brachiocephalic veins

III. **VEINS OF UPPER EXTREMITIES**

  A. **Superficial Veins**
1. Cephalic veins drain lateral hand and forearm
2. Basilic veins drain medial hand and forearm
3. Cephalic and basilic veins empty into axillary veins in shoulder
4. Superficial median cubital connects cephalic and basilic veins

  B. **Deep Veins**
1. Drain blood from superficial and deep palmar venous arches
    a. radial veins drain on lateral side of forearm
    b. ulnar veins drain on medial side of forearm
2. Radial and ulnar veins unite to form brachial vein
3. Brachial veins drain into axillary veins
4. Axillary veins become subclavian vein
5. Subclavian veins unite with external jugular veins and form brachiocephalic veins

IV. **VEINS OF THE THORAX**

  A. **The Azygos System**
1. Azygos vein is single
    a. drains right side of thorax wall
    b. drains ascending lumbar veins from abdomen
2. Hemiazygos vein drains left posterior thorax and abdomen
    a. crosses to right and empties into azygos vein
3. Accessory hemiazygos vein drains upper left thorax
    a. crosses to right and empties into azygos vein

V. Hepatic Portal System

A. Hepatic Portal Vein

1. Formed by junction of superior mesenteric and splenic veins
2. Superior mesenteric veins drain stomach, small intestine, and portion of ascending colon
3. Splenic vein drains spleen, stomach, and pancreas
4. Inferior mesenteric vein joins splenic vein
   a. drains rectum, sigmoid colon, and descending colon

VI. Veins of Abdominal and Pelvic Regions

A. Inferior Vena Cava

1. Drains blood from abdominopelvic viscera
2. Formed by right and left common iliac veins
   a. internal and external iliac veins form common iliac veins
3. Pelvic viscera drained by internal iliac veins
4. Tributaries of inferior vena cava:
   a. right gonadal veins
   b. renal veins (left gonadal veins drain into renal vein)
   c. inferior phrenic
   d. right suprarenal

VII. Veins of the Lower Extremities

A. Superficial Veins

1. Great saphenous vein is body's longest vein
   a. originates from medial dorsal venous arch
   b. ascends on medial side of foot, leg, and thigh
   c. drains into femoral vein
2. Small saphenous vein originates from lateral dorsal venous arch
   a. ascends on posterior leg and empties into popliteal vein

B. Deep Veins

1. Anterior and posterior tibial veins originate from veins in foot
2. Ascend deep in posterior muscles of leg
3. Merge to form popliteal vein posterior to knee
4. Superior to knee, popliteal vein becomes femoral vein
5. Femoral vein ascends in thigh and becomes external iliac vein under the inguinal ligament

LABORATORY EXERCISE 17.1
The Veins of the Heart, Head, Neck, and Upper Extremities
Part I

1. Veins join or unite as _____ and arteries branch.

2. To ensure that blood flows toward the heart, veins of the extremities have

   _____ to prevent the backflow of blood.

3. Which major veins deliver venous blood to the right atrium?

   a. _____

   b. _____

   c. _____

4. Which vessel drains venous blood from the heart musculature?

   _____

5. The _____ returns blood from the lower body

   regions to the heart, and the _____ returns venous

   blood from the upper body regions.

6. The vessels that bring oxygenated blood from the lungs to the left atrium of

   the heart are called the _____.

7. Veins that drain the scalp, face, and superficial neck regions are called

   _____, and veins that drain blood from the brain and

   deep regions of the neck are _____.

8. The major veins from the head and neck regions deliver blood to the

   _____. Union of these two vessels then forms the

   brachiocephalic veins, which unite to form the major vein of the upper

   region, the _____.

9. Which are the main superficial veins of the upper extremities?

   a. _____

b. _____

10. Which are the veins that drain the medial and lateral sides of the forearm?

a. _____

b. _____

Part II

Veins of the Thorax, Abdomen, Pelvis, and Lower Extremities

1. Which veins constitute the azygos system?

a. _____

b. _____

c. _____

2. Which vein drains the right side of the abdominal and thoracic cavities?

_____

3. Which vein drains the left side of the abdominal and thoracic cavities?

_____

4. Which vein drains the upper left region of the thorax via the intercostal veins?_____

5. Venous blood from the digestive organs, spleen, pancreas, and gallbladder enter the liver via the _____.

6. From which two sources does the liver receive blood?

a. _____

b. _____

7. The hepatic portal vein is normally formed by the junction of the

a. _____

b. _____

8. Which veins drain the following organs:

    a. spleen, stomach, and pancreas _____

    b. stomach, small intestine, cecum, and ascending and transverse colon

    _____

    c. rectum, sigmoid colon, and descending part of colon _____

    d. stomach only (four veins) _____

    e. descending and sigmoid colon _____

9. Which veins unite to form the inferior vena cava?

    _____

10. Which veins drain the major and deep pelvic organs?

    _____

Part III

Using the listed terms, label the following major veins of the head and neck and then color them.

Right external jugular   Left brachiocephalic
Right brachiocephalic   Subclavian
Superior vena cava   right axillary
Right internal jugular

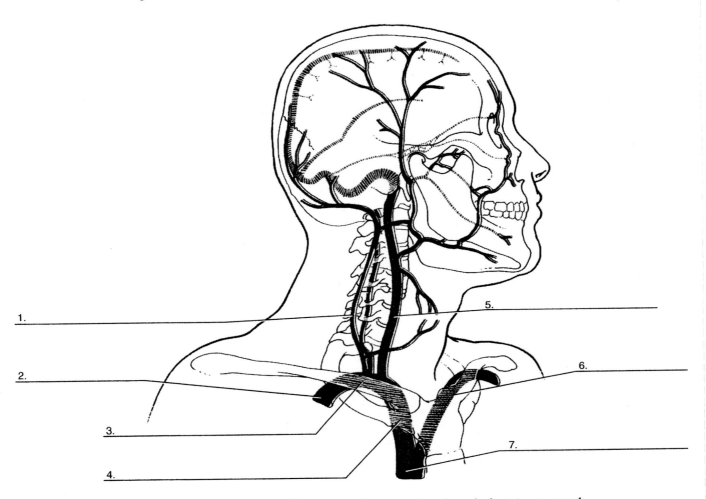

Figure 17.7 Right lateral view of the veins of the head and neck draining toward the heart.

LABORATORY EXERCISE 17.2

Using the listed terms, label the veins of the right upper arm and heart and then color them.

Digital   Brachiocephalic
Palmar venous arch   Basilic
Brachial   Ulnar
Superior vena cava   Axillary
Cephalic   Subclavian
Radial   Median cubital

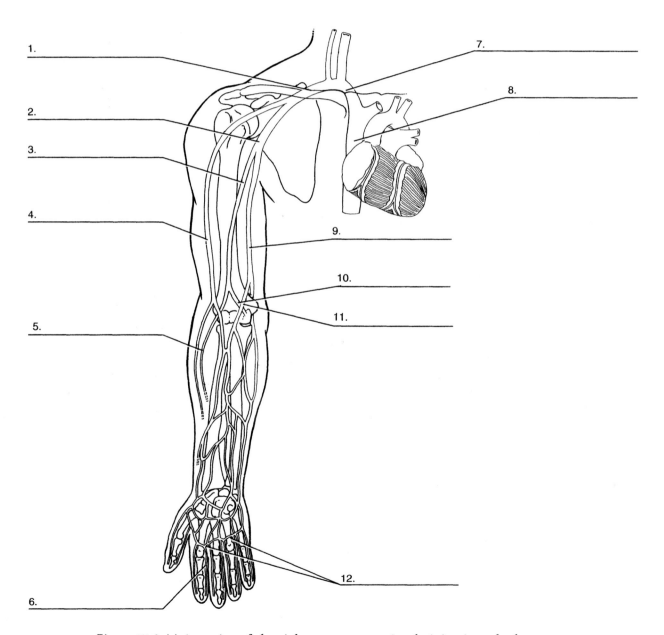

1. _____   7. _____

                                       8. _____

2. _____

3. _____

4. _____   9. _____

                                       10. _____

5. _____   11. _____

6. _____   12. _____

Figure 17.8 Major veins of the right upper extremity draining into the heart.

LABORATORY EXERCISE 17.3

Using the listed terms, label the veins in the thoracic, abdominal, and pelvic regions of the body and then color them.

Azygos
Superior vena cava
Hemiazygos
Inferior vena cava
Gonadal (right and left)
Right femoral
Left internal iliac

Intercostal (right and left)
Renal (right and left)
Accessory hemiazygos
Ascending lumbar (right and left)
Right common iliac
Left external iliac
Hepatic

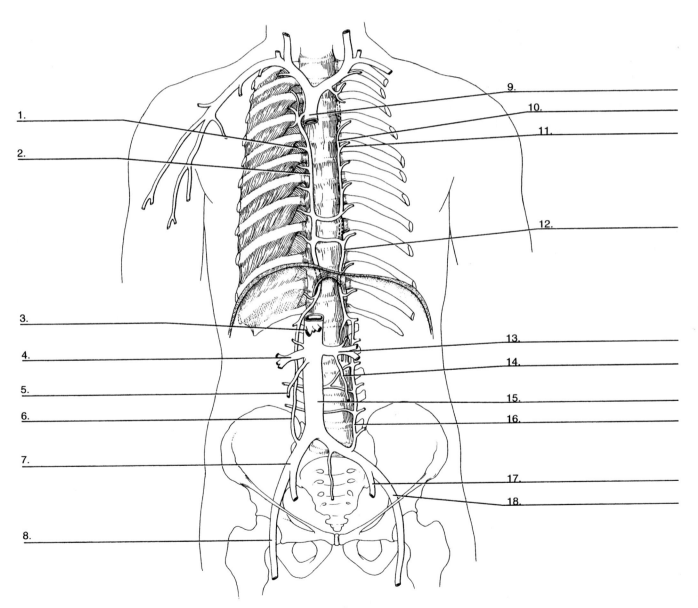

1.
2.
3.
4.
5.
6.
7.
8.
9.
10.
11.
12.
13.
14.
15.
16.
17.
18.

Figure 17.9 Anterior view of the veins of the thorax, abdomen, and the pelvic region.

LABORATORY EXERCISE 17.4

Using the listed terms, label the veins in the portal system and abdominal region of the body and then color them.

Hepatic veins
Left gastric
Superior mesenteric
Right colic
Superior rectal
Inferior mesenteric
Hepatic portal
Left gastroepiploic

Jejunal branches
Middle colic
Sigmoid
Splenic
Right gastric
Ileocolic
Left colic
Right gastroepiploic

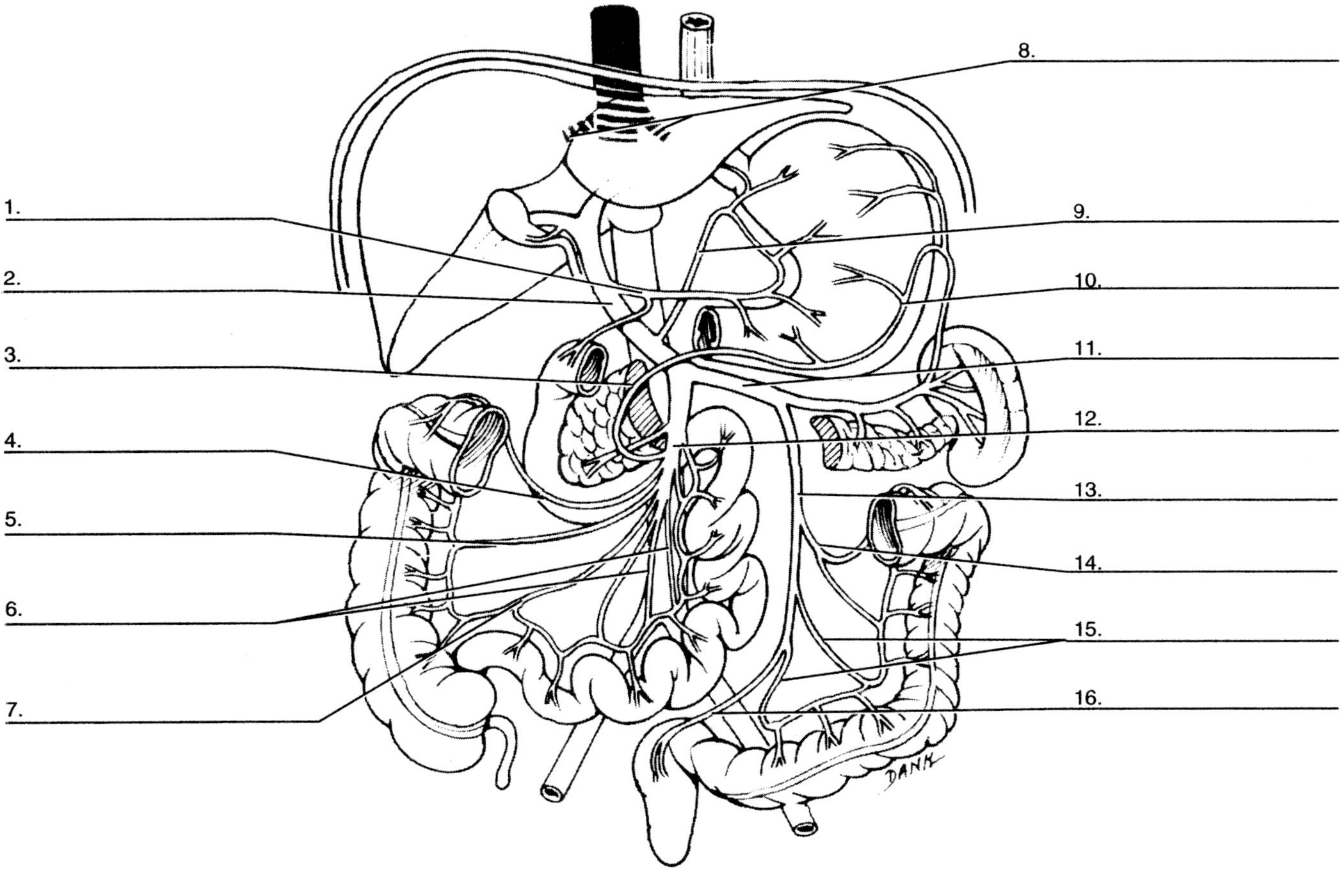

Figure 17.10 The hepatic portal system and the major gastrointestinal veins.

18

The Respiratory System

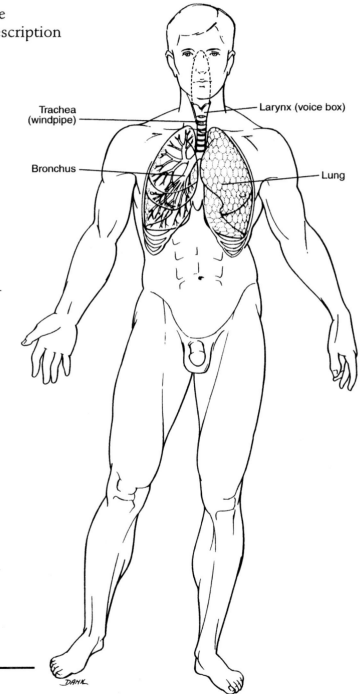

Trachea (windpipe)

Larynx (voice box)

Bronchus

Lung

DANK

Objective

The objective of Chapter 18, "The Respiratory System," is for you to be able to identify and compare the following structures in the human respiratory system:

1. **Pharynx, larynx, and trachea**
2. **Membranes of lung and thorax**
3. **Passages and vessels in right and left lungs**

Description [Figure 18.1]

The human respiratory system consists of the **oral and nasal cavities, pharynx, larynx, trachea, bronchi,** and **bronchioles,** all of which lead to and from the **lungs.** The **mouth** and **pharynx** are **common pathways** for solid food, fluids, and air (the respiratory and digestive systems), and the **larynx, trachea, bronchi,** and **bronchioles** are the passageway to the respiratory system. These structures filter, moisten, and warm the air as it enters the lungs, where gaseous exchange with blood occurs in the microscopic alveoli.

During inspiration, the intercostal muscles and the diaphragm contract. This increases thoracic volume so that air is able to enter the lungs. During expiration, the process is passive; the inspiratory muscles relax, the elastic fibers in the expanded lungs recoil, thoracic volume is decreased, and the air leaves the lungs.

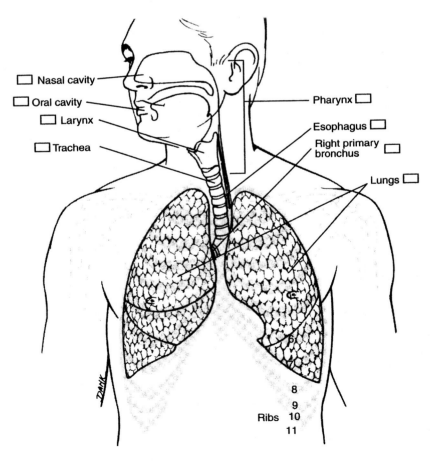

Nasal cavity
Oral cavity
Larynx
Trachea

Pharynx
Esophagus
Right primary bronchus
Lungs

8
9
Ribs 10
11

Figure 18.1 The location of the respiratory system in the head, neck, and thorax.

The Respiratory Passages in the Head and Upper Neck—Description *[Figures 18.1 and 18.2]*

The Nasal Cavity

When the mouth is closed, air enters the **nasal cavity** and the respiratory system through the **external nares**, or **nostrils**. This nasal cavity extends from the external nares anteriorly to the **internal nares** (singular, **naris**) or **choanae**, posteriorly. The nasal cavity opens into the **nasopharynx** through the internal nares. The lateral walls of the nasal cavity are characterized by irregular projections called **conchae** (superior, middle, and inferior). Beneath the conchae are grooves or recesses called the **meatuses**. The floor of the nasal cavity is formed by the bony anterior **hard palate**, which is the **palatine process of the maxillary bone**, and a more posterior muscular structure, **the soft palate**. The hard palate forms the roof of the **oral cavity**.

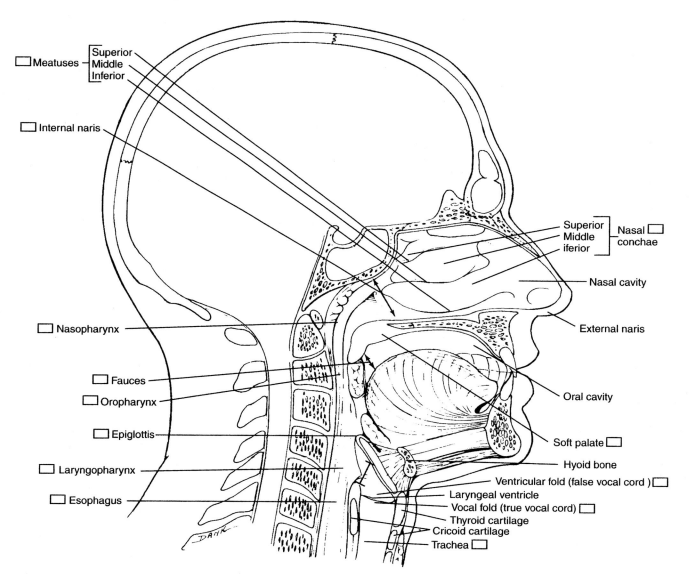

Figure 18.2 A sagittal plane of the head and upper neck illustrates the respiratory passages.

The Pharynx [Figures 18.1 and 18.2]

The **pharynx** is a muscular tube that is the common pathway for food and air. It is divided into **nasopharynx**, **oropharynx**, and **laryngopharynx**.

The **nasopharynx** is the most superior portion of the pharynx. It lies immediately posterior to the nasal cavity, with which it communicates through the **internal nares**, and extends down to the **soft palate**.

The **oropharynx** is a direct continuation of the nasopharynx; it is the middle portion of the pharynx. It lies posterior to the oral cavity and extends inferiorly from the **soft palate** to the superior border of the **epiglottis**, which is the beginning of the **laryngopharynx**. The oropharynx communicates anteriorly with the oral cavity through an opening, the **fauces**. The oropharynx receives food and fluids from the oral cavity and air from the nasopharynx.

The **laryngopharynx** is the most inferior portion of the pharynx; it extends from the oropharynx superiorly and is continuous with the **esophagus** inferiorly. The laryngopharynx is a passageway for food, fluids, and air.

The Larynx [Figures 18.1–18.3]

The **larynx** is a tubular structure that connects the inferior portion of the **pharynx** to the **trachea**. The common carotid arteries and jugular veins, vagus nerve, and the sternocleidomastoid muscles lie on each side of the larynx. The larynx allows uninterrupted air flow into and out of the trachea for vocalization and prevents solid objects and liquids from entering the trachea.

The wall of the larynx consists of nine pieces of **cartilage**, three of which are unpaired and three paired. The paired cartilages are the **arytenoid**, **cuneiform**, and **corniculate**. The unpaired cartilages are the **thyroid cartilage**, **the cricoid cartilage**, and the **epiglottis**. The **thyroid cartilage** is triangular and the largest of all. It forms the anterior portion of the larynx. In adult males, this cartilage forms the laryngeal projection called the **Adam's apple**. The **cricoid cartilage**, shaped like a ring, is located below the thyroid cartilage. This cartilage attaches to the thyroid cartilage superiorly and the trachea inferiorly. The leaf-shaped **epiglottis** is located posterior to the thyroid cartilage. Its narrow inferior end is attached to the inner surface of the thyroid cartilage, and its unattached superior portion projects behind the base of the tongue. During swallowing, the larynx is elevated and the unattached edge of the epiglottis closes the entrance to the larynx, deflecting solid and liquid material from the trachea and into the esophagus.

The interior of the larynx is lined with folds that extend into the pharyngeal passageway. Within the laryngeal aperture is the **vestibule**; at its inferior border it exhibits two folds called the **vestibular folds**, or **false vocal cord**. These cords do not produce sound, but below them is a second pair of folds, the **true vocal cords**. These cords vibrate and produce sound as air passes over them from the lungs.

Nasal Cavity, Pharynx, and Larynx— Identification *[Figures 18.1–18.3]*

Examine a midsagittal section of the head and the upper neck in a model. Examine the larynx in a model and the diagrams, and then color the parts. Identify the following structures in the model and/or cadaver.

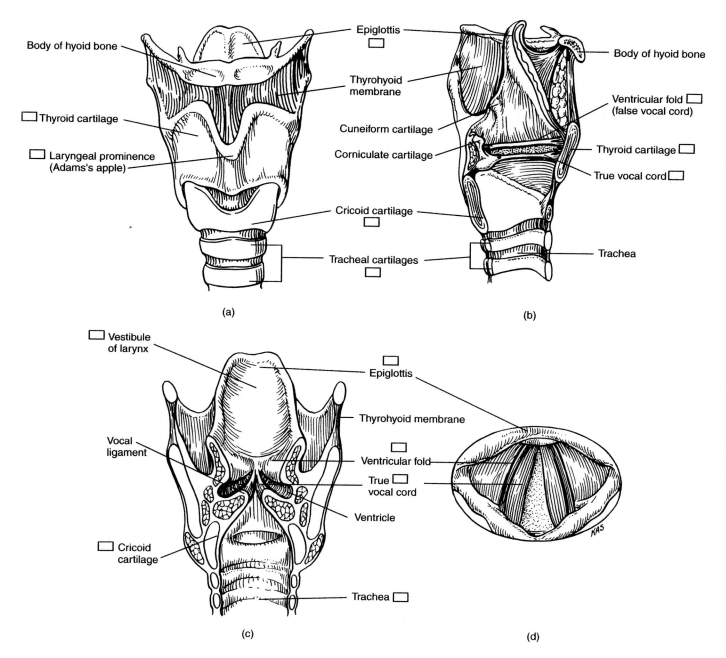

Body of hyoid bone

Epiglottis ☐

Thyroid cartilage ☐

Laryngeal prominence ☐
(Adams's apple)

Thyrohyoid
membrane

Cuneiform cartilage

Corniculate cartilage

Cricoid cartilage ☐

Tracheal cartilages ☐

(a)

Body of hyoid bone

Ventricular fold ☐
(false vocal cord)

Thyroid cartilage ☐

True vocal cord ☐

Trachea

(b)

☐ Vestibule
of larynx

Vocal
ligament

☐ Cricoid
cartilage

☐
Epiglottis

Thyrohyoid membrane

☐
Ventricular fold

True ☐
vocal cord

Ventricle

Trachea ☐

(c)

(d)

Figure 18.3 (a) Anterior, (b) sagittal, (c) posterior, and (d) superior views of the larynx from the pharynx.

1. **external nares**
2. **nasal cavity**
3. **conchae**
4. **meatuses**
5. **internal nares (choanae)**
6. **nasopharynx**
7. **oral cavity**
8. **hard palate**
9. **soft palate**
10. **fauces**

11. oropharynx
12. laryngopharynx
13. esophagus
14. epiglottis
15. larynx
16. true vocal cord
17. false vocal cord
18. thyroid cartilage
19. cricoid cartilage

The Respiratory Passages and Structures in the Thoracic Cavity—Description [Figures 18.1–18.4]

The Trachea and Bronchi

The **trachea** is a tube about 12–14 cm long and 2.5 cm wide. It descends directly from the **larynx** into the thoracic cavity to the level of the fourth or fifth thoracic vertebrae. The trachea then passes posterior to the **arch** of the **aorta** and bifurcates (divides) into two smaller **left** and **right primary bronchi** (singular, **bronchus**). The bronchi then enter the left and right lungs. At the bifurcation of the trachea is a prominent, semilunar internal ridge called the **carina**.

The wall of the **trachea** contains 15 to 20 C-shaped **hyaline cartilage (tracheal) rings** that prevent the trachea from collapsing and keep its air passage patent (open). The cartilage rings encircle the trachea anteriorly but do not completely

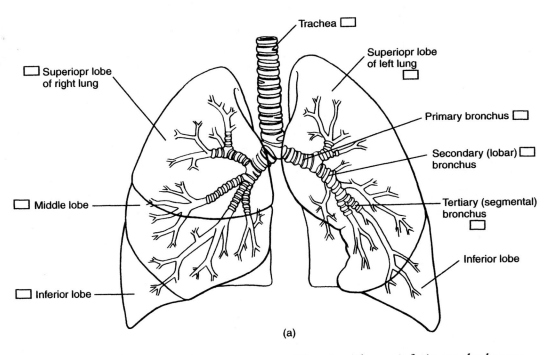

Trachea ☐

Superiopr lobe of left lung ☐

☐ Superiopr lobe of right lung

Primary bronchus ☐

Secondary (lobar) ☐ bronchus

☐ Middle lobe

Tertiary (segmental) bronchus ☐

☐ Inferior lobe

Inferior lobe

(a)

Figure 18.4a Conducting respiratory passages. The air pathway inferior to the larynx consists of the trachea and primary, secondary, and tertiary bronchi. These branch into smaller and smaller bronchioles.

close on its posterior surface, which faces the **esophagus**. The **trachealis muscle** and connective tissue connect the opening of the tracheal rings.

In the neck, the **thyroid gland** covers part of the anterior and lateral sides of the trachea. In the thoracic cavity, the trachea lies posterior to the heart and the great vessels. Posterior to the trachea is the esophagus, and posterior to the esophagus are the vertebrae.

The **trachea** and **primary bronchi** lie outside the lungs. The **right primary bronchus** is shorter, more vertical, and wider than the **left primary bronchus**. For these reasons, foreign objects inhaled into the air passages almost always lodge in the right primary bronchus.

The walls of the primary bronchi are similar to those of trachea. In the lungs, however, the cartilage rings are replaced by small cartilage plates that completely

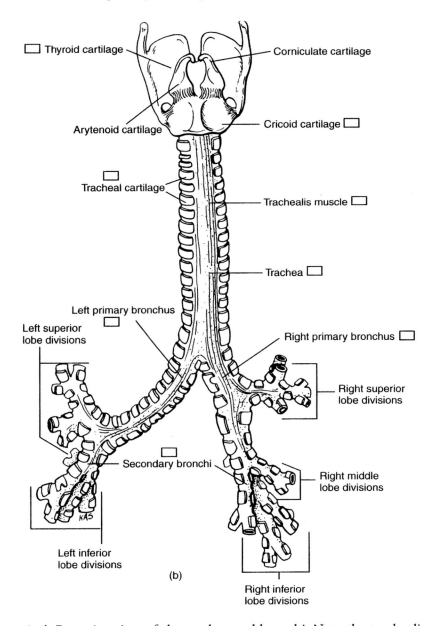

(b)

Figure 18.4b Posterior view of the trachea and bronchi. Note the trachealis muscle spanning the opening of the C-shaped cartilage rings.

encircle the bronchi. As the bronchi continue to branch, the cartilage plates become smaller and disappear in the walls of the **bronchioles**, which are about 1 mm in diameter.

Upon entering the lungs, the primary bronchi bifurcate to form smaller **secondary bronchi**, one for each **lobe** of the lung. The right lung consists of three lobes, and the left lung two. All five lobes are readily visible externally. In the lungs, the secondary bronchi divide further, forming many smaller bronchi, the **tertiary bronchi**. Each lung lobe is further subdivided by connective tissue partitions into numerous smaller functional units called **bronchopulmonary segments**. Each bronchopulmonary segment is supplied by a tertiary bronchus. These bronchi branch further, finally giving rise to smaller, microscopic **terminal** and **respiratory bronchioles**.

The Respiratory Passages and Thoracic Cavity

The Trachea and Bronchi—Identification

[Figures 18.1–18.4]

Examine models of the trachea and then color its parts in the illustration. If possible, examine the trachea in the neck as it descends into the thorax of a prepared cadaver. Identify the following parts of the trachea in the thorax:

1. **larynx**
2. **tracheal rings**
3. **bifurcation of the trachea**
4. **carina**
5. **right primary bronchus**
6. **left primary bronchus**
7. **secondary bronchi**

The Lungs—General Description *[Figures 18.5–18.7]*

The lungs are cone-shaped, spongy organs that lie against the ribs anteriorly, laterally, and posteriorly in the thorax. Each lung has an **apex** and a **base; costal, mediastinal**, and **diaphragmatic surfaces**; and a **hilus**. The lung surfaces are named for their associations with adjacent structures. The **apex** of the lung is the round, superior tip that extends superiorly in thorax, above the first rib and posterior to the clavicle. The **base** is the broad, concave portion whose **diaphragmatic surface** lies on the convex surface of the **diaphragm**. The **hilus** is the region on the **mediastinal surface** of the lung where structures that form the **root of the lung** (bronchi, pulmonary vessels, lymphatic vessels, and nerves) enter and leave the lung. The **costal surface** of the lung lies against the rib cage and has rounded anterior, lateral, and posterior surfaces.

Lateral and Anterior Surfaces—Specific Description

[Figures 18.5–18.7]

The lungs are different in anatomy. The right and left lungs are divided into **lobes** by **fissures**, but the division is not identical in both lungs.

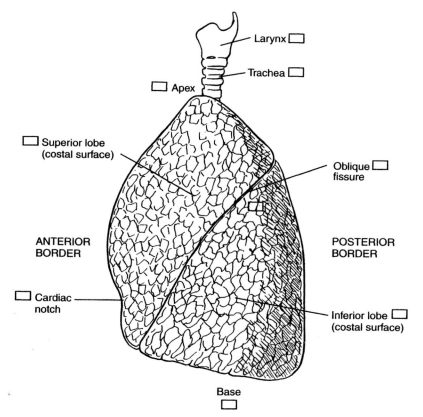

Figure 18.5 Lateral view of the left lung illustrates the oblique fissure, two lobes, and borders.

In both lungs, the **oblique fissure** extends inferiorly and anteriorly through the **costal, diaphragmatic,** and **mediastinal surfaces**. In the left lung, the oblique fissure separates the lung into a **superior lobe** and **inferior lobe**. The superior lobe has a wide **cardiac notch** on its anterior border to accommodate the bulging heart. The cardiac notch produces a tonguelike projection of the left lung called the **lingula**. The lingula is embryologically homologous to the middle lobe of the right lung. The right lung, in addition to possessing an oblique fissure, also has a **horizontal fissure**. This fissure begins laterally at the oblique fissure and continues horizontally across the costal surface of the lung, subdividing the superior lobe of the right lung to form a small **middle lobe**.

Thus, the three lobes of the right lung are supplied by the **three secondary** or **lobar bronchi**, and the two lobes of the left lung are supplied by **two secondary bronchi**. The secondary bronchi are the subdivisions of the primary bronchi. The right lung is also wider and shorter than the left lung due to the presence of the liver below the diaphragm.

Mediastinal (Medial) Surfaces—Specific Description

[Figure 18.7]

The **bronchus** in the **left lung** lies posteriorly in the **hilus**, the **pulmonary artery** lies superiorly, and the **pulmonary veins** lie anteriorly and inferiorly. In the right lung, the bronchus to the upper lobe is superior, and the pulmonary artery slightly inferior.

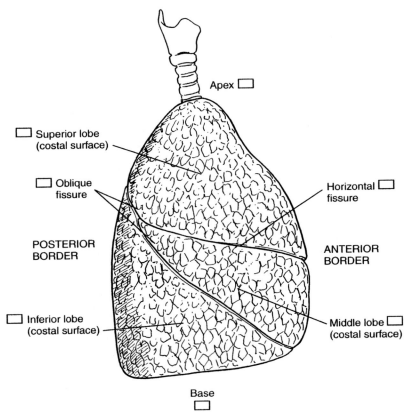

Apex ☐

Superior lobe ☐
(costal surface)

Oblique ☐
fissure

POSTERIOR
BORDER

Inferior lobe ☐
(costal surface)

Horizontal ☐
fissure

ANTERIOR
BORDER

Middle lobe ☐
(costal surface)

Base
☐

Figure 18.6 Lateral view of the right lung illustrates the fissures, three lobes, and borders.

Distinct impressions of the heart, great vessels, ribs, and other associated structures are readily apparent in hardened cadaver lungs and models. A hollow depression, called the **cardiac impression**, in the lower medial side of the left lung is even more obvious and represents the location and projection of the heart to the left side. Also seen in the left lung is the impression of the **aortic arch** and **descending aorta**. The medial surface of the right lung exhibits a less prominent **groove** for the **esophagus, arch** of the **azygos vein**, and **superior vena cava**. The anterior, lateral, and posterior surfaces of both lungs also exhibit impressions of the **ribs**.

The Lungs—Identification of Structures and/or Impressions *[Figures 18.5–18.7]*

Examine the lungs of a model and in the diagrams, and then color them. Note the association of the lungs to other organs in the mediastinum (heart and great vessels). Identify the following structures in a model or cadaver:

BOTH LUNGS

1. **apex**
2. **base**
3. **diaphragmatic surface**
4. **mediastinal surface**
5. **costal surface**
6. **hilus**

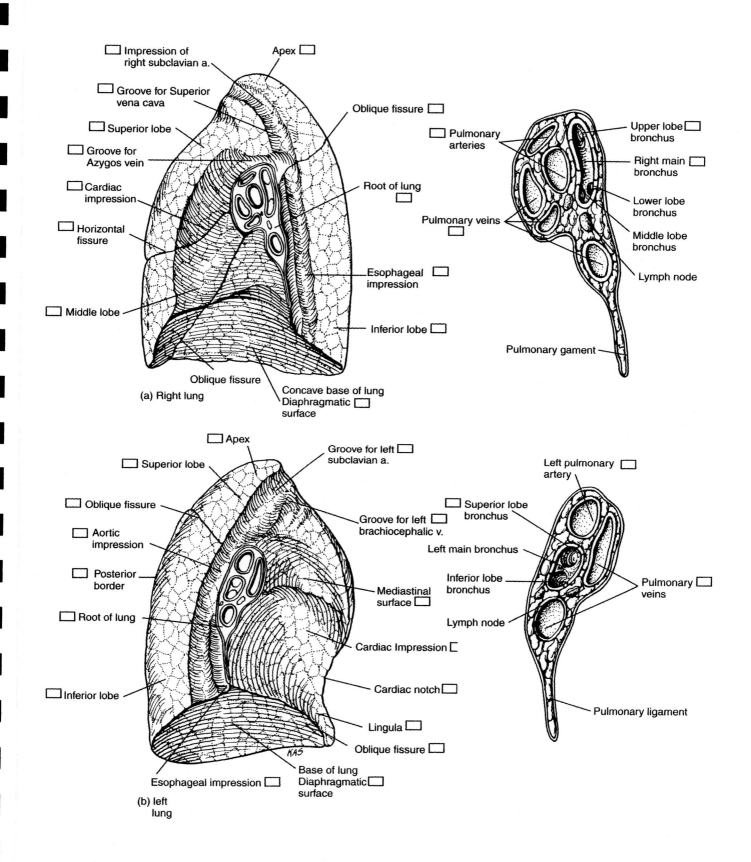

□ Impression of right subclavian a.

Apex □

□ Groove for Superior vena cava

Oblique fissure □

□ Superior lobe

□ Pulmonary arteries

Upper lobe □ bronchus

□ Groove for Azygos vein

Right main □ bronchus

□ Cardiac impression

Root of lung □

Lower lobe bronchus

□ Horizontal fissure

Pulmonary veins □

Middle lobe bronchus

□ Middle lobe

Esophageal □ impression

Lymph node

Inferior lobe □

Oblique fissure

Pulmonary gament

(a) Right lung

Concave base of lung Diaphragmatic □ surface

□ Apex

Groove for left □ subclavian a.

Left pulmonary □ artery

□ Superior lobe

□ Oblique fissure

Groove for left □ brachiocephalic v.

□ Superior lobe bronchus

□ Aortic impression

Left main bronchus

□ Posterior border

Inferior lobe bronchus

Pulmonary □ veins

□ Root of lung

Mediastinal surface □

Lymph node

Cardiac Impression □

□ Inferior lobe

Cardiac notch □

Pulmonary ligament

Lingula □

KAS

Oblique fissure □

Esophageal impression □

Base of lung Diaphragmatic □ surface

(b) left lung

Figure 18.7 Medial aspect of the (a) right lung and (b) left lung. Details of the root of each lung are illustrated on the right.

RIGHT LUNG

1. **superior lobe**
2. **middle lobe**
3. **inferior lobe**
4. **oblique fissure**
5. **horizontal fissure**
6. **pulmonary artery**
7. **pulmonary vein**
8. **primary bronchus**
9. **Secondary lobar bronchi**
10. **groove for superior vena cava**
11. **esophageal impression**
12. **groove for azygos vein**
13. **groove for right subclavian artery**

LEFT LUNG

1. **superior lobe**
2. **inferior lobe**
3. **oblique fissure**
4. **pulmonary artery**
5. **pulmonary vein**
6. **primary bronchus**
7. **Secondary lobar bronchi**

The Thoracic Cage, Mediastinum, and Pleura— Description [*Figures 18.8 and 18.9*]

The thoracic cage consists of ribs, bones, and muscle. Superiorly, the thorax is bordered by the **first ribs** and **manubrium**, laterally by the **ribs**, anteriorly by the **sternum**, posteriorly by the **vertebral bodies**, and inferiorly by the **diaphragm**. Intercostal muscles are located between the ribs; the diaphragm, a sheet of skeletal muscles, separates the thoracic cavity from the abdominal cavity (Figure 18.8).

The thorax contains two **pleural cavities** and a **mediastinum**. The right and left pleural cavities are each filled with a lung. The mediastinum is a median partition that separates the lungs and pleural cavities from each other. The mediastinum contains the **heart, pericardial cavity, aorta, venae cavae, pulmonary vessels, esophagus**, a portion of the **trachea** and **bronchi**, and the **thymus gland**. The central structure in the mediastinum is the heart, which is surrounded the connective tissue **pericardium** or **pericardial sac**.

The mediastinum is subdivided into **superior** and **inferior mediastinum** by an imaginary plane that passes from the sternal angle anteriorly to the lower border of the fourth thoracic vertebra posteriorly. The inferior mediastinum is further subdivided into the **anterior, middle**, and **posterior mediastinum**. The heart and pericardium occupy the middle mediastinum. In front of the pericardium is the anterior mediastinum, a narrow, unimportant region that contains fat, connective tissue, and, in adults, remnants of the thymus gland. The area

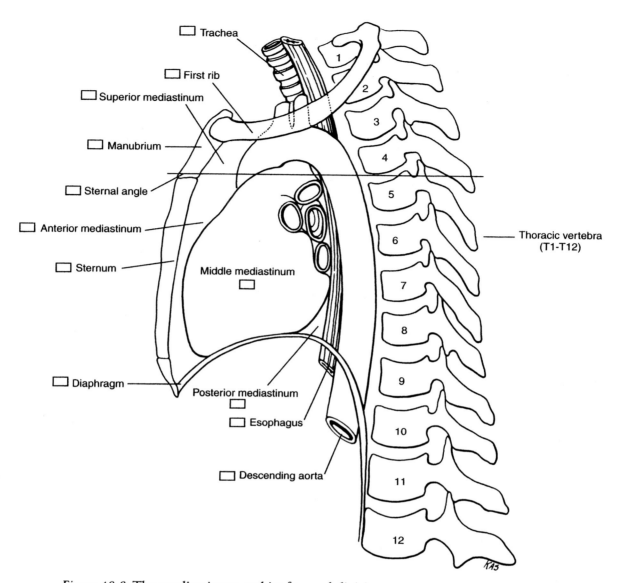

□ Trachea

□ First rib

□ Superior mediastinum

□ Manubrium

□ Sternal angle

□ Anterior mediastinum

□ Sternum

Middle mediastinum
□

□ Diaphragm

Posterior mediastinum
□

□ Esophagus

□ Descending aorta

1
2
3
4
5
6
7
8
9
10
11
12

Thoracic vertebra
(T1-T12)

Figure 18.8 The mediastinum and its four subdivisions.

above the pericardium is the superior mediastinum, and the area posterior to the pericardium and anterior to the vertebral column is the posterior mediastinum. Thus, the mediastinum extends from the superior aperture of the thorax inferiorly to the diaphragm and anteriorly from the sternum to the thoracic vertebrae posteriorly.

Each pleural cavity of the thorax is lined by serous membranes, or **pleura**. The pleura are divided into **parietal pleura** and **visceral pleura**. The outer, or parietal, pleura line the inner wall of the thorax. The inner, or visceral, pleura enclose each lung and are directly attached to the lung surfaces. When the lung lobes are gently pulled apart at the fissures, the visceral pleura are visible as a thin membrane that lines the deep fissures. Each lung with its pleura is attached to the mediastinum at the **lung root**, an area where the bronchi, blood vessels, lymphatic vessels, and nerves enter and leave the lungs (Figures 18.7a and 18-9b).

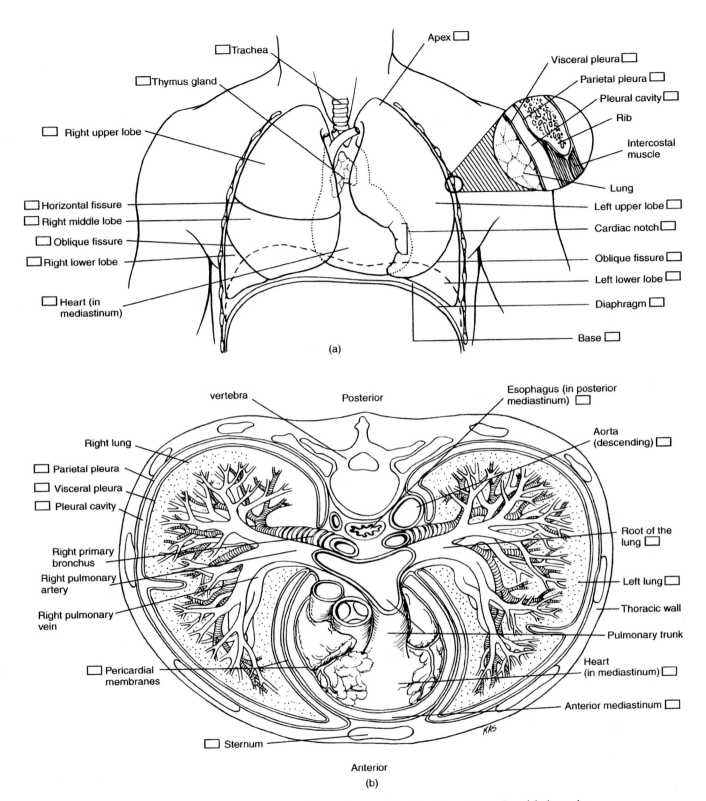

Figure 18.9 Anatomical relationships of organs in the thoracic cavity. (a) Anterior view of the thoracic cavity shows the position of the lungs, which surround the mediastinal structures. (b) Transverse section through the thorax shows the lungs, pleural membranes, and major organs.

(a)

- Trachea
- Thymus gland
- Apex
- Visceral pleura
- Parietal pleura
- Pleural cavity
- Rib
- Intercostal muscle
- Right upper lobe
- Lung
- Horizontal fissure
- Right middle lobe
- Oblique fissure
- Right lower lobe
- Left upper lobe
- Cardiac notch
- Oblique fissure
- Left lower lobe
- Diaphragm
- Heart (in mediastinum)
- Base

(b)

- vertebra
- Posterior
- Esophagus (in posterior mediastinum)
- Aorta (descending)
- Right lung
- Parietal pleura
- Visceral pleura
- Pleural cavity
- Right primary bronchus
- Right pulmonary artery
- Right pulmonary vein
- Root of the lung
- Left lung
- Thoracic wall
- Pulmonary trunk
- Heart (in mediastinum)
- Pericardial membranes
- Anterior mediastinum
- Sternum
- KAS
- Anterior

At the root of each lung, the visceral pleura become continuous with the parietal pleura. The pleura that line the mediastinum are the **mediastinal pleura**, and the pleura that line the surface of the diaphragm are the **diaphragmatic pleura**. Between the two pleura layers is the **pleural cavity**, a space that contains a small amount of pleural fluid that lubricates the lung surfaces and reduces friction during respiratory movement.

CHAPTER SUMMARY AND CHECKLIST

I. **THE MAJOR RESPIRATORY PASSAGES**

A. **Interconnected Air Pathways in the Respiratory System**
   1. Mouth
   2. Nose
   3. Nasal cavity
   4. Pharynx
   5. Larynx
   6. Trachea
   7. Bronchi

B. **Common Pathways of the Respiratory and Digestive Systems**
   1. Mouth
   2. Pharynx

C. **Pathways Exclusively of the Respiratory System**
   1. Larynx
   2. Trachea
   3. Bronchi and bronchioles

II. **THE RESPIRATORY PASSAGES AND RELATED STRUCTURES IN THE HEAD AND THE UPPER NECK**

A. **Nasal Cavity**
   1. Nose, external nares, internal nares, and nasal cavity
      a. conchae—three lateral projections into cavity
      b. meatuses—grooves or recesses under the conchae
   2. Hard palate
      a. forms floor of nasal cavity
      b. forms roof of oral cavity

B. **Pharynx**
   1. Common pathway for digestive and respiratory systems
   2. Anatomically subdivided into three regions
      a. nasopharynx is the uppermost region
      b. oropharynx is the middle portion
      c. laryngopharynx is the lowest region

C. **Larynx**
   1. Connects pharynx to trachea
   2. Lies in the middle of the neck
   3. Important structure for uninterrupted air flow to lungs

    4. Three unpaired cartilages
       a. thyroid cartilage with laryngeal prominence (Adam's apple in males)
       b. cricoid cartilage is below the thyroid cartilage
       c. epiglottis deflects fluids and solids into esophagus
    5. Contains true and false vocal cords

III. THE RESPIRATORY PASSAGES AND STRUCTURES IN THE THORACIC CAVITY

A. Trachea
    1. Open, rigid tube that descends from larynx into thorax
    2. Passes behind aortic arch and bifurcates into primary bronchi
    3. Lined with incomplete cartilage rings (C-shaped)
    4. Lies in front of the esophagus

B. Bronchi
    1. Wall of primary bronchi similar to trachea
    2. Right primary bronchus shorter, wider, and more vertical than left
    3. Intrapulmonary bronchi lined by cartilage plates
    4. In lungs, bronchi bifurcate repeatedly to form bronchioles

IV. THE LUNGS

A. Regions and Surfaces
    1. Apex—superior pointed portion in the upper thorax
    2. Base—broad, concave portion on the diaphragm
    3. Hilus—region on the mediastinal surface
    4. Lung root—structures in the hilus
    5. Costal surface lies against the rib cage
    6. Diaphragmatic surface lies on the diaphragm

B. Lateral Surfaces
    1. Fissures partition both lungs into lobes
    2. Oblique fissure seen in both lungs
    3. Left lung is partitioned by oblique fissure into:
       a. superior lobe
       b. inferior lobe
    4. Right lung partitioned by oblique and horizontal fissures into:
       a. superior lobe
       b. middle lobe
       c. inferior lobe
    5. Lobes of the lungs supplied by secondary bronchi

C. The Lungs: Mediastinal (Medial) Surfaces
    1. Hilus contains bronchi and pulmonary vessels
    2. In hilus of left lung:
       a. bronchus is posterior
       b. pulmonary artery is superior
       c. pulmonary veins are inferior and anterior
    3. In the hilus of the right lung, the bronchus to upper lobe is the highest structure
    4. Impressions in left cadaver lung:
       a. cardiac impression

           b. aortic arch and descending aorta
      5. Impressions in right cadaver lung:
           a. groove for esophagus
           b. arch of azygos
           c. superior vena cava

V. THE THORACIC CAGE, MEDIASTINUM, AND PLEURAL CAVITIES
   A. Boundaries
      1. First ribs, superiorly
      2. Sternum, anteriorly
      3. Vertebral bodies, posteriorly
      4. Diaphragm, inferiorly
   B. Pleural Cavities and Mediastinum
      1. Pleural cavities filled with lungs
      2. Mediastinum contains:
           a. heart
           b. aorta
           c. venae cavae
           d. pulmonary vessels
           e. esophagus
           f. trachea and bronchi
           g. thymus gland
      3. Regions of mediastinum
           a. anterior
           b. middle
           c. superior
           d. posterior
      4. Lining of pleural cavities
           a. visceral pleura
           b. parietal pleura

Laboratory Exercises 18

NAME _____

LAB SECTION _____ DATE _____

LABORATORY EXERCISE 18.1

The Respiratory System

Part I

The Upper Respiratory Passages

1. From the external nares, where does the nasal cavity extend?

   _____

2. Into what does the nasal cavity open?

   _____

3. What form the floor of the nasal cavity?

   a. _____

   b. _____

4. Into which three regions is the pharynx divided?

   a. _____

   b. _____

   c. _____

5. For what does the pharynx serve as a passageway?

   a. _____

   b. _____

6. What structures make up the wall of the larynx and what is the prominent laryngeal projection in males called?

   a. _____

   b. _____

7. What structure closes larynx when food is swallowed?

   _____

8. What structures in the larynx vibrate, producing sound, when air passes over them?

   _____

9. What are the subdivisions of the trachea?

a. _____

b. _____

c. _____

d. _____

10. What structures keep the trachea open and prevent it from collapsing?

_____

Part II

The Lungs

Match the description in the left column with the answer in the right. Some answers may be correct for more than one description.

1. Superior tip of the lung _____

2. Where vessels, nerves, and bronchi

   enter the lung _____

3. Lies against the ribs _____

4. Contacts the diaphragm _____

5. Contains the cardiac notch _____

6. Exhibits a horizontal fissure _____

7. Homolog to middle lobe _____

8. Contains three lobes _____

a. Base
b. Left lung
c. Right lung
d. Costal surface
e. Apex
f. Hilus
g. Lingula
h. Diaphragmatic surface

Part III

Mediastinum and Pleura—Thoracic Cage

1. What fills the pleural cavities in the thorax?

   _____

2. What are the subdivisions of the mediastinum?

   a. _____

b. _____

3. What is found in the middle mediastinum?

   a. _____

   b. _____

4. The outer pleura that lines the inner wall of the thorax is the _____ pleura, and inner pleura that closely enclose each lung is the

   _____ pleura.

5. What attaches each lung to the mediastinum?

   _____

6. The space containing fluid that lubricates each lung is called

   _____.

Part IV

Using the listed terms, label the respiratory passages in the illustration and then color them.

Meatuses
Internal nares
Nasopharynx
Soft palate
Palatine bone
Trachea
Nasal cavity

Laryngopharynx
Nasal conchae
External nares
Oropharynx
Maxilla
Epiglottis

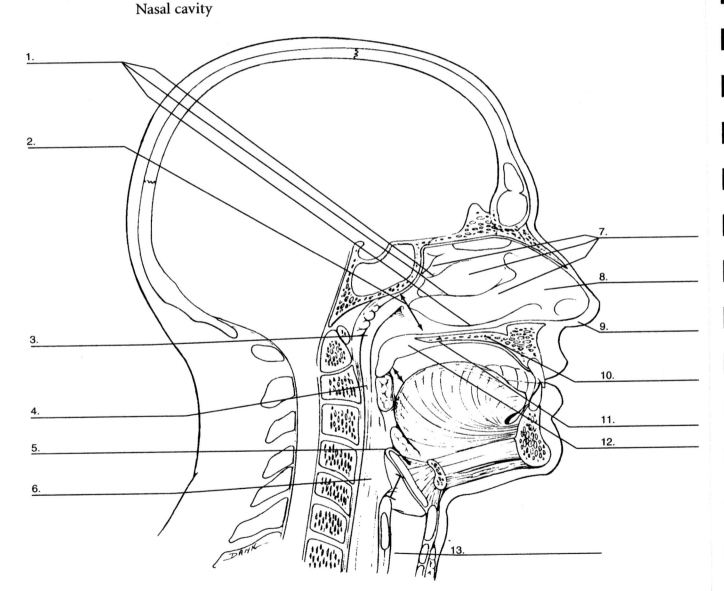

Figure 18.10 A sagittal plane illustrates the respiratory passages in the head and neck.

LABORATORY EXERCISE 18.2

Using the listed terms, label the parts of the lung and/or the impressions of different organs in the lungs, and then color them.

Lung Structures

Apex
Middle lobe
Oblique fissure
Base
Superior lobe
Root of the lung
Inferior lobe
Horizontal fissure
Lingula

Impressions or Grooves in Lungs

Subclavian artery (right and left)
Cardiac notch
Aortic
Esophageal
Cardiac impression
Azygos vein

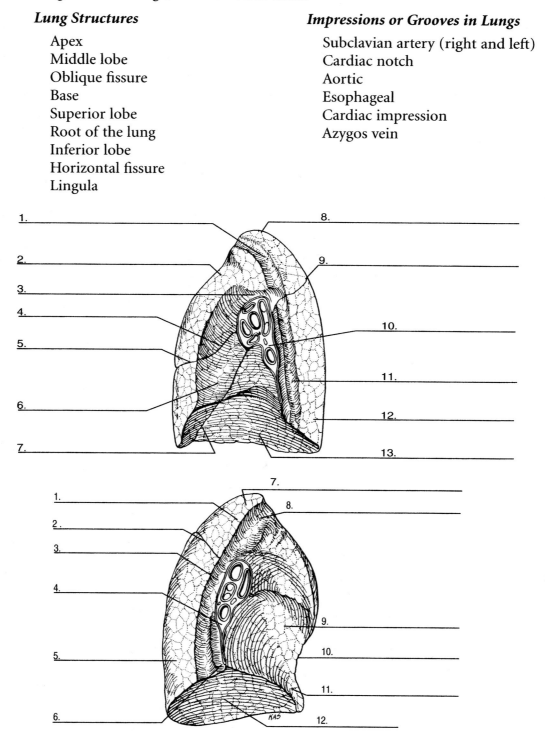

Figure 18.11 Medial views of the right lung (top) and left lung (bottom).

19

The Digestive System

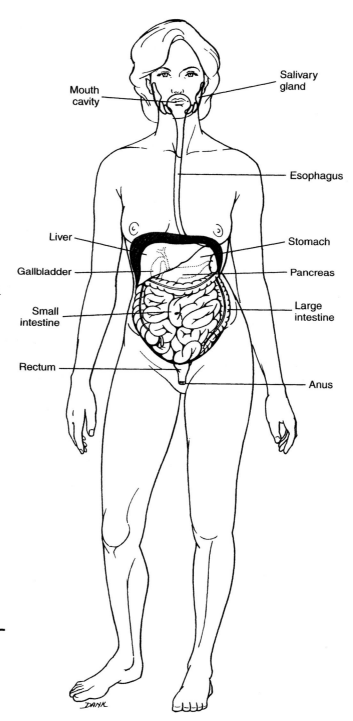

Objective

The objective of Chapter 19, "The Digestive System," is for you to be able to identify the location and structure of the following digestive organs:

1. **The mouth, pharynx, and esophagus**
2. **The stomach, small intestine, and large intestine**
3. **The liver, pancreas, and gallbladder**

Description *[Figure 19.1]*

The digestive system, or **gastrointestinal (GI) tract** is a long tube that starts at the **mouth** and terminates at the **anus**. In its course through the body, the GI tract passes through the **head, neck, thorax, abdomen,** and **pelvis.** The GI tract is divided into different regions: the **mouth, pharynx, esophagus, stomach, small intestine, large**

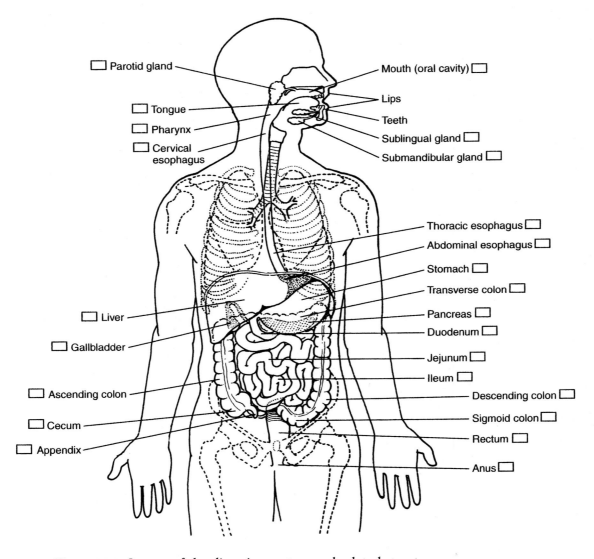

□ Parotid gland
Mouth (oral cavity) □
□ Tongue
Lips
□ Pharynx
Teeth
□ Cervical esophagus
Sublingual gland □
Submandibular gland □
Thoracic esophagus □
Abdominal esophagus □
Stomach □
Transverse colon □
□ Liver
Pancreas □
Duodenum □
□ Gallbladder
Jejunum □
Ileum □
□ Ascending colon
Descending colon □
□ Cecum
Sigmoid colon □
□ Appendix
Rectum □
Anus □

Figure 19.1 Organs of the digestive system and related structures.

intestine, and **rectum**. Associated with the GI tract are the **salivary glands, liver,** and **pancreas**, which are outside the tract but connected to it by ducts (Figure 19.1).

The Mouth and Its Cavity—
Description *[Figure 19.2]*

The mouth, or **oral cavity**, extends from the lips to the **oropharynx**. The roof of the mouth is formed by the **hard palate** anteriorly and the **soft palate** posteriorly. The **uvula** is a small muscular extension that hangs down from the soft palate. The tongue forms the floor of the mouth and the cheeks the lateral walls. Posterior to the oral cavity is the **pharynx**, a **common passageway** for both the digestive and respiratory systems. Food enters the **esophagus** and air enters the **larynx**. During **deglutition** (swallowing), the epiglottis moves down and seals off the larynx, diverting solids or liquids from the respiratory tube into the esophagus.

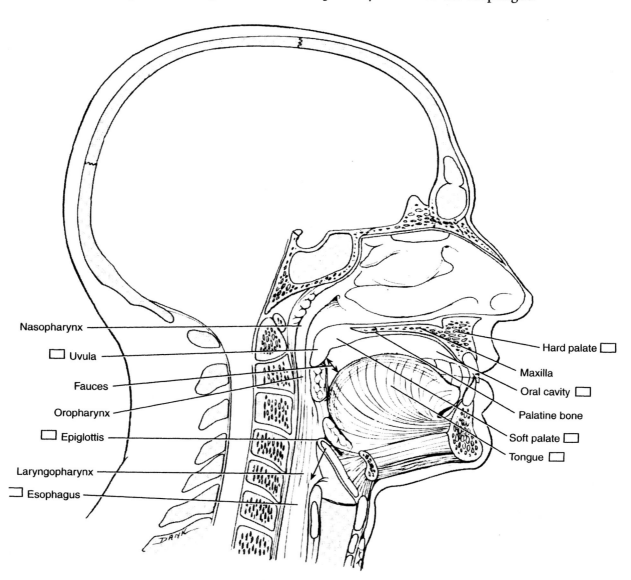

Figure 19.2 Sagittal section of the head and neck shows the upper parts of the digestive tract.

The Mouth and Its Cavity—
Identification [Figure 19.2]

Examine a midsagittal section of a model, chart, or human head. Identify and color the following structures in the illustration:

1. **oral cavity**
2. **hard palate**
3. **soft palate**
4. **tongue**
5. **uvula**
6. **epiglottis**
7. **esophagus**

The Mouth and Accessory Organs—
Description [Figures 19.1 and 19.3]

There are three pairs of salivary glands: the **parotid, submandibular (sub-maxillary)**, and **sublingual**. These are located outside the oral cavity but they deliver **saliva**, their secretory product, into the oral cavity via excretory ducts. The largest salivary gland is the **parotid gland**, located inferior and anterior to the ear on each side of the face. The duct of each parotid gland crosses the **masseter** muscle, pierces the **buccinator** muscle, and opens into the oral cavity as the **parotid duct orifice**.

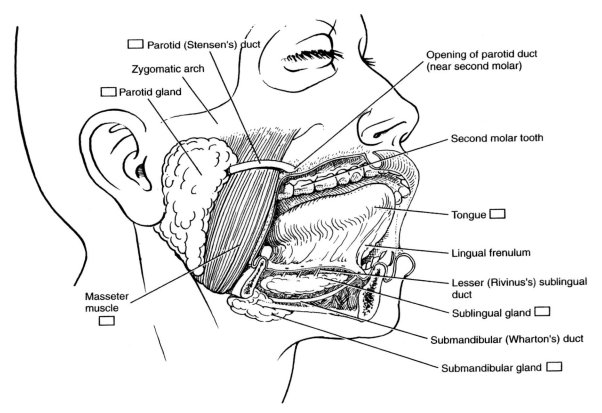

Figure 19.3 Location of the parotid, sublingual, and submandibular salivary glands.

(Buccinator and masseter muscles assist in mastication.) The **submandibular glands** are smaller and located medial and inferior to the body of the mandible. The **sublingual glands** lie in the floor of the oral cavity on either side of the tongue. Their ducts join medially to form a pair of ducts that can be seen under the tongue.

The Mouth and the Accessory Organs— Identification *[Figure 19.3]*

Examine the head and neck of a model and/or prepared cadaver. Identify the following structures and color them in the illustration:

1. **masseter muscle**
2. **temporalis muscle**
3. **buccinator muscle**
4. **parotid gland**
5. **parotid duct**
6. **submandibular gland**
7. **sublingual gland**

The Esophagus—Description
[Figures 19.1, 19.2, and 19.4]

A soft, collapsible muscular tube about 25 cm long, the **esophagus** is continuous superiorly with the laryngeal part of the pharynx and inferiorly with the stomach. The esophagus lies anterior to the vertebral column and posterior to the trachea. The esophagus passes through the cervical region and thorax and then the **diaphragm** through an opening called the **esophageal hiatus**. Thus, about 2 cm of

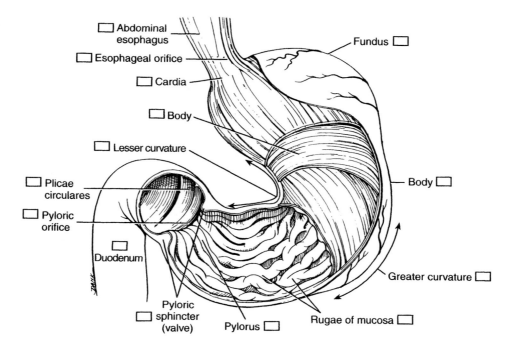

Figure 19.4 External and internal anatomy of the esophagus, stomach, and duodenum.

the esophagus is in the abdomen (**abdominal**), where it terminates in the superior portion of the **stomach**.

The Esophagus— Identification

[Figures 19.1, 19.2, and 19.4]

Examine the esophagus as it extends from the cervical region of the head and through the thorax and stomach in a model and/or cadaver. Identify the following structures and then color them in the illustration:

1. **cervical esophagus**
2. **thoracic esophagus**
3. **diaphragm**
4. **esophageal hiatus**
5. **abdominal esophagus**
6. **stomach**

The Stomach—Description *[Figures 19.1 and 19.4]*

The stomach is a pouchlike portion of the GI tract. Superiorly, it is connected to the **abdominal esophagus** and inferiorly to the **duodenum**. No true anatomical sphincter separates the esophagus from the stomach. Inferiorly, the stomach opening into the duodenum is guarded by a **pyloric sphincter**.

The stomach is divided into the **cardia, fundus, body**, and **pylorus**. The **cardia** is the superior, narrow region inferior to the **esophageal orifice**. The **fundus** is the dome-shaped region of the stomach that lies inferior to the diaphragm and projects superiorly and left of the esophagus. The **body** is the large central portion, and the **pylorus** is the funnel-shaped part of the stomach that ends at the **pyloric sphincter**. The pylorus communicates with the duodenum through the **pyloric orifice** in the pyloric sphincter.

The right, concave border of the stomach, the **lesser curvature**, extends from the esophageal orifice to the pylorus. The lesser curvature of the stomach and the first 2 cm of the duodenum are attached to the inferior surface of the liver by a double layer of peritoneum (a mesentery) called the **lesser omentum**. On the right side, the lesser omentum has a free border. In this region, the **portal vein**, **hepatic artery**, and **bile duct** course between the two layers of the lesser omentum. The left, convex border of the stomach is the **greater curvature**. Two layers of the lesser omentum separate at the lesser curvature and rejoin at the greater curvature of the stomach to form the **greater omentum**, an apronlike, fat-filled structure that hangs down from the stomach posterior to the anterior abdominal wall and anterior to the intestinal coils.

In an empty stomach, the interior surface exhibits longitudinal and circular folds called **rugae**. These folds are temporary and disappear as the stomach is filled.

The Stomach—Identification *[Figures 19.1 and 19.4]*

Study the diagrams and a model of the human abdominal cavity. If possible, examine the abdominal cavity in a prepared cadaver. Identify the following structures in the abdominal cavity and then color them in the illustration.

1. esophageal orifice
2. cardia of stomach
3. fundus of stomach
4. body of stomach
5. pyloric antrum
6. pyloric orifice
7. pyloric sphincter
8. rugae
9. greater curvature
10. lesser curvature
11. greater omentum
12. lesser omentum

The Small Intestine—Description

[Figures 19.1, 19.4, 19.5, and 19.6]

The **small intestine** is a coiled portion of the GI tract situated between the **pyloric sphincter** of the stomach and the ileocecal valve in the large intestine, the **ileocecal junction**. The small intestine is about 6 meters long and is subdivided into the **duodenum, jejunum,** and **ileum.**

The **duodenum** is the first part of the small intestine. Shaped like the letter C, it is a tube about 25 cm long that curves around the **head** of the **pancreas.** The duodenum has four parts: the first, or **superior duodenum,** runs transversely to the right; the second, or **descending duodenum** runs inferiorly; the third, or **horizontal duodenum,** runs transversely to the left on the anterior surface of the inferior vena cava and aorta; and the fourth, or **ascending duodenum,** turns superiorly to become the jejunum at the duodenal-jejunal junction. This junction is supported and fixed to the posterior wall by a fibromuscular band called the suspensory muscle of the duodenum (ligament of Treitz).

The duodenum receives secretions from the **liver** and the **pancreas.** The **common bile duct** from the liver descends in the lesser omentum and courses behind the first part of the duodenum. Before entering the duodenum, the common bile duct from the liver and **main pancreatic duct** from the pancreas join and open as one duct into the descending or second portion of the duodenum in the **major duodenal papilla.**

The first centimeter of the duodenum is suspended by the **lesser omentum.** The rest of the duodenum is **retroperitoneal,** that is, attached to the posterior abdominal wall (discussed at the end of this chapter).

There is no sharp division between the jejunum and ileum; the ileum makes up the final two or more meters of the small intestine. The terminal portion of the ileum empties into the medial side of the **cecum** through the **ileocecal valve.** Both ileum and jejunum are suspended by a **mesentery.** As a result, the small intestine is freely movable and more motile during peristaltic contractions. The jejunum is normally in the superior left portion and the ileum in the inferior right portion of the abdominal cavity.

The inner wall of the small intestine is marked by permanent deep circular folds called **plicae circulares.** These structures increase the absorptive surface area of the small intestine. Because most of the absorption of nutrient material in the

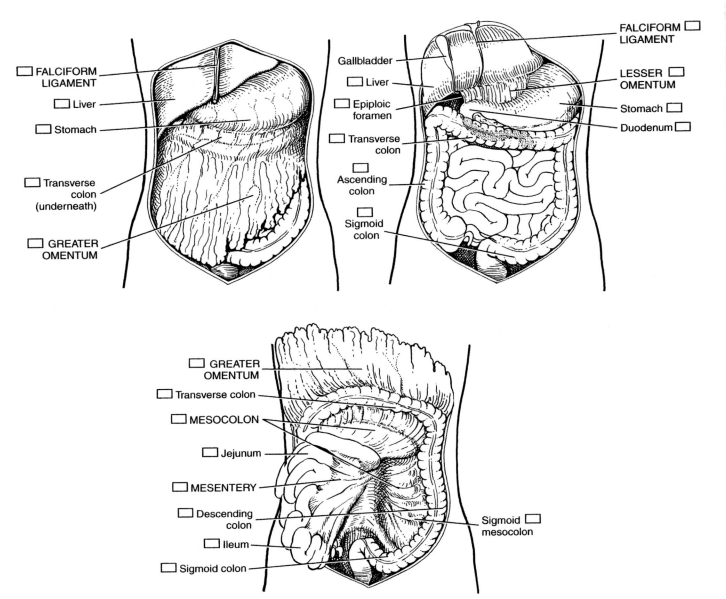

Figure 19.5 The digestive organs in the abdominal cavity and the extensions of the peritoneal lining.

small intestine takes place in the duodenum-jejunum region, the plicae circulares are most developed in these regions. They gradually decrease in size in the ileum (Figure 19.4).

Small Intestine—Identification

[Figures 19.1, and 19.4, 19.5, and 19.6]

Examine the small intestine and associated structures in a model and/or cadaver. Identify the following structures and then color them in the illustrations.

1. **duodenum**
   a. **major duodenal papilla**
   b. **common bile duct**

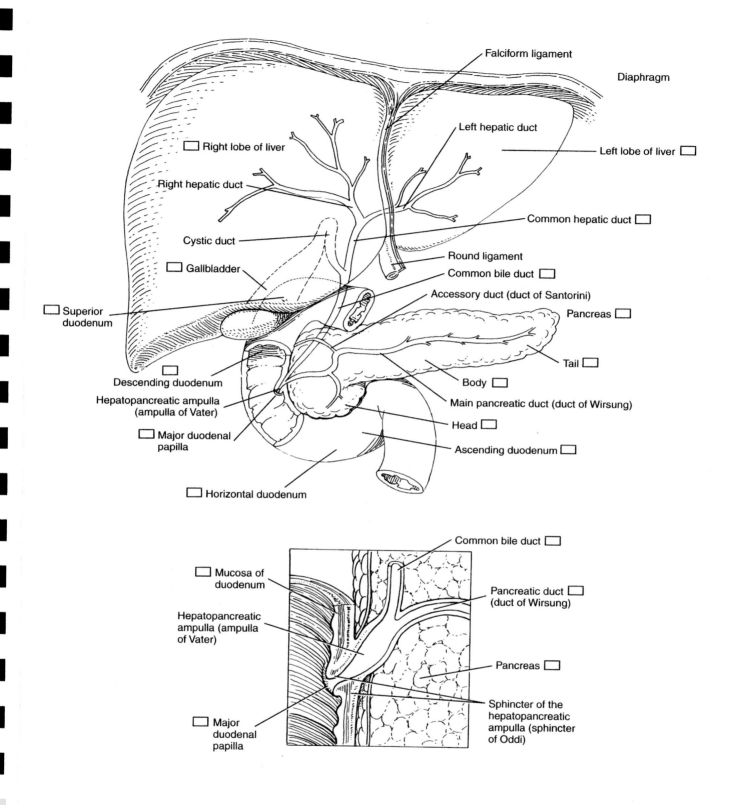

Figure 19.6 Diagram shows the relation of the pancreas to the liver, gallbladder, and duodenum. The inset shows details of the common bile duct and pancreatic duct that form the hepatopancreatic ampulla (of Vater) and empty into the duodenum via the major duodenal papilla.

2. jejunum
3. ileum
4. plicae circulares
5. ileocecal valve
6. ileocecal junction
7. cecum
8. lesser omentum
9. intestinal mesentery

The Large Intestine (Colon)—Description

[Figures 19.1, 19.5, and 19.7]

Extending from the ileum to the anus, the large intestine is about 1.5 meters long and 6.0 cm wide. It forms a perimeter of three and a half sides around the coils of

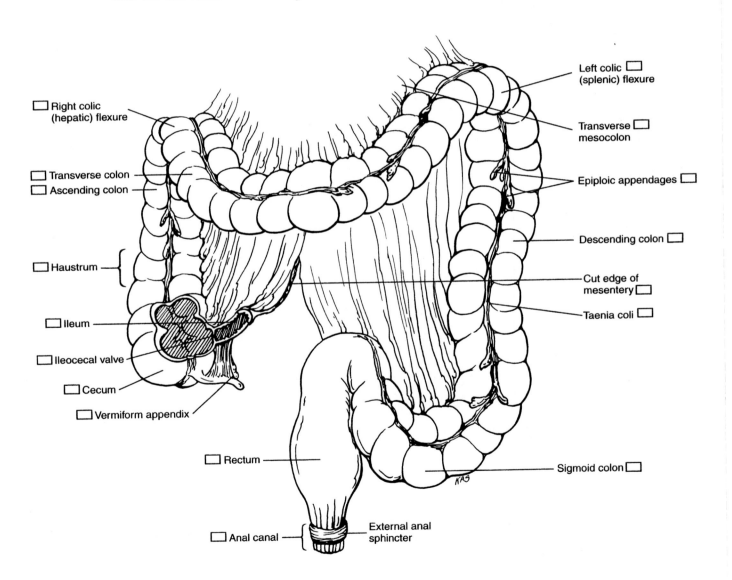

Right colic (hepatic) flexure

Transverse colon

Ascending colon

Haustrum

Ileum

Ileocecal valve

Cecum

Vermiform appendix

Rectum

Anal canal

External anal sphincter

Left colic (splenic) flexure

Transverse mesocolon

Epiploic appendages

Descending colon

Cut edge of mesentery

Taenia coli

Sigmoid colon

Figure 19.7 The large intestine.

the small intestines. The large intestine begins in the blind pouch **cecum** and continues superiorly on the right side to the inferior surface of the liver as the **ascending colon**. The **vermiform appendix** is a worm-shaped structure attached to the cecum. Inferior to the liver, the ascending colon bends left at the **hepatic flexure** and then passes to the left as the **transverse colon**. Anterior to the **spleen**, the transverse colon turns inferiorly at the **splenic flexure** and continues on the left side as the **descending colon**.

Near the pelvis, the descending colon passes over the pelvic brim and curves to the midplane as the S-shaped **sigmoid colon**. Past the sigmoid colon, the large intestine continues into the pelvic bowl as the **rectum** for about the last 20 cm. The last 2 to 3 cm of the rectum comprise the **anal canal**. The exterior opening of the anal canal is the **anus**. Except for the appendix, rectum, and anal canal, the various parts of the large intestine have no differentiating characteristics.

In the large intestine, the longitudinal smooth muscle forms three distinct longitudinal strips spaced equally around the external circumference of the tube. These muscles strips, called **taenia coli**, run the length of the colon and are easily seen with the naked eye. The taenia coli are shorter than the colon, causing the wall of the large intestine to form pouches called **haustra** (singular, **haustrum**). Attached to the wall of the large intestine near the taenia coli are numerous fat-filled pouches called the **epiploic appendages** (or appendices). The rectum is similar in structure to the large intestine but it does not have taenia coli, haustra, or epiploic appendages on its surface.

The ascending and descending colon and the rectum are attached directly to the posterior body wall; these organs are retroperitoneal. The transverse and sigmoid colon are suspended by mesenteries, the **transverse mesocolon** and the **sigmoid mesocolon** (Figure 19.5).

Large Intestine—Identification

[Figures 19.1, 19.5, and 19.7]

Examine the large intestine in a model and/or cadaver, identify the following parts, and then color them in the illustrations:

1. **cecum**
2. **appendix (if present)**
3. **ascending colon**
4. **hepatic flexure of colon**
5. **transverse colon**
6. **splenic flexure of colon**
7. **descending colon**
8. **sigmoid colon**
9. **taenia coli**
10. **haustra**
11. **epiploic appendages**
12. **transverse mesocolon**
13. **sigmoid mesocolon**
14. **rectum**
15. **anus**

The Liver and Gallbladder—Description

[Figures 19.1, 19.5, 19.6, and 19.8]

The liver lies in the right superior part of the abdominal cavity, directly inferior to the diaphragm. Part of the liver extends to the left side and lies superior to the cardia of the stomach. Viewed anteriorly, the liver is divided into the larger **right lobe** and the smaller **left lobe**. On the posterior and inferior surfaces of the right lobe are the smaller **caudate** and **quadrate lobes**. The caudate lobe is next to (right of) the **inferior vena cava**, and the quadrate lobe lies more inferiorly next to the **gallbladder**. The right and left lobes are separated by a **fissure**. On the anterior side of the fissure, the **falciform ligament** attaches the liver to the anterior abdominal wall. In the free margin of the falciform ligament is the **round ligament** of the liver, which in adults is the remnant of the fetal umbilical vein.

The falciform ligament, after it attaches to the anterior and superior surfaces of the liver, splits into two layers. The right layer forms the **coronary ligament** on the superior surface of the liver. The right extremity of the coronary ligament is called the **right triangular ligament**. The left layer of the falciform ligament forms the **left triangular ligament**. The coronary ligament attaches the liver to the diaphragm and surrounds the **bare area** of the liver. (The bare area has no peritoneal lining on its surface because the peritoneal layers forming the coronary ligaments are separated in this area.)

The fissure on the inferior surface of the liver attaches to the lesser curvature of the stomach via the **lesser omentum**. Within the right free margin of the lesser omentum are the **hepatic artery**, **portal vein**, and **common bile duct**. Also on the inferior surface of the liver is the **porta hepatis**, a horizontal fissure between the caudate and quadrate lobes. Here the portal vein and hepatic artery enter the liver and the hepatic ducts exit.

The **gallbladder** is a sac located on the inferior surface of the liver. It is drained by a **cystic duct** that joins the **hepatic duct** from the liver. The cystic and hepatic ducts form the **common bile duct**. This duct then passes in the lesser omentum to the duodenum behind the head of the pancreas, where it joins the major pancreatic duct (Figure 19.6).

The Liver and Gallbladder—Identification

[Figures 19.1, 19.5, 19.6, and 19.8]

Examine the liver and gallbladder in a model and/or cadaver, identify the following structures, and then color them in the illustrations:

1. right lobe
2. left lobe
3. caudate lobe
4. quadrate lobe
5. falciform ligament
6. round ligament
7. coronary ligament
8. left triangular ligament
9. right triangular ligament

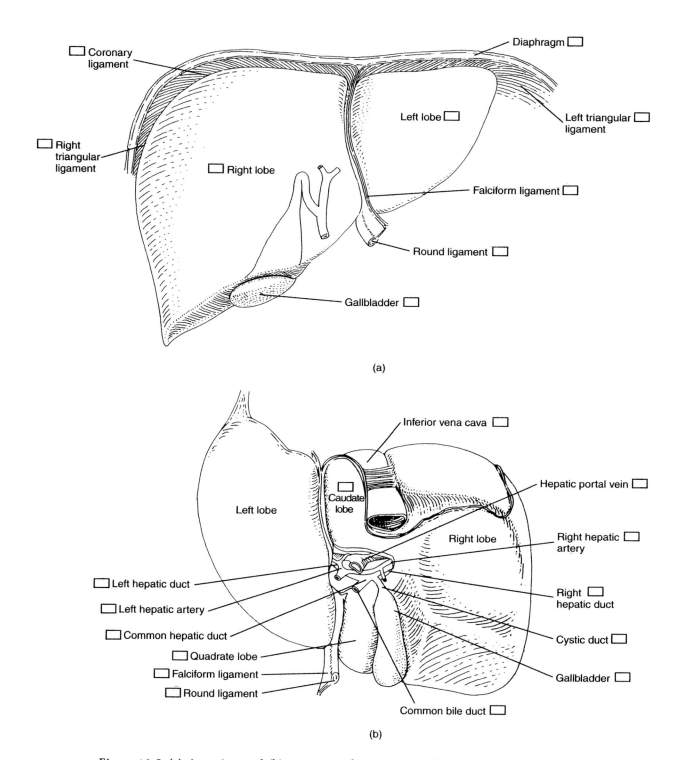

Figure 19.8 (a) Anterior and (b) postero-inferior views of the liver.

10. bare area
11. impression for inferior vena cava
12. porta hepatis
13. hepatic artery
14. portal vein
15. hepatic duct

16. common bile duct
17. gallbladder
18. cystic duct

The Pancreas—Description [Figures 19.1 and 19.6]

The pancreas is located inferior and posterior to the stomach. The gland is divided into a **head, neck, body**, and **tail**. The head lies in the curve of the duodenum, whereas the neck, body, and tail extend left and contact the **spleen**. The pancreas is drained by two ducts. The **main pancreatic duct** extends from the tail and through the length of the gland to the head. Near its termination, the main pancreatic duct joins the **common bile duct** (see preceding description of the liver and gallbladder). Together the main pancreatic duct and the common bile duct share a common passageway (**hepatopancreatic ampulla**) into the duodenum. The common duct pierces the muscular medial wall of the descending duodenum and opens at the **major duodenal papilla**. An accessory pancreatic duct, when present, lies superior to the main duct in the head of the pancreas. It empties into the duodenum by the **minor duodenal papilla**, located superior to the major papilla (Figure 19.6).

The Pancreas—Identification [Figures 19.1 and 19.6]

Examine the pancreas and its association with the duodenum, liver, and spleen in a model and/or cadaver. Identify the following structures and color them in the illustrations:

1. head
2. neck
3. body
4. tail
5. major pancreatic duct
6. spleen

The Peritoneum—Description [Figures 19.9 and 19.10]

The **peritoneum** is a **serous membrane** that lines the walls of the abdominal cavity and viscera. The peritoneum that lines the abdominal walls, the inferior surface of the diaphragm, and the walls of the pelvic cavity is the **parietal peritoneum**. The peritoneum that covers the abdominal organs is called the **visceral peritoneum**. Most of the abdominal and pelvic organs are suspended from the posterior wall by double-layered membranes of the peritoneum called **mesenteries**, which bind the small intestine, transverse colon, and sigmoid colon to the abdominal wall. The **greater omentum** is a double-layered fold of the peritoneum that attaches to the greater curvature of the stomach and hangs down from it like an apron. It is located in the space between the anterior abdominal wall and the coils of the small intestine. The **lesser omentum** attaches the lesser curvature of the stomach to the inferior surface of the liver.

During embryological development, the developing intestines rotate and come to lie near the abdominal wall of the fetus. As a result, most of the duodenum,

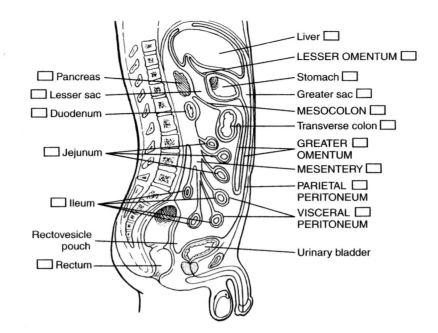

Figure 19.9 A sagittal section of the body shows the peritoneal lining of the abdominal cavity.

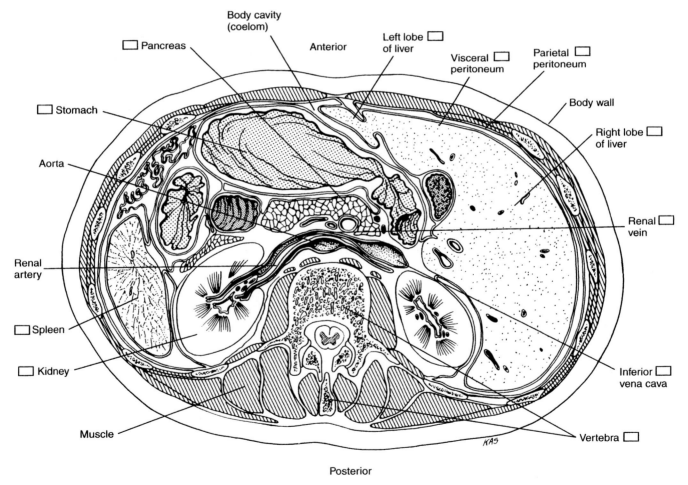

Figure 19.10 A transverse section of the abdomen shows the peritoneal lining and the retro-peritoneal organs, the pancreas and kidneys.

ascending colon, and descending colon closely adhere to the posterior abdominal wall and are covered only by the peritoneum on their anterior surfaces. Such organs, including the rectum and pancreas, are **retroperitoneal**.

The space between the parietal and visceral peritoneum is called the **peritoneal cavity**. It is divided into the **greater sac** and the **lesser sac**. The greater sac is the main compartment of the peritoneal cavity. It extends from the diaphragm to the pelvis. The lesser sac is a smaller compartment behind the lesser omentum and stomach. The right margin of the lesser sac opens into the greater sac, the peritoneal cavity, through an opening of the lesser sac, the **epiploic foramen** (Figure 19.5). The free border of the lesser omentum forms the anterior boundary of the opening of the lesser sac (epiploic foramen).

CHAPTER SUMMARY AND CHECKLIST

I. THE MOUTH AND ITS CAVITY

A. The Oral Cavity
1. Hard palate forms the roof of the oral cavity
2. Soft palate forms the posterior boundary of the roof
3. Uvula hangs from the soft palate
4. The tongue is composed of skeletal muscles
5. The pharynx is posterior to oral cavity
6. The epiglottis diverts material into the esophagus

B. The Accessory Organs
1. The three pairs of salivary glands
   a. parotid glands located on side of face
   b. submandibular glands located under the mandible
   c. sublingual glands located under the tongue
2. Salivary glands enter oral cavity via ducts

II. THE ESOPHAGUS

A. Parts
1. Cervical part is located in the neck
2. Thoracic part is located in the thoracic cavity
3. Abdominal part is short and located in the abdomen

III. THE STOMACH

A. Location and Structural Associations
1. Inferior to diaphragm and partially hidden by it
2. Superiorly attached to abdominal esophagus
3. Inferiorly attached to duodenum

B. Parts of the Stomach
1. Cardia is upper, narrow region inferior to esophagus
2. Fundus is the dome-shaped region to left of esophagus
3. Body is the large central portion
4. Pyloric region ends at pyloric sphincter
5. Lesser curvature is the concave border attached to lesser omentum

6. Greater curvature is the convex border
7. Greater omentum hangs down from the stomach
8. Rugae are temporary folds in stomach wall

IV. SMALL INTESTINE

A. Location and Regions
 1. Situated between pyloric sphincter and large intestine
 2. Subdivided into duodenum, jejunum, and ileum
 3. Duodenum
  a. first part of the small intestine
  b. C-shaped tube that curves around the pancreas
  c. common bile and major pancreatic duct enter at major duodenal papilla
  d. most of the organ is retroperitoneal
 4. No sharp line of demarcation between jejunum and ileum
 5. Ileum empties into cecum through ileocecal valve
 6. Plicae circulares prominent in duodenum and jejunum

V. LARGE INTESTINE

A. Location and Parts
 1. Begins at the cecum
 2. Ascends on right side as ascending colon
 3. At the liver exhibits hepatic flexure
 4. Passes to left as transverse colon
 5. Exhibits splenic flexure at the spleen
 6. Descends inferiorly as descending colon
 7. Near pelvic brim curves as the sigmoid colon
 8. In pelvic bowl it becomes the rectum
 9. Terminal 2–3 cm of the rectum is the anal canal

B. Characteristic Features
 1. Longitudinal strips of muscle are taenia coli
 2. Pouches in the colon are haustra
 3. Fat-filled pouches are the epiploic appendages
 4. Ascending colon, descending colon, and rectum are retroperitoneal

VI. THE ACCESSORY DIGESTIVE ORGANS

A. Liver
 1. Divided into larger right and smaller left lobes
 2. On posterior side are caudate and quadrate lobes
 3. Lobes are separated by fissures
 4. Falciform ligament attaches liver to anterior abdominal wall and forms coronary ligament which attaches liver to diaphragm and surrounds bare area
 5. Round ligament is the remnant of fetal umbilical vein
 6. Attached by lesser omentum to the stomach

B. Gallbladder
 1. Located on the inferior surface of the liver

      2. Cystic duct drains it

      3. Joins the hepatic duct from liver

      4. Cystic and hepatic ducts join to form common bile duct

C. Pancreas

      1. Located inferior and posterior to stomach

      2. Is divided into head, neck, body, and tail

      3. Neck, body, and tail extend to the spleen

      4. Main pancreatic duct joins the common bile duct

VII. PERITONEUM

A. Location and Characteristic Features

      1. Serous membrane that lines the abdominopelvic cavity

      2. Parietal peritoneum lines the abdominal and pelvic walls

      3. Visceral peritoneum lines the abdominal organs

      4. Most of duodenum, ascending and descending colon, and rectum are retroperitoneal

      5. Peritoneal cavity is located between visceral and parietal peritoneum

      6. Greater sac extends from diaphragm to pelvis

      7. Lesser sac is behind the stomach and lesser omentum

LABORATORY EXERCISE 19.1

Part I

The Organs of the Digestive System

1. What structure serves as a common pathway for digestive and respiratory systems?

   _____

2. What are the three pairs of salivary glands?

   a. _____

   b. _____

   c. _____

3. Which salivary gland is the largest?

   _____

4. Through what opening does the esophagus enter the abdominal cavity?

   _____

5. The stomach is connected superiorly to the _____

   and inferiorly to the _____.

6. What structure separates the stomach from the duodenum?

   _____

7. What are the four parts of the stomach?

   a. _____

   b. _____

   c. _____

   d. _____

8. What is the upper narrow region of the stomach inferior to the esophagus?

   _____

9. What are the two curvatures of the stomach?

a. _____

b. _____

10. What structure attaches the inferior surface of the liver to the stomach?

    _____

11. What structures are found in the lesser omentum?

    a. _____

    b. _____

    c. _____

12. What is the apronlike structure that hangs from the stomach?

    _____

13. What is the organ located between the pyloric sphincter and ileocecal junction?

    _____

14. What are the major subdivisions of the small intestine?

    a. _____

    b. _____

    c. _____

15. The small intestine is located between which two structures?

    a. _____

    b. _____

16. What are the three parts of the colon as it passes from the ileum to the anus?

    a. _____

    b. _____

    c. _____

Part II

Match the description on the left with the structures on the right.

1. Temporary folds in the stomach _____
2. Circular folds in the small intestine _____
3. There common bile duct enters duodenum _____
4. Duodenum curves around it _____
5. Receives secretion from liver and pancreas _____
6. Suspends small part of duodenum _____
7. Location of most of the duodenum _____
8. Terminal portion of ileum empties into _____
9. Valve between ileum and cecum _____
10. Worm-shaped structure of cecum _____
11. Colon on the left side of body _____
12. Three muscle strips on colon _____
13. Fat-filled pouches on colon _____
14. S-shaped portion of colon _____
15. Colon in pelvis _____

A. Plicae circulares
B. Duodenum
C. Pancreas
D. Major duodenal papilla
E. Rugae
F. Retroperitoneal
G. Lesser omentum
H. Cecum
I. Taenia coli
J. Epiploic appendages
K. Ileocecal
L. Appendix
M. Descending colon
N. Sigmoid
O. Rectum

Part III

1. The liver is divided into which lobes?

   a. _____

   b. _____

   c. _____

   d. _____

2. What structure attaches the liver to the anterior abdominal wall?

   _____

3. The round ligament of the liver is a remnant of what fetal structure?

   _____

4. What structure attaches the liver to the stomach?

   _____

5. What two structures join to form the common bile duct?

   a. _____

   b. _____

6. The porta hepatis of the liver is the entrance and exit of which structures?

   a. _____

   b. _____

   c. _____

7. The pancreas is divided into what parts?

   a. _____

   b. _____

   c. _____

   d. _____

8. What pancreatic structure joins the common bile duct?

   _____

9. What does the tail of the pancreas comes into contact with when it extends to the left?

   _____

10. The peritoneum that lines the abdominal walls and pelvic cavity is called the _____ peritoneum, and the peritoneum that covers the abdominal organs is the _____ peritoneum.

Part IV

Using the listed terms, label the organs of the digestive system and then color them.

Anus	Pancreas
Mouth	Jejunum
Sublingual gland	Transverse colon
Cervical esophagus	Cecum
Abdominal esophagus	Parotid gland
Duodenum	Pharynx
Ascending colon	Stomach
Descending colon	Liver
Appendix	Gallbladder
Tongue	Ileum
Submandibular gland	Rectum
Thoracic esophagus	Sigmoid colon

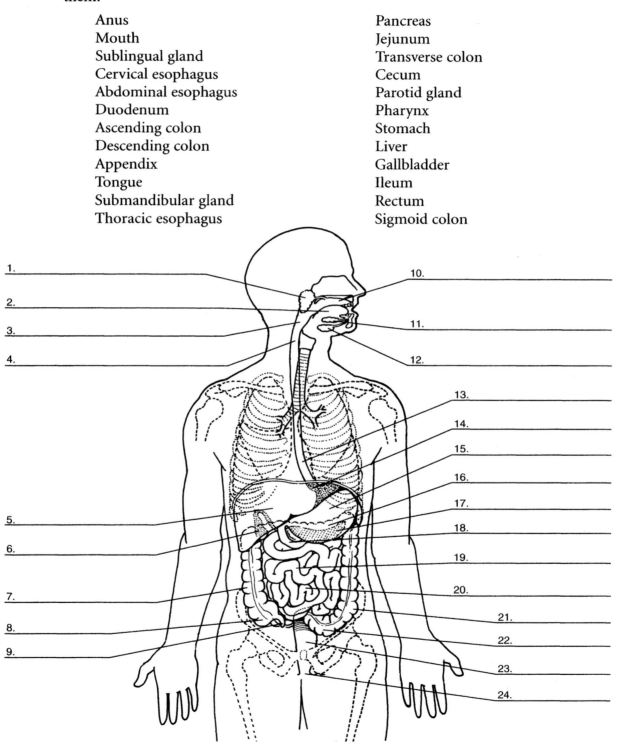

1. _____

2. _____

3. _____

4. _____

5. _____

6. _____

7. _____

8. _____

9. _____

10. _____

11. _____

12. _____

13. _____

14. _____

15. _____

16. _____

17. _____

18. _____

19. _____

20. _____

21. _____

22. _____

23. _____

24. _____

Figure 19.11 The organs of the digestive system.

LABORATORY EXERCISE 19.2

Using the listed terms, label the esophagus, stomach, and duodenum and then color them.

Abdominal esophagus
Cardia
Lesser curvature
Plicae circulares
Pyloric orifice
Esophageal orifice
Body

Greater curvature
Pyloric sphincter
Fundus
Duodenum
Rugae
Pylorus

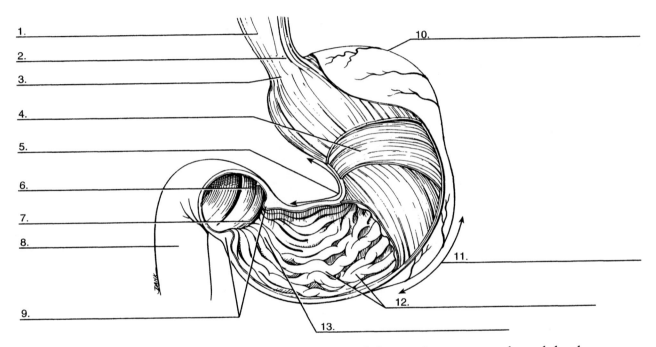

1. _____
2. _____
3. _____
4. _____
5. _____
6. _____
7. _____
8. _____
9. _____
10. _____
11. _____
12. _____
13. _____

Figure 19.12 External and internal anatomy of the esophagus, stomach, and duodenum.

LABORATORY EXERCISE 19.3

Using the listed terms, label the liver, pancreas, and gallbladder, and then color them.

Right lobe of liver
Common hepatic duct
Common bile duct
Head of pancreas
Main pancreatic duct
Major duodenal papilla
Ascending duodenum
Gallbladder
Body of pancreas

Left lobe of liver
Falciform ligament
Cystic duct
Tail of pancreas
Descending duodenum
Horizontal duodenum
Superior duodenum
Round ligament
Pancreas

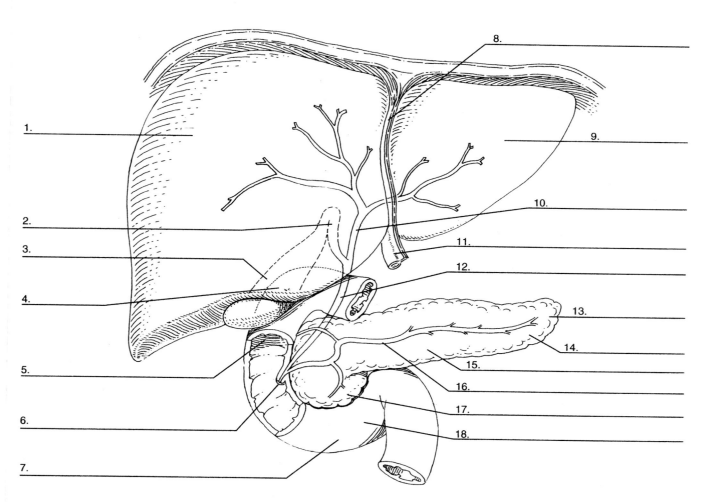

Figure 19.13 External and internal anatomy of the liver, pancreas, and gallbladder.

LABORATORY EXERCISE 19.4

Using the listed terms, label the large intestine and then color it.

Appendix
Transverse colon
Ileocecal valve
Sigmoid colon
Epiploic appendages
Anal canal
Hepatic flexure
Ascending colon
Descending colon

Splenic flexure
Taenia coli
Transverse mesocolon
Ileum
Haustrum
Cecum
Mesentery
Rectum

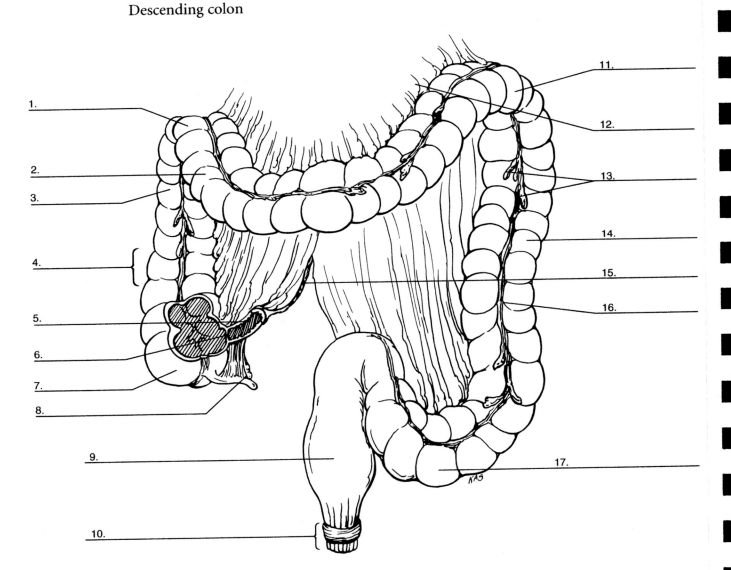

Figure 19.14 The large intestine.

20

The Urinary System

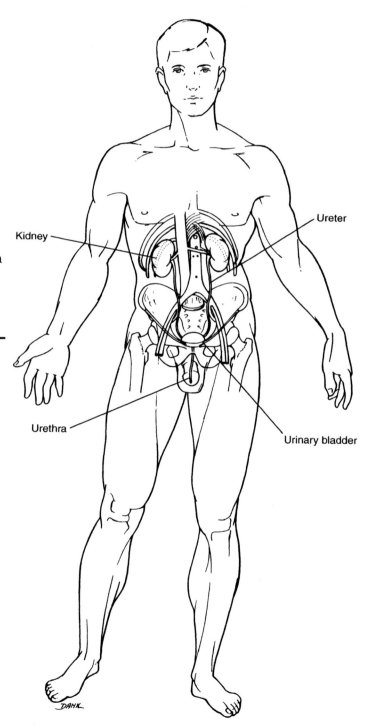

Kidney

Ureter

Urethra

Urinary bladder

Objective

The objective of Chapter 20, "The Urinary System," is for you to learn about the following organs in the human urinary system:

1. **External and internal anatomy of the kidneys**
2. **Ureters, bladder, and urethra**

Description *[Figure 20.1]*

The urinary system consists of two **kidneys** and two **ureters**, one **bladder**, and one **urethra**. The kidneys filter the blood and produce urine. The urine leaves the kidneys through the ureters, which convey it to the bladder. It is stored here until voided via the **urethra**.

The External Anatomy of the Kidneys

[Figures 20.1–20.3]

The kidneys, ureters, and bladder are **retroperitoneal**. The left kidney is higher than the right kidney because the liver lies above the right kidney. The superior region of each kidney is covered by an endocrine organ, the **suprarenal (adrenal) gland**. The adrenal glands do not bear any morphological connection with the organs of the urinary system.

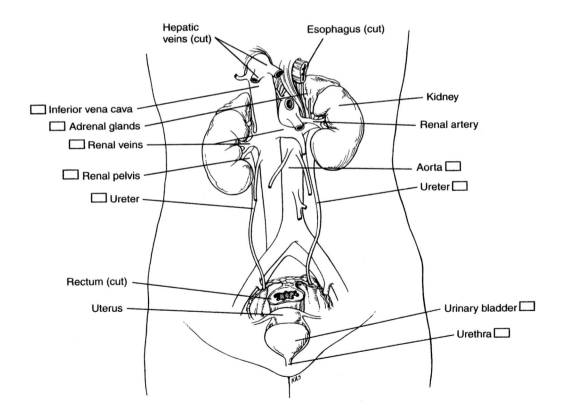

Figure 20.1 The urinary system in the female.

Three layers of connective and fatty tissue enclose each kidney. The innermost layer around the kidney is the fibrous **renal capsule.** External to the renal capsule is the **adipose capsule,** or **perirenal fat.** The third layer around the adipose capsule and kidneys is the **renal fascia,** which is divided into anterior and posterior layers. Outside the renal fascia is the **pararenal fat,** covered anteriorly by the **parietal peritoneum** (Figure 20.2).

The kidneys are bean-shaped organs that face the vertebral column. In the center of the medial concavity is an indentation called the **hilus.** Here the **renal arteries, renal veins,** lymph vessels, nerves, and **ureters** enter and leave the kidneys. Extending from the hilus into the kidney proper is a cavity called the **renal sinus.** This cavity contains fat and loose connective tissue that surround the renal blood vessels and the dilated end of the ureter, the **renal pelvis** (Figure 20.3).

Leaving each kidney hilus is the **ureter.** The ureters descend toward the pelvis between the parietal peritoneum and the posterior abdominal wall muscles. In the pelvis, the ureters enter the **urinary bladder.** Also leaving each kidney hilus is the **renal vein.** The **left renal vein** is longer than the right and passes in front of the **aorta** to drain into the **inferior vena cava.** The **right renal vein** is shorter and drains into the right side of the **inferior vena cava.** The **renal arteries** arise from each side of the aorta. The **left renal artery** passes behind the left renal vein to supply the left kidney. The longer **right renal artery** passes behind the inferior vena cava to supply the right kidney (Figure 20.1).

In the kidney hilus, the renal vein is most **anterior,** the renal artery **intermediate,** and the renal pelvis most **posterior.** The disposition of these structures in the kidney hilus allows us to distinguish between isolated right and left kidneys. In the hilus, the renal arteries and veins divide into numerous smaller branches anterior and posterior to the renal pelvis (Figures 20.1 and 20.3).

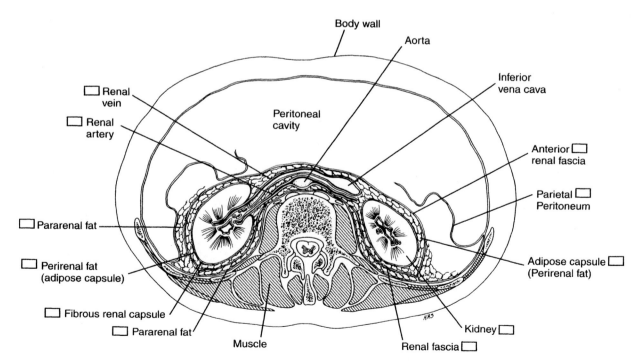

Figure 20.2 Transverse section illustrates the retroperitoneal location of the kidneys and the fat layers.

The External Anatomy of the Kidneys—
Identification *[Figures 20.1–20.3]*

Examine the urinary organs in a model and/or prepared cadaver. Identify the following structures and then color them in the illustrations:

1. right kidney
2. left kidney
3. renal arteries (right and left)
4. renal veins (right and left)
5. aorta
6. inferior vena cava
7. suprarenal (adrenal) gland
8. ureter
9. hilus
10. renal pelvis
11. renal capsule
12. renal fascia
13. renal fat (pararenal, adipose capsule

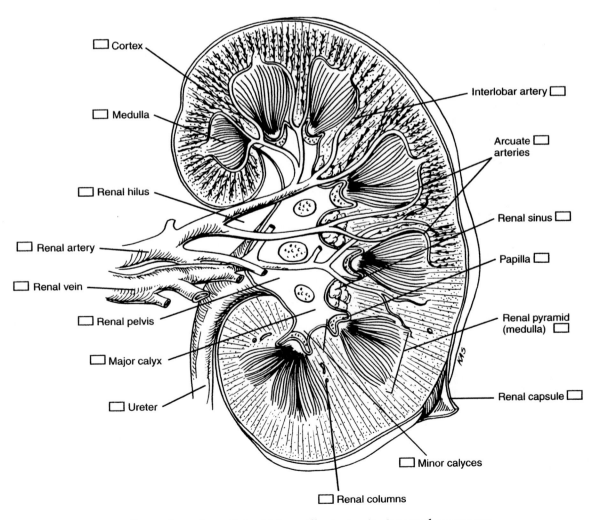

Figure 20.3 Frontal section of the kidney illustrates its internal anatomy.

The Internal Anatomy of the Kidney—
Description [Figure 20.3]

The outer, darker area of the kidney under the **renal capsule** is called the **cortex**, and adjacent to the cortex is the inner and lighter-colored **medulla**. The medulla is characterized by several distinct **renal**, or **medullary**, **pyramids**. The bases of the pyramids are broad and situated adjacent to the cortex. Their pointed apices, called the **renal papillae** (singular, **papilla**), project inward into the kidney. The cortex of each kidney extends between the pyramids as the **renal columns**.

The **minor calyces** (four to 13 in number) are cup-shaped expansions that surround the renal papillae. The minor calyces merge to form two or three short tubes called the **major calyces**. These join to form a larger, funnellike structure, the **renal pelvis**. The minor calyces collect urine from the tips of the papillae and direct it into the major calyces, the pelvis, and the ureter, which carries it to the bladder. (The renal pelvis exits the renal hilus and becomes a narrow ureter outside the kidney.)

In the hilus of each kidney, the **renal artery** divides into several anterior and posterior branches. These vessels continue into the kidney between the renal pyramids and the renal columns as the **interlobar arteries**. Between the medulla and the cortex, the interlobar arteries arch over the base of the pyramids and become the **arcuate arteries**. These arteries then branch further into smaller microscopic interlobular arteries. Leaving the kidneys, the interlobular veins empty into **arcuate veins** and **interlobar veins**. Venous blood from the kidneys is then carried by the **renal veins** into the **inferior vena cava**. The renal veins generally follow the same course as the arteries of the same name.

The Internal Anatomy of the Kidneys—
Identification [Figure 20.3]

Examine the internal anatomy of the kidney in a model and/or prepared cadaver. Identify the following internal structures and then color them in the illustration:

1. medulla
2. cortex
3. renal artery
4. renal vein
5. interlobar artery
6. arcuate artery
7. renal pelvis
8. major calyces (calyx)
9. minor calyces (calyx)
10. renal pyramid
11. renal papilla
12. renal columns
13. renal sinus

The Ureters, Urinary Bladder, and Urethra—
Description *[Figures 20.1, 20.4, 20.5 and 20.6]*

The ureters descend toward the bladder retroperitoneally and enter the bladder on the postero-inferior side. In males, the bladder lies anterior to the **rectum** and is separated from it by the **rectovesical pouch** or **space**. In females, the bladder is anterior to the **uterus** and is separated from it by the **vesicouterine pouch** or **space** (Figures 20.4 and 20.5).

The **urethra** is a small tube that exits the bladder to the exterior of the body. In females, the urethra is a short structure (about 4 cm long); it passes along the anterior surface of the **vagina** and opens to the exterior as the **external urethral orifice**. This orifice is located between the clitoris and vagina (Figures 20.4 and 20.6).

In males, the urethra is much longer (about 20 cm long) and serves as a common passageway for urine and semen. The male urethra passes to the exterior of the body through the tip of the penis as the **external urethral orifice**. In the male, the urethra has three different regions: the **prostatic, membranous,** and **penile (cavernous) urethra,** named according to the regions through which the urethra passes (Figures 20.5 and 20.6).

The internal anatomy of the bladder is wrinkled when empty. At the inferioposterior wall of the bladder is a smooth area called the **trigone**. As its name implies, this region has a triangular outline, with a **ureteral orifice** located at each corner superiorly and the **urethral orifice** inferiorly (Figure 20.6).

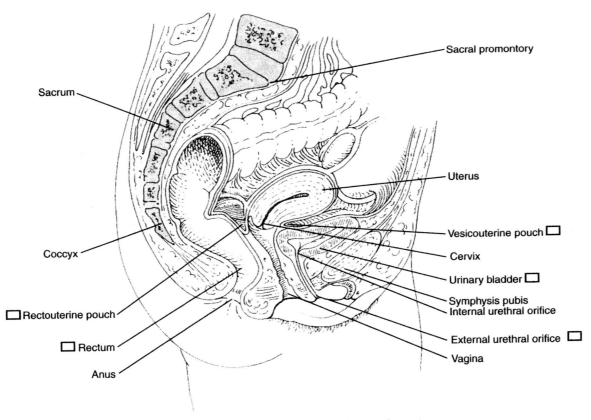

Figure 20.4 Sagittal section of the female pelvis shows the urinary organs.

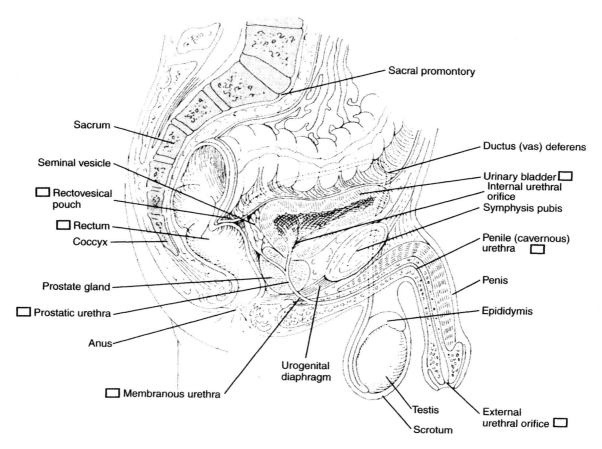

Figure 20.5 Sagittal section of the male pelvis shows the urinary organs.

In the area where urethral leaves the bladder, bundles of circular smooth muscle fibers surround the urethra and form the involuntary **internal urethral sphincter**. Below the internal urethral sphincter is an **external urethral sphincter**, which is voluntary and composed of skeletal muscles. It is located within the muscles of the **urogenital diaphragm**.

The Ureters, Urinary Bladder, and Urethra— Identification *[Figures 20.4–20.6]*

Examine the location of the ureters, bladder, and urethra in models of both genders. If possible, examine the abdominopelvic cavity in a prepared cadaver, identify the following structures, and color them in the illustrations.

1. **ureters**
2. **urinary bladder**
3. **trigone**
4. **ureteral orifice in bladder**
5. **male urethra (prostatic, membranous, and cavernous)**
6. **external urethral orifice (male and female)**
7. **uterovesical pouch (female)**
8. **rectovesical pouch (male)**
9. **urethra**

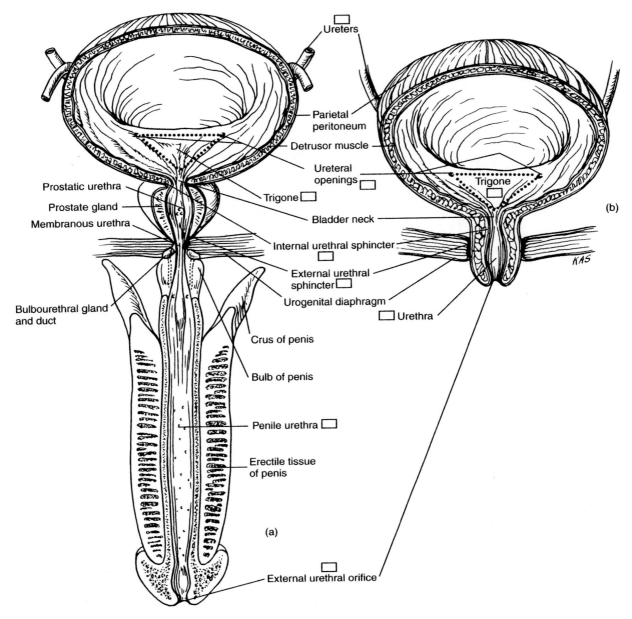

Figure 20.6 Structure of the urinary bladder and urethra. The anterior wall of the bladder has been cut to reveal the trigone. (a) The bladder and urethra of the male. (b) The bladder and urethra of the female.

CHAPTER SUMMARY AND CHECKLIST

I. URINARY SYSTEM

A. Retroperitoneal Organs
1. One pair of kidneys
2. One pair of ureters enter the bladder
3. One bladder
4. One urethra leaves the bladder

B. **Surrounding Tissues**
   1. Connective tissue, or renal capsule, is the innermost layer
   2. Adipose tissue capsule (perirenal fat)
   3. Renal fascia is the outermost layer
   4. Pararenal fat is covered by peritoneum

II. **EXTERNAL ANATOMY OF THE KIDNEY**

A. **Associated Structures**
   1. Renal sinus is filled with fat and connective tissue
   2. Renal pelvis is the most posterior structure in hilus
   3. Hilum is medial slit in the kidney
   4. Renal vein located most anterior
   5. Renal artery located intermediate
   6. Suprarenal (adrenal) gland situated superior to kidney

III. **THE INTERNAL ANATOMY OF THE KIDNEY**

A. **Characteristic Structures**
   1. Renal capsule surrounds the kidney
   2. Cortex is outer region, located inside the capsule
   3. Medulla is inner region, inside the cortex
   4. Medullary pyramids are in medulla
   5. Renal papillae are bases of pyramids
   6. Renal columns are cortical regions between pyramids
   7. Renal pelvis is expanded portion of ureter in renal sinus
   8. Major calyx is a subdivision of renal pelvis
   9. Minor calyx is a small tube for each pyramid; it collects urine
   10. Interlobar arteries run between renal pyramids
   11. Arcuate arteries arch over base of pyramids
   12. Veins follow arteries out of kidney

IV. **THE URETERS, BLADDER, AND URETHRA**

A. **Associated Structures in Both Sexes**
   1. Ureteral orifices seen at the base of bladder
   2. Trigone is the smooth triangular area at the base of bladder
   3. Urethral orifice is the exit of urethra from bladder
   4. Rectum is located posterior to bladder in males, uterus in females
   5. Rectovesical pouch separates bladder from rectum in males
   6. Uterovesical pouch separates bladder from uterus in females
   7. Urethra in females is short
   8. Urethra in males has three regions (prostatic, membranous, penile)
   9. External urethral orifice (male and female)

Laboratory Exercises 20

NAME _____

LAB SECTION _____ DATE _____

LABORATORY EXERCISE 20.1

Part I

Organs of the Urinary System

1. What are the three layers of connective tissue that surround each kidney?

   a. _____

   b. _____

   c. _____

2. What major structures enter and leave the kidney at the hilus?

   a. _____

   b. _____

   c. _____

3. What is the connection between the kidneys and the bladder called?

   _____

4. In the hilus of the kidney, which structure is located:

   a. most anteriorly _____

   b. most posteriorly _____

   c. intermediate _____

5. What are the outer and an inner regions of the kidney?

   a. _____

   b. _____

6. What structures surround the renal papillae?

   _____

7. What forms the major calyces?

   _____

8. What forms the renal pelvis?

   _____

9. Which blood vessels in the kidney are described below?

   a. in the kidney hilus _____

   b. between the renal pyramids and renal columns _____

   c. between the medulla and cortex _____

10. In males, the bladder is located anterior to the

    _____ and in females the bladder is anterior

    to the _____.

11. Which are the different regions of the male urethra?

    a. _____

    b. _____

    c. _____

Part II

Using the listed terms, label the structures and then color them in the illustrations.

Adrenal glands   Renal artery
Renal veins   Bladder
Renal pelvis   Aorta
Kidney   Ureter
Inferior vena cava   Urethra

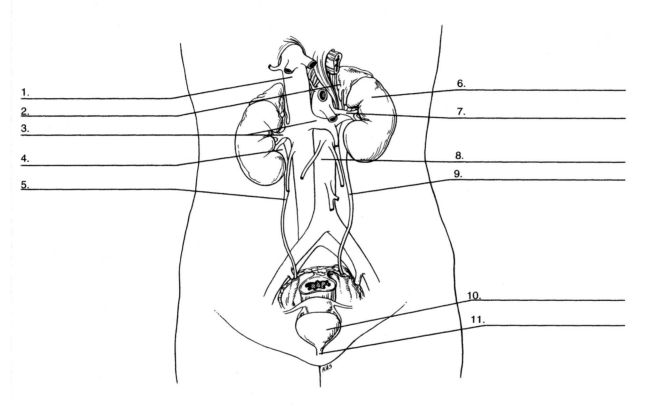

1.
2.
3.
4.
5.
6.
7.
8.
9.
10.
11.

Figure 20.7 The urinary organs and blood vessels in the female.

LABORATORY EXERCISE 20.2

Using the listed terms, label the structures in the kidney and then color them in the illustration.

Cortex
Renal hilus
Renal pelvis
Major calyx
Renal columns
Interlobar artery
Medulla
Renal artery

Ureter
Arcuate arteries
Renal sinus
Renal capsule
Renal vein
Minor calyx
Renal pyramid
Papilla

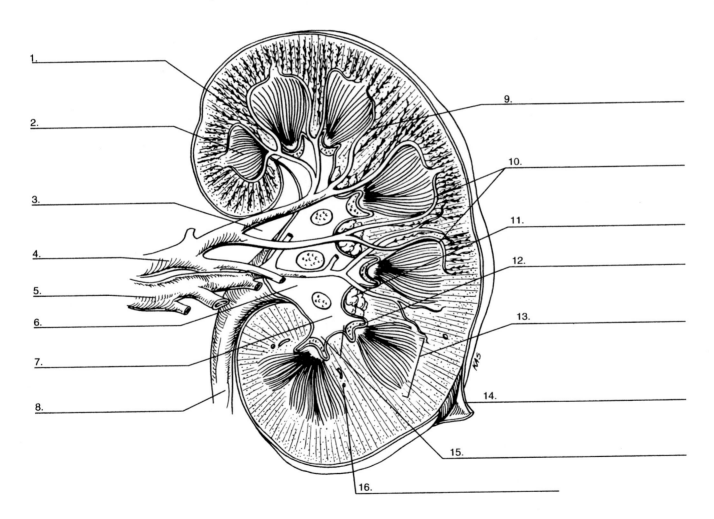

1.
2.
3.
4.
5.
6.
7.
8.
9.
10.
11.
12.
13.
14.
15.
16.

Figure 20.8 Frontal section of the kidney.

LABORATORY EXERCISE 20.3

Using the listed terms, label the structures of the urinary organs of both genders, and color them in the illustration.

External urethral orifice
Internal urethral sphincter
Prostatic urethra
Membranous urethra
Trigone

External urethral sphincter
Ureters
Penile urethra
Ureteral orifice
Urethra

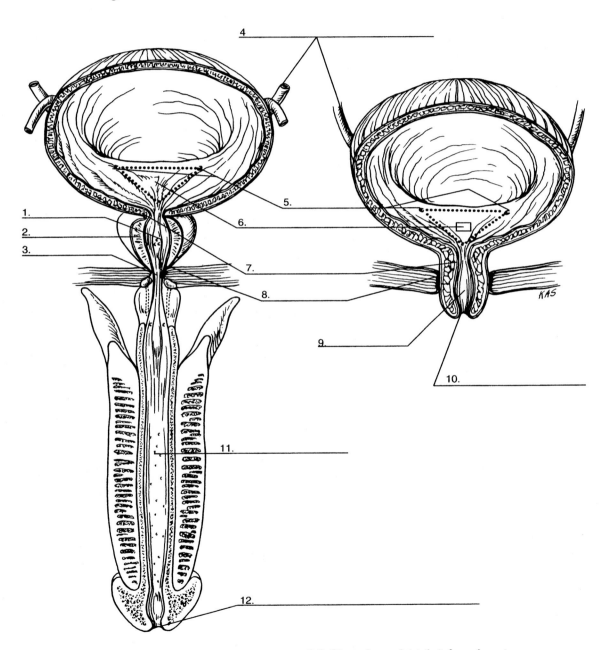

Figure 20.9 Comparative cross sections of (left) male and (right) female urinary organs.

Part Six

The Continuity of Human Life

21

The Male Reproductive System

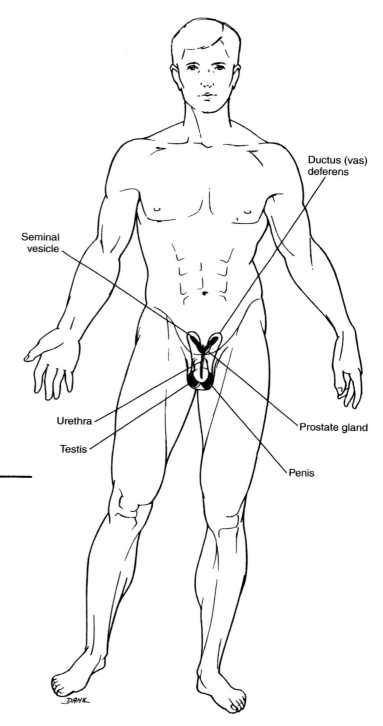

Seminal
vesicle

Ductus (vas)
deferens

Urethra

Testis

Prostate gland

Penis

Objective

The objective of Chapter 21, "The Male Reproductive System," is for you to be able to identify the structures, contents, and location of the following organs in the male reproductive system:

1. **Scrotum, testes, and epididymis**
2. **Ductus (vas) deferens and spermatic cord**
3. **Accessory sex glands and penis**

Description *[Figure 21.1]*

The reproductive system of the human male consists of primary or essential sex organs, accessory sex glands, a system of ducts, and penis. The **testes** are the **essential sex organs**; they produce germ cells, or **sperm**, and a male sex hormone, **testosterone**. The **accessory organs** or **glands** are the paired **seminal vesicles**, a single **prostate gland**, and paired **bulbourethral glands**. A system of excurrent ducts bring sperm from the testes to outside the body. Efferent ducts convey sperm from the

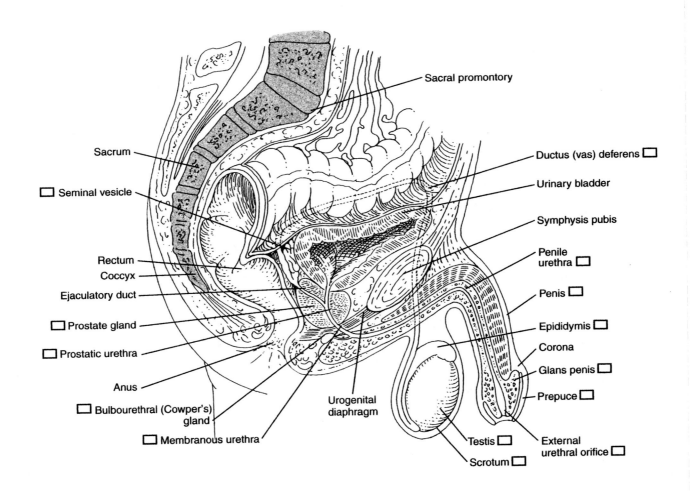

Figure 21.1 Sagittal section of the male reproductive tract.

testes to the epididymis, where they are stored and mature. From the epididymis, a mixture of sperm and accessory gland secretions, called **semen**, is carried to the exterior via the **ductus (vas) deferens** and **urethra**. The penis is an organ of both copulation and urination. The urethra, which conveys both urine and semen, is located in the body of the penis.

The Scrotum *[Figures 21.1 and 21.2]*

The primary male sex organs are the **testes**, located outside the body in a pouch called the **scrotum**. Externally, the scrotum appears to be a single pouch, but internally, it is divided by a **scrotal septum** into two chambers that each contain a single **testis, epididymis**, and the lower portion of the **spermatic cord**.

The skin of the scrotum is thin and contains hairs. Deep to the skin is the **dartos (tunic)** consisting of a thin layer of smooth muscle. When the outside temperature is cold, the dartos muscle contracts and wrinkles the skin. This action thickens the scrotal wall and raises the testes closer to the body and warmer temperature. Warm temperature reverses the process. Also elevating the testes are the **cremaster muscles**. These skeletal muscles surround each testis and are derived from the **internal oblique muscles** of the abdominal wall (Figure 21.2).

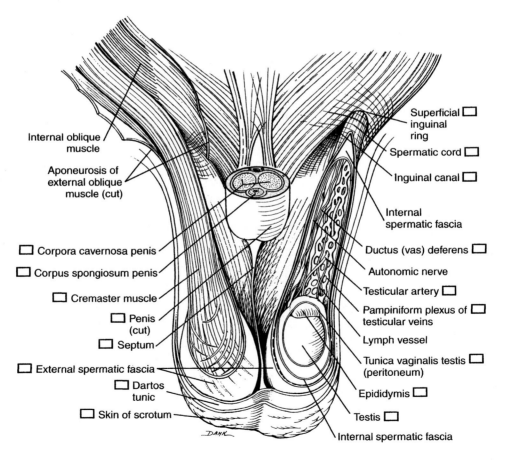

Figure 21.2 Anterior view of spermatic cord, scrotum, and inguinal canal. The left spermatic cord has been opened to expose its contents.

The Testes [*Figures 21.2 and 21.3*]

The testes are paired oval organs located in individual compartments in the scrotum. Each testis is surrounded by a dense, white, glistening connective tissue capsule called the **tunica albuginea**. External to the tunica albuginea is a thin layer of peritoneum, the **tunica vaginalis**. The connective tissue septa from the tunica albuginea extend inward and divide each testis into numerous internal compartments called **lobules**. Each lobule contains several highly coiled **seminiferous tubules**. The epithelium lining these tubules, the **germinal epithelium**, gives rise to cells that become sperm. The sperm then move from the seminiferous tubules through the **mediastinum testis**, **rete testis**, and **efferent ducts (ductuli efferentes)** into the **epididymis** (Figure 21.3).

The Scrotum and Testes—Identification

[*Figures 21.1–21.3*]

Examine a model and/or the prepared male reproductive organs in a cadaver. Identify the following structures and then color them in the illustrations.

1. **scrotum**
2. **scrotal septum**
3. **dartos tunic**

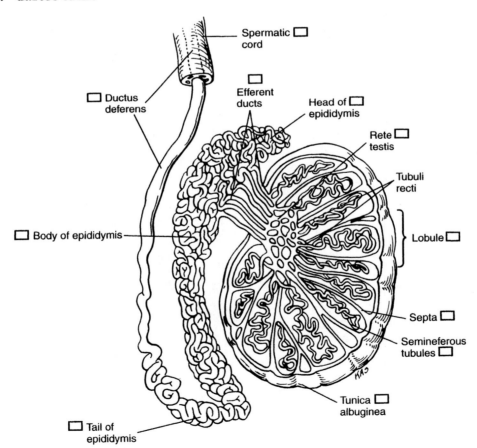

Figure 21.3 Dissected testis shows internal tubules, lobules, and the epididymis, which leads into the ductus deferens.

4. testis
5. epididymis
6. tunica albuginea
7. seminiferous tubules

The Epididymis *[Figures 21.2 and 21.3]*

The epididymis is a single, coiled, elongated, crescent-shaped mass on the superior surface of each testis. Numerous efferent ducts from the testis enter the epididymis. The epididymis is divided into the superior **head**, where the efferent ducts enter; a central **body**; and an inferior **tail**. The tail gradually becomes less coiled and then continuous with the straight **ductus (vas) deferens** (Figure 21.3).

The Ductus (Vas) Deferens
[Figures 21.1, and 21.3, 21.4, and 21.5]

As the **ductus deferens** ascends lateral to the testis, it becomes part of the **spermatic cord** (see next paragraph) that passes superiorly through the **inguinal canal** into the pelvic cavity; the inguinal canal is an oblique passageway in the anterior abdominal wall. The **superficial inguinal ring** is located above and lateral to the pubic tubercle, and constitutes the external opening of the inguinal canal (Figures 21.2 and 21.4).

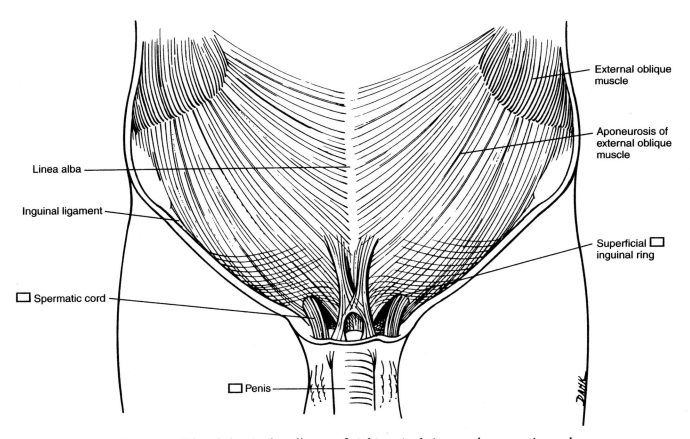

Figure 21.4 The abdominal wall, superficial inguinal ring, and spermatic cord.

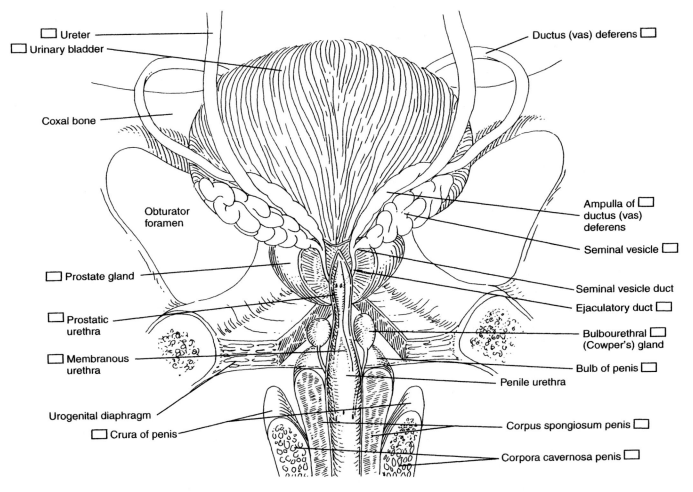

Labels on the figure:
- Ureter
- Urinary bladder
- Coxal bone
- Obturator foramen
- Prostate gland
- Prostatic urethra
- Membranous urethra
- Urogenital diaphragm
- Crura of penis
- Ductus (vas) deferens
- Ampulla of ductus (vas) deferens
- Seminal vesicle
- Seminal vesicle duct
- Ejaculatory duct
- Bulbourethral (Cowper's) gland
- Bulb of penis
- Penile urethra
- Corpus spongiosum penis
- Corpora cavernosa penis

Figure 21.5 Male reproductive organs and glands seen from postero-superior view.

Because of its thick, muscular wall, the ductus deferens can be identified as a firm, cordlike tube in the spermatic cord. The ductus deferens continues under the peritoneum of the pelvic cavity, crosses over the ureter, and descends along the posterior surface of the **bladder**, where its terminal portion dilates into an **ampulla**. On the postero-inferior surface of the urinary bladder, the ampulla of each ductus deferens is joined by the duct from the accessory sex gland, the **seminal vesicle** (described in the next section under Accessory Sex Glands) to form a short **ejaculatory duct**. This duct passes through the **prostate gland** into the **prostatic urethra** (Figure 21.1).

The Epididymis, Ductus (Vas) Deferens, and Associated Sex Glands—Identification

[Figures 21.1, 21.3, 21.4, and 21.5]

Examine the male reproductive organs in a model and/or prepared cadaver. Identify the following organs or structures and then color them in the illustrations.

1. **epididymis (head, body, and tail)**
2. **ductus deferens**

3. superficial inguinal ring
4. inguinal canal
5. ampulla of ductus deferens
6. seminal vesicle
7. prostate gland
8. ejaculatory duct

The Spermatic Cord [Figures 21.2 and 21.4]

The spermatic cord extends from the deep inguinal ring through the inguinal canal to the testes. The sheaths or layers around the cord and testes are derived from the anterior abdominal wall during the descent of the testes (in the fetus) into the scrotum. The outermost layer of the spermatic cord is the **external spermatic fascia**, derived from the external oblique muscle. Internal to this layer is the **cremaster muscle** and **fascia**, derived from the internal oblique muscle. Inside this layer is the **internal spermatic fascia**, derived from the transversalis fascia. Enclosed within these layers of the spermatic cord are the **testicular artery**, **testicular vein**, nerves, lymphatic vessels, and **ductus deferens**. As the veins leave the testes and enter the spermatic cord, they form a convoluted anastomosis of veins called the **pampiniform plexus**. This plexus serves as a heat exchange mechanism that lowers the temperature of the blood going to the testes and warms the blood as it returns to the body (Figure 21.2).

The Spermatic Cord—Identification
[Figures 21.2 and 21.4]

Examine a model and/or cadaver and identify the following structures of the spermatic cord:

1. spermatic cord
2. ductus deferens
3. testicular artery
4. testicular vein
5. cremaster muscle
6. pampiniform plexus

The Accessory Sex Glands

Seminal Vesicles and the Bulbourethral Glands
[Figures 21.3, 21.5, and 21.6]

The **seminal vesicles** are a pair of convoluted, pouchlike accessory sex glands situated on the lateral side of the **ductus deferens** on the postero-inferior surface of the **bladder**. Short ducts from each seminal vesicle join the ductus deferens and form the ejaculatory ducts that immediately enter the **prostate gland**. The **bulbourethral glands** are a pair of small glands that situated inferior to the prostate gland and on each side of the membranous **urethra**. These glands embedded in the **urogenital diaphragm** (a sheet of skeletal muscle) and their ducts empty into the **cavernous (penile) urethra**, a short distance from the glands.

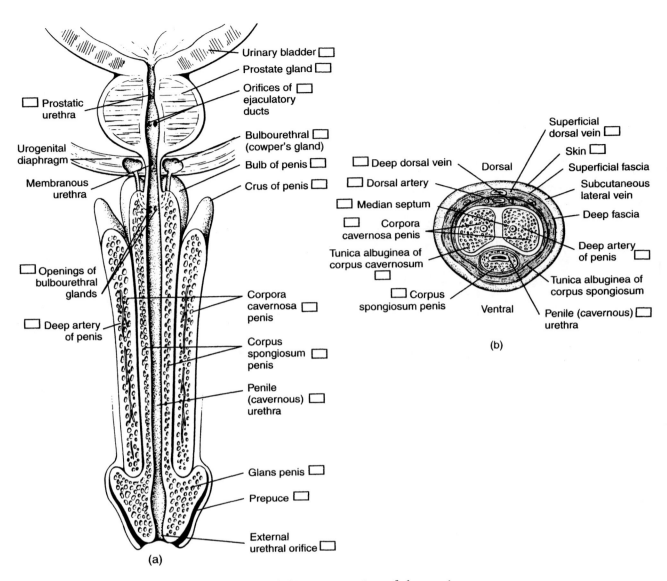

Figure 21.6 (a) Longitudinal and (b) cross section of the penis.

The Prostate Gland [Figures 21.1, 21.5, and 21.6]

The **prostate gland** is a single, walnut-sized accessory sex gland located inferior to the **bladder** and superior to the urogenital diaphragm and **bulb** of the **penis**. The prostate gland surrounds the superior portion of the urethra (the **prostatic urethra**) that passes through it. During ejaculation, the ejaculatory ducts deliver semen into the prostatic urethra. Opening into the prostatic urethra are numerous ducts that deliver secretions from the interior of the prostate gland to mix with the semen.

The Penis [Figures 21.6 and 21.7]

The penis is a long **shaft** composed of three cylindrical bodies of erectile tissue. These are surrounded by dense fibrous capsule, the **tunica albuginea**. Superficial to this capsule is the deep fascia, surrounded by superficial fascia and skin. Two of these erectile bodies, the **corpora cavernosa penis**, are paired and situated adjacent

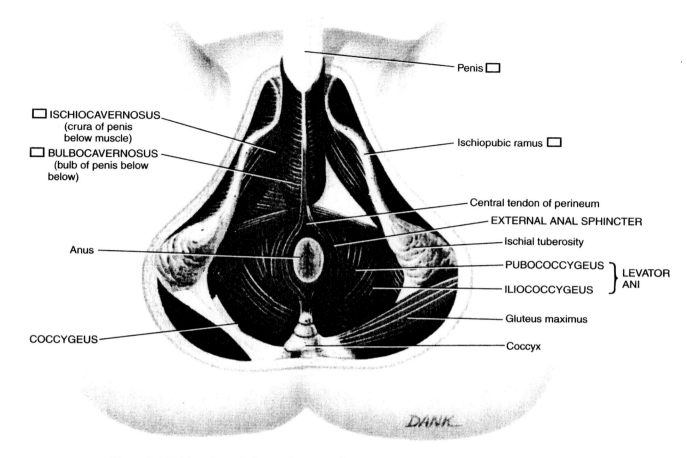

Penis ☐

☐ISCHIOCAVERNOSUS
(crura of penis
below muscle)

☐ BULBOCAVERNOSUS
(bulb of penis below
below)

Ischiopubic ramus ☐

Central tendon of perineum

EXTERNAL ANAL SPHINCTER

Anus

Ischial tuberosity

PUBOCOCCYGEUS
} LEVATOR
ANI
ILIOCOCCYGEUS

Gluteus maximus

COCCYGEUS

Coccyx

DANK

Figure 21.7 Muscles of the male reproductive organs and perineum.

to each other in the dorsal part of the penis. A fibrous internal connective tissue between the two cylinders forms a **median septum**. A single **corpus spongiosum penis**, located ventrally to the corpora cavernosa penis, expands at its distal end to form the **glans penis**. The skin of the penis continues over the glans as the **prepuce**, or **foreskin**. The entire length of the corpus spongiosum penis is transversed by the **urethra**, which opens at the end of glans penis as the **external urethral orifice**. The urethra in the male is long and consists of three sections: the **prostatic, membranous**, and **penile urethra** (Figure 21.6).

The penis has a rich blood supply. The **dorsal arteries** of the penis supply the **tunica albuginea** and the **glans**. The **deep arteries** of the penis supply the corpora cavernosa penis and are located in the center of each corpus cavernosum. The two major dorsal veins of the penis are the **superficial dorsal vein** under the skin and the **deep dorsal vein** in the deep fascia (Figure 21.6).

The corpora cavernosa extend back to the perineum and separate into **crura** (singular, **crus**) of the penis. The crura attach the penis to the rami of the ischial and pubic bones. Each crus is covered by a skeletal **ischiocavernosus muscle**. The corpus spongiosum enlarges proximally to form the **bulb** of the penis, which lies medially between the two crura. The bulb is attached to the inferior surface of the urogenital diaphragm and is covered by the skeletal **bulbospongiosum (bulbocavernosus) muscle** (Figure 21.7).

The Penis—Identification [Figures 21.6 and 21.7]

Examine longitudinal and cross sections of the penis in a model and/or prepared cadaver. Identify the following structures and then color them in the illustrations:

1. **corpora cavernosa penis**
2. **corpus spongiosum penis**
3. **tunical albuginea**
4. **penile urethra**
5. **deep arteries of penis**
6. **dorsal arteries of penis**
7. **superficial dorsal vein**
8. **deep dorsal vein**
9. **glans penis**
10. **prepuce**
11. **crura of penis**
12. **bulb of penis**
13. **ischiocavernosus muscle**
14. **bulbospongiosum muscle**

CHAPTER SUMMARY AND CHECKLIST

I. **THE SCROTUM AND TESTES**

   A. **Scrotum**
   1. Abdominal wall sac
   2. Partitioned by a septum
   3. Each partition contains one testis
   4. Spermatic cord extends from scrotum to superficial ring
   5. Scrotal wall contains dartos tunic

   B. **Testes**
   1. Tunica albuginea surrounds the testicular contents
   2. Connective tissue partitions testes into testicular lobules
   3. Located in the lobules are numerous seminiferous tubules
   4. Seminiferous tubules produce sperm
   5. Efferent ducts connect testes with epididymis

II. **THE EPIDIDYMIS, DUCTUS (VAS) DEFERENS, AND SPERMATIC CORD**

   A. **Epididymis and Ductus Deferens**
   1. Epididymis is located superiorly and lateral to testes
   2. Consists of head, body, and tail
   3. Epididymis is the storage site for maturing sperm
   4. Epididymis is continuous with ductus deferens
   5. Ductus deferens transports sperm during ejaculation
   6. Ductus deferens is located in the spermatic cord
   7. Terminal part of ductus deferens enlarges into ampulla behind the bladder

B. The Spermatic Cord
   1. Extends from the testis to the deep inguinal ring
   2. Enters pelvis through inguinal canal
   3. Surrounded by numerous tissue layers
      a. external spermatic fascia
      b. internal spermatic fascia
      c. cremaster muscle
   4. Contains the prominent ductus deferens, the testicular artery and vein
   5. Pampiniform plexus serves as heat exchange mechanism to cool blood entering testes

III. ASSOCIATED GLANDS

A. Seminal Vesicles
   1. Paired glands lie posterior to and at base of bladder, lateral to ductus deferens
   2. Produce important fluids for semen
   3. Ducts from glands join ampulla of ductus deferens to form ejaculatory ducts
   4. Ejaculatory ducts deliver sperm into prostatic urethra

B. Prostate Gland
   1. A single gland located inferior to the bladder
   2. Surrounds prostatic urethra (first part of urethra)
   3. Ejaculatory ducts open into prostatic urethra

IV. THE PENIS

A. Erectile Tissues
   1. The paired corpora cavernosa penis lie dorsally
   2. A single corpus spongiosum lies ventrally
      a. anteriorly, enlarges into glans penis
      b. posteriorly, enlarges into bulb of the penis
      c. bulb of penis covered by bulbospongiosum muscle
   3. Connective tissue tunica albuginea surrounds the erectile tissues
   4. Urethra passes through corpus spongiosum penis
      a. opens at the glans as external urethral orifice
   5. Posteriorly, corpora cavernosa penis divide into crura
      a. attach penis to rami of ischial and public bones
      b. crura covered by ischiocavernosus muscle
   6. Rich blood supply in the penis
      a. dorsal arteries supply tunica albuginea and glans
      b. deep arteries supply corpora cavernosa penis
      c. superficial and deep dorsal veins drain the penis

Laboratory Exercises 21

NAME _____

LAB SECTION _____ DATE _____

LABORATORY EXERCISE 21.1

Part I

Organs of the Male Reproductive System

1. What one essential and three accessory glands or organs comprise the male reproductive system?

   a. _____

   b. _____

   c. _____

   d. _____

2. Inside what structure outside the body are the testes located?

   _____

3. What structures are associated with the testes?

   a. _____

   b. _____

4. What two muscles are involved in temperature regulation of the testes?

   a. _____

   b. _____

5. What are the highly coiled tubules contained in the lobules of the testes?

   _____

6. What is the path (in order) of mature sperm from the testes?

   a. _____

   b. _____

   c. _____

7. What are the parts of the epididymis?

   a. _____

b. _____

c. _____

8. What does the duct of the epididymis become after it straightens out?

_____

9. What male reproductive structure passes through the inguinal canal?

_____

10. What structures/organs form the ejaculatory duct?

a. _____

b. _____

11. What organ does the ejaculatory duct pass through before it reaches the urethra?

_____

12. What structures associated with the testes are found in the spermatic cord?

a. _____

b. _____

c. _____

13. What structure in the spermatic cord is involved in regulating the temperature of blood that reaches the testes?

_____

14. Where are the bulbourethral glands located and into what structure do their ducts empty?

a. _____

b. _____

Part II

Match the description on the left with the correct term on the right.

1. Located inferior to bladder _____

2. Delivers semen to urethra _____

3. Surrounds shaft of penis _____

4. Paired erectile bodies in penis _____

5. Divides erectile bodies in penis _____

6. Single erectile body _____

7. Expanded terminal portion of penis _____

8. Urethra in prostate gland _____

9. Arteries of corpora cavernosa _____

10. Corpora cavernosa separates into _____

11. Muscle covering crura of penis _____

12. Muscle covering bulb of penis _____

A. Crura
B. Prostatic urethra
C. Median septum
D. Deep arteries
E. Tunica albuginea
F. Bulbospongiosum
G. Ischiocavernosus
H. Ejaculatory ducts
I. Corpora cavernosa
J. Prostate gland
K. Glans penis
L. Corpus spongiosum

Using the listed terms, label the organs and/or structures of the male reproductive system, and then color them in the illustration.

Seminal vesicle
Prostate gland
Penis
External urethral orifice
Urogenital diaphragm
Membranous urethra
Bladder
Prostatic urethra
Glans penis

Testis
Ductus deferens
Epididymis
Ejaculatory duct
Penile urethra
Prepuce
Scrotum
Bulbourethral gland

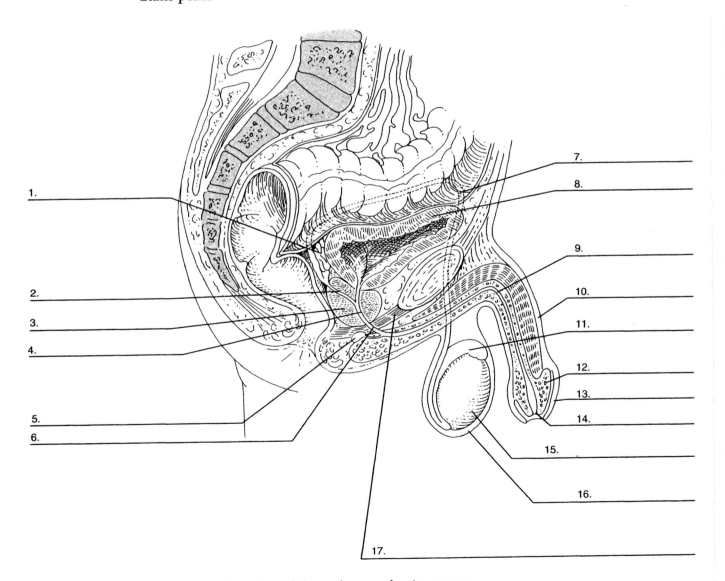

Figure 21.8 Sagittal section of the male reproductive system.

LABORATORY EXERCISE 21.2

Using the listed terms, label the organs and/or structures of the male reproductive system, and then color them in the illustration.

Ductus deferens
Seminal vesicle
Bulbourethral gland
Bulb of penis
Corpora cavernosa
Prostate gland
Bladder
Ejaculatory duct

Membranous urethra
Penile urethra
Urogenital diaphragm
Ampulla of ductus
Prostatic urethra
Crura of penis
Corpus spongiosum

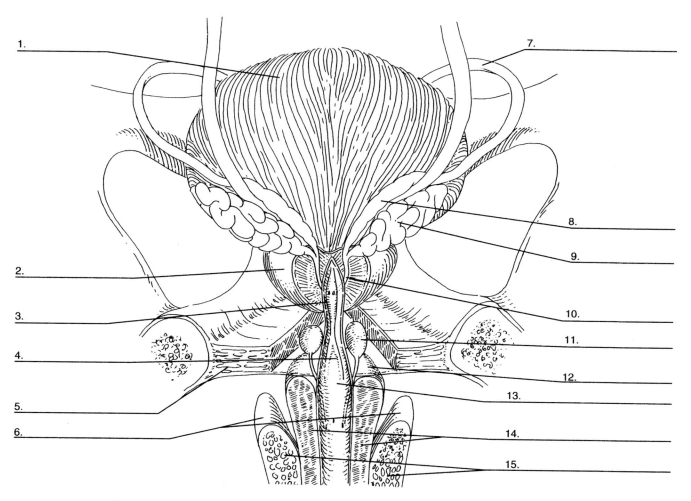

1.
2.
3.
4.
5.
6.

7.
8.
9.
10.
11.
12.
13.
14.
15.

Figure 21.9 Male reproductive organs and glands seen from posterior view.

22

The Female Reproductive System

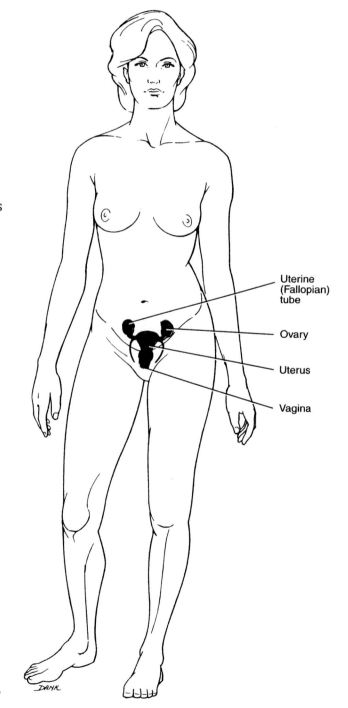

Uterine
(Fallopian)
tube

Ovary

Uterus

Vagina

Objective

The objective of Chapter 22, "The Female Reproductive System," is for you to be able to identify the structure and location of the following organs of the female reproductive system:

1. Ovaries, uterus, uterine tubes, and vagina
2. External genitalia
3. Mammary glands

Description [Figures 22.1 and 22.2]

The reproductive system of the human female consists of **primary sex organs**, **accessory sex organs**, and **external genitalia**. The primary sex organs are the paired **ovaries**. These produce **oocytes** that develop into mature **ova** upon fertilization (union with sperm) and the female sex hormones, **estrogen** and **progesterone**. The **accessory organs** are the unpaired **vagina** and **uterus**, and the paired **uterine (Fallopian) tubes**, or **oviducts**. These organs are located in the pelvis.

The reproductive organs located outside the pelvis are the external genitalia or, collectively, the **vulva**, or **pudendum**. These include the **mons pubis, labia majora, labia minora, clitoris, vestibule of the vagina, bulb of the vestibule**, and **greater vestibular glands**.

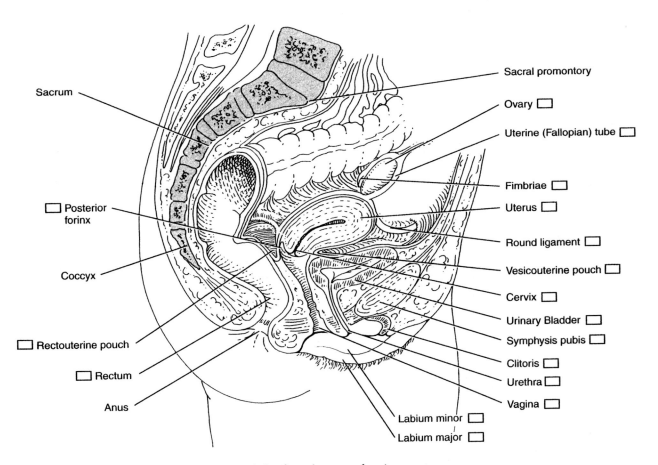

Figure 22.1 Sagittal section of the female reproductive system.

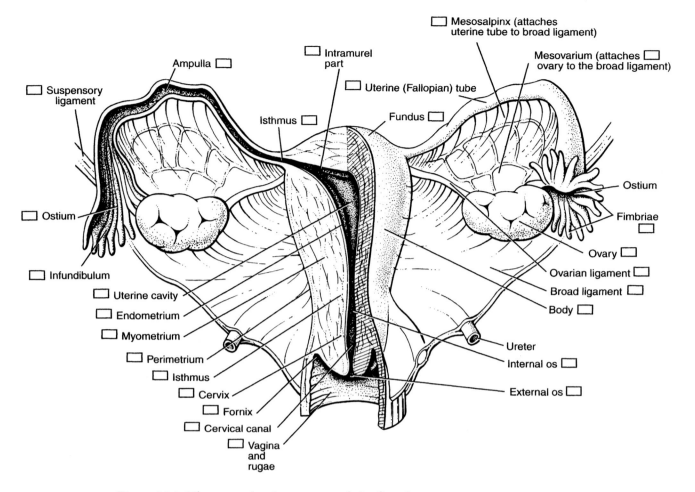

Ampulla ☐

☐ Intramurel part

☐ Mesosalpinx (attaches uterine tube to broad ligament)

Mesovarium (attaches ☐ ovary to the broad ligament)

☐ Suspensory ligament

☐ Uterine (Fallopian) tube

Isthmus ☐

Fundus ☐

Ostium

☐ Ostium

Fimbriae ☐

Ovary ☐

☐ Infundibulum

Ovarian ligament ☐

☐ Uterine cavity

Broad ligament ☐

☐ Endometrium

Body ☐

☐ Myometrium

Ureter

☐ Perimetrium

Internal os ☐

☐ Isthmus

External os ☐

☐ Cervix

☐ Fornix

☐ Cervical canal

☐ Vagina and rugae

Figure 22.2 The reproductive organs of the female.

The Internal Reproductive Organs— Description *[Figures 22.1 and 22.2]*

The Ovaries [Figures 22.1 and 22.2]

The ovoid **ovaries** are flattened and located close to the lateral wall in the pelvic cavity. Each ovary sits in a shallow depression on the posterior body wall called the **ovarian fossa** and is secured in place by a series of ligaments.

The largest of these ligaments is the **broad ligament**, part of the parietal peritoneum. The peritoneal layers that cover the anterior and posterior uterus fuse along its lateral borders and extend into the pelvic cavity as the broad ligament. The broad ligament supports the ovaries, uterine tubes, uterus, and vagina.

The anterior surface of each ovary, the **hilus**, is attached to the posterior extension of the broad ligament, called the **mesovarium**, which is a short fold of the peritoneum; blood vessels and nerves pass through the mesovarium to and from the hilus of the ovary. Each ovary is also attached by an **ovarian ligament** that extends medially to the lateral wall of the uterus near the entrance of the uterine tube. The lateral margins of the broad ligament form the **suspensory ligament** of the ovary. This ligament attaches each ovary to the pelvic wall; ovarian vessels, nerves, and lymphatic vessels reach the ovaries via this ligament.

The Uterine (Fallopian) Tubes [Figures 22.1 and 22.2]

The paired **uterine (Fallopian) tubes**, or **oviducts**, open into the abdominopelvic cavity near the ovaries. Each uterine tube extends from the ovary to the uterus. Each tube is also attached to the broad ligament by the **mesosalpinx**, a portion of the broad ligament that extends inferiorly as far as the attachment of the ovary. The distal end of each uterine tube is a funnel-shaped **infundibulum**. The opening of the infundibulum, the **ostium**, is surrounded by ciliated, fingerlike projections called **fimbriae**. One fimbria is attached to the ovary, and the rest extend over the surface of the ovary without attachment.

From the infundibulum, the uterine tube continues medially and inferiorly as the **ampulla**, the longest and widest part of the uterine tube. The uterine tube continues toward the uterus and narrows into an **isthmus**, a short, constricted medial one-third of the tube located lateral to the uterus. The **intramural** portion of the uterine tube passes through the wall of the uterus and opens into the uterine cavity.

The Uterus [Figures 22.1 and 22.2]

An unpaired, thick-walled, pear-shaped, muscular organ located inferiorly in the pelvic cavity, the uterus lies posterior to the urinary bladder and anterior to the rectum. The uterus is normally divided into the **fundus, body, cervix,** and **isthmus.**

The **fundus** is the dome-shaped portion above the entrance of the uterine tubes into the uterus. The main portion is the **body.** Below the body, the uterine cavity narrows into the **isthmus.** The inferior, constricted portion of the uterus that opens into the vagina is the **cervix.**

The **uterine cavity** is the interior space within the body and fundus of the uterus. The **cervical canal** is a small opening that extends through the cervix into the vagina. The junction of the isthmus of the uterus with the cervical canal is the **internal os.** The opening of the uterine cavity through the cervix into the vagina is the **external os.**

The peritoneum covering the urinary bladder folds backward to form the **vesicouterine pouch.** Similarly, the peritoneum continues over the posterior surface of the uterus and upper part of the posterior vagina to form the deep **rectouterine pouch.** The peritoneum then continues over the anterior part of the rectum (Figure 22.1).

Several ligaments support the uterus. These include the **broad ligament** that attaches the uterus to either side of the pelvic cavity and the **round ligaments** that extend from the uterus to the skin of the **labia majora.** The other ligaments are **uterosacral** and **cardinal (lateral cervical) ligaments.**

The thick uterus wall has three layers. Lining the uterine cavity is the **endometrium,** the inner mucous layer. The thick, middle layer, the **myometrium,** forms the bulk of the uterus, and is composed of smooth muscle. The outer layer is the **perimetrium,** a thin outer covering of the peritoneum.

The Vagina [Figures 22.1 and 22.2]

The vagina is an elastic, muscular tube anterior to the rectum and posterior to the urethra and urinary bladder. The vagina extends from the cervix to the exterior of the body; the vaginal canal is continuous with the cervical canal of the uterus. The terminal part of the vagina surrounds and forms a continuous recess around the cervix called the **fornix**. The cervix projects into the vaginal lumen from above. The deepest vaginal recess lies posterior to the cervix and is called the **posterior fornix**. Also formed around the cervix are the **anterior fornix** and two **lateral fornices**. The vagina is normally collapsed. The vaginal wall is much thinner than the uterine wall and bears numerous transverse folds or ridges called **rugae**.

The Ovaries, Uterine Tubes, Uterus, and Vagina—Identification *[Figures 22.1 and 22.2]*

Examine the female reproductive organs on a model and/or in a prepared cadaver. Identify the following structures and then color them in the illustrations:

1. uterus
2. body of the uterus
3. fundus of the uterus
4. uterine tubes (oviducts)
5. fimbriae
6. infundibulum
7. ostium
8. ampulla
9. isthmus
10. ovaries
11. broad ligament
12. ovarian ligament
13. mesosalpinx
14. mesovarium
15. suspensory ligament

Midsagittal Anatomy of the Uterus and Vagina—Identification *[Figures 22.1 and 22.2]*

Examine the midsagittal section of the uterus and vagina in a model and/or prepared cadaver. Identify the following structures and then color them in the illustrations.

1. uterine cavity
2. endometrium
3. myometrium
4. perimetrium
5. isthmus of uterus
6. cervix
7. internal os
8. cervical canal
9. external os
10. vaginal fornix
11. posterior fornix
12. anterior fornix
13. lateral fornices
14. vaginal rugae
15. vesicouterine pouch
16. rectouterine pouch

The External Genitalia—Description *[Figure 22.3]*

Mons Pubis

A mound of subcutaneous fat tissue, the mons pubis is located over the bony **pubic symphysis**. In females who have reached puberty, the skin over the mons pubis is covered with coarse pubic hair.

Labia Majora

Two longitudinal folds of skin that extend inferiorly and posteriorly from the **mons pubis**, the labia majora are located laterally on either side of the **labia minora** and are also covered with pubic hair.

Labia Minora

Two thin folds of skin located medial to the labia majora, the labia minora lack pubic hair. Anteriorly, the labia minora merge over the **glans** of the **clitoris** to form a fold of skin called the **prepuce**.

The Vestibule

The vestibule of the vagina is the space or cleft between the labia minora. Found in the vestibule are the posterior **vaginal orifice**, anterior **urethral orifice**, and **openings of numerous secretory glands** that moisten the vaginal walls and vagina.

The Clitoris

The clitoris is a small body of erectile tissue located at the anterior border of the labia minora. Most of the clitoris is covered by the foreskin, or **prepuce**, formed by the junction of the labia minora. The exposed portion of the clitoris is called the

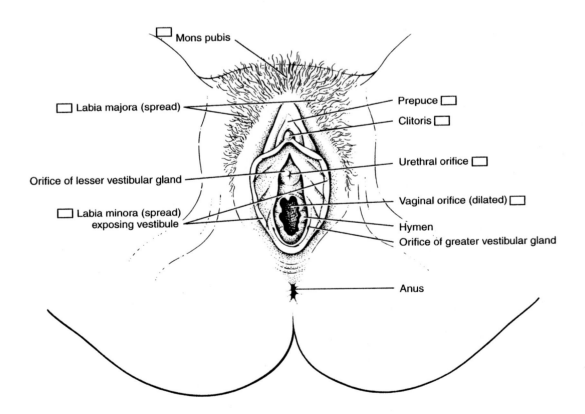

Figure 22.3 The external genitalia of the female.

glans. The skin-covered portion of the clitoris consists of two erectile tissue bodies called the **corpora cavernosa clitoris.** Posteriorly, these bodies diverge to form **crura** that attach to the pubic bones.

Unlike the penis, the clitoris does not have a true corpus spongiosum and does not incorporate the **urethra** into its body. As a result, in the female the urethra is a passageway for urine only.

Bulb of the Vestibule and the Greater Vestibular Glands [Figure 22.4]

Located deep to the labia (minora and majora) and lying on both sides of the vaginal orifice are two elongated masses of erectile tissue, each called the **bulb** of the **vestibule.** In addition, ducts of mucous-secreting glands empty into the vestibule. The **greater vestibular glands** lie on each side of the vaginal orifice and posterior to the bulb of the vestibule. During intercourse, these glands secrete mucus and deliver it through the ducts and into the vaginal orifice to lubricate the passageway.

The External Genitalia—Identification *[Figure 22.3]*

Examine the external genitalia of a female model and/or prepared cadaver. Identify the following structures and then color them in the illustration.

1. **mons pubis**
2. **pubic symphysis**
3. **prepuce of the clitoris**
4. **glans of the clitoris**
5. **urethral orifice**
6. **vaginal orifice**
7. **vestibule**
8. **labia majora**

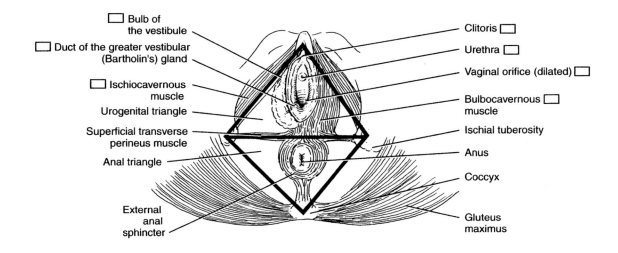

Figure 22.4 Deep structures of the female reproductive system.

Additional Deep Structures of the External Genitalia—Identification [Figure 22.4]

Examine the deep structures of the external genitalia in a female model and/or a specially prepared dissection of the cadaver. Identify the following structures:

1. corpora cavernosa clitoris
2. crura of clitoris
3. bulbs of the vestibule
4. greater vestibular glands
5. ischiocavernosus muscle
6. bulbospongiosum (bulbocavernosus) muscle

Homologous Female and Male Reproductive Organs

Portions of the genitalia of the two genders correspond or match each other in structure. They are said to be homologous. The **labia majora** are homologous to the male **scrotum**, and the **clitoris** to the **glans penis**. The **bulb** of the **vestibule** is homologous to the **corpus spongiosum penis** and **bulb** of the **penis**. The **greater vestibular glands** are homologous to the **bulbourethral glands** in the male.

The **crura** of the **clitoris**, like the crura of the penis, are covered by the skeletal **ischiocavernosus muscles**; the **bulb** of the **vestibule**, like the bulb of the penis, is covered by the skeletal **bulbospongiosum** (bulbocavernosus) **muscles** (Figure 22.4).

The Mammary Glands

Mammary glands are present in both genders. In males, they remain rudimentary, or undeveloped. In females, the mammary glands develop and function to nourish the newborn. The mammary glands are located inferior to the skin of the thorax and superior to the pectoralis major and serratus anterior muscles. The glands attach to the fascia of these muscles via connective tissue. Each mammary gland is contained within the breast, which has a small conical **nipple** surrounded by a pigmented area of skin called the **areola**.

Internally, each female mammary gland consists of 15 to 20 lobes that radiate from the nipple. The lobes are separated from each other by fat and connective tissue bands that form the suspensory ligaments of the breast. In each lobe are several smaller compartments called **lobules**; these contain the **alveolar glands** that produce milk during lactation. The **mammary gland ducts** that reach the nipple expand into **lactiferous sinuses**, which continue as **lactiferous ducts** that open to the outside of the nipple.

The Mammary Glands—Identification [Figure 22.5]

Examine the external anatomy of the human mammary gland on a model and/or prepared cadaver. Identify the following structures:

1. mammary gland lobe
2. lactiferous ducts
3. areola
4. nipple

CHAPTER SUMMARY AND CHECKLIST

I. INTERNAL ORGANS

A. The Ovaries
1. Paired organs located laterally in the pelvic cavity
2. Anterior surface is the hilus
3. Attached to mesovarium
4. Attached to the uterus by ovarian ligament
5. Attached to pelvic wall by suspensory ligament

B. The Uterine (Fallopian) Tubes
1. Paired structures that open near the ovaries
2. Attached to mesosalpinx
3. Distal end is the infundibulum
4. Infundibulum opening is the ostium
5. Ostium surrounded by fimbriae
6. Infundibulum continues medially as ampulla
7. Isthmus attaches to uterus

C. The Uterus
1. Unpaired organ located at bottom of pelvic cavity
2. Subdivided into fundus, body, isthmus, and cervix
3. Fundus is domeshaped, above entrance of uterine tubes
4. Body is the enlarged main portion
5. Isthmus is the narrow cavity below body
6. Cervix is constricted portion that opens into vagina
7. Cervical canal extends through the cervix into vagina
8. Peritoneum folds around uterus to form vesicouterine and rectouterine pouches
9. Attached by the broad ligament to pelvic walls
10. Round ligament attaches uterus to skin
11. Endometrium is the innermost layer
12. Myometrium is the thick, smooth-muscle middle layer
13. Perimetrium is the outer peritoneal lining

D. The Vagina
1. Extends from the uterus to exterior of body
2. Initial portion forms the fornix around cervix
3. Deepest fornix is the posterior
4. Shallower recesses around cervix are the anterior and lateral fornices
5. Interior wall bears folds or rugae

II. THE EXTERNAL GENITALIA

A. Mons Pubis
1. Mound located over the pubic symphysis

B. Labia Majora
1. Two longitudinal folds of skin located around vestibule of vagina

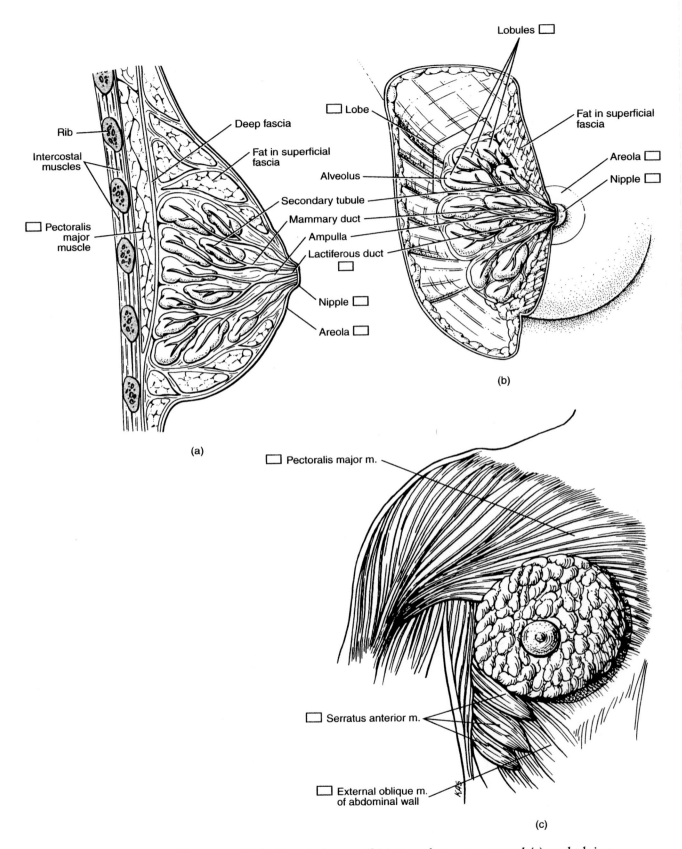

Figure 22.5 (a) Structure of the female breast, (b) internal structures, and (c) underlying structures.

C. **Labia Minora**
  1. Two small longitudinal folds medial to labia majora

D. **Vestibule**
  1. Space between the labia minora

E. **Clitoris**
  1. Small body of erectile tissue located at anterior border of labia minora
  2. Most is covered by prepuce
  3. Exposed portion is the glans
  4. Unexposed portion has two corpora cavernosa clitoris
  5. Does not contain urethra
  6. Crura covered by ischiocavernosus muscle

F. **Bulb of the Vestibule and the Greater Vestibular Glands**
  1. Erectile tissues
  2. Located deep to the labia on both sides of the vagina
  3. Greater vestibular glands are secretory, lie on each side of vaginal orifice and lubricate the canal
  4. Bulb of vestibule covered by bulbospongiosum (bulbocavernosus) muscle

G. **Homologous Male and Female Organs**
  1. Labia majora is homologous to scrotum
  2. Clitoris is homologous to glans penis
  3. Bulb of vestibule is homologous to corpus spongiosum penis and bulb of penis
  4. Greater vestibular glands are homologous to bulbourethral glands

III. **MAMMARY GLANDS**

A. **Description**
  1. Modified sweat glands present in both sexes
  2. In females, become functional for nutrition of offspring
  3. Location is beneath skin over pectoralis major and serratus anterior muscles
  4. Each gland exhibits a nipple surrounded by pigmented areola
  5. Internally, 15–20 lobes divide into lobules
  6. Lobules contain alveolar glands that secrete milk
  7. Lactiferous ducts from glands open at nipple to outside

Laboratory Exercises 22

NAME _____

LAB SECTION _____ DATE _____

LABORATORY EXERCISE 22.1

Part I

Organs of the Female Reproductive System

1. What are the one primary and three accessory internal organs of the female reproductive system?

   a. _____

   b. _____

   c. _____

   d. _____

2. Which ligaments attach to the ovary?

   a. _____

   b. _____

3. Which are the three parts of the uterine tube?

   a. _____

   b. _____

   c. _____

4. What surrounds the ostium of the uterine tube?

   _____

5. The uterus is subdivided into which four regions?

   a. _____

   b. _____

   c. _____

   d. _____

6. What structure connects the uterine cavity with the vagina?

   _____

7. The external os is located in which organ of the female?

   _____

8. What are the three layers of the wall of the uterus?

   a. _____

   b. _____

   c. _____

9. What is the recess formed by vagina around the cervix?

   _____

10. What structures are found in the vestibule?

    a. _____

    b. _____

    c. _____

11. What is the name of the exposed part of the clitoris?

    _____

12. What erectile bodies compose the clitoris?

    _____

13. Which glands are located on each side of the vagina?

    _____

Part II

Match the description on the left with the correct terms on the right. Some terms may be used more than once.

1. Fold of skin over the glans clitoris_____

2. Erectile tissue on each side of the vagina_____

3. Homologous to scrotum_____

4. Homologous to bulb of penis_____

5. Covered by bulbospongiosum

    (bulbocavernosus) muscle_____

6. Homologous to the glans penis_____

7. Pigmented skin around nipple_____

A. Bulb of the vestibule
B. Labia majora
C. Prepuce
D. Clitoris
E. Areola
F. Crura of the clitoris

Part III

Using listed terms, label the organs of the female reproductive system and then color them in the illustration.

Ovary
Uterine tube
Clitoris
Labia majora
Uterus
Round ligament

Vagina
Posterior fornix
Fimbria
Cervix
Labia minora

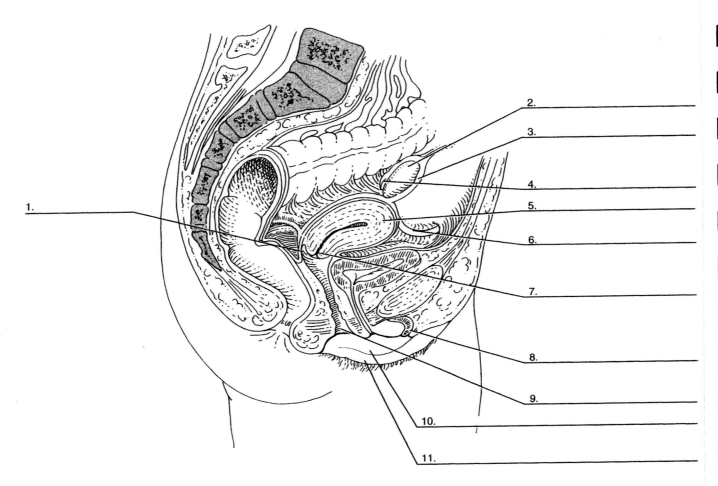

Figure 22.6 Sagittal section of the female reproductive system.

LABORATORY EXERCISE 22.2

Using the listed terms, label the organs of the female reproductive system, and then color them in the illustration.

Ostium
Infundibulum
Uterine cavity
Myometrium
Fornix
Ovarian ligament
Ovary
Broad ligament
Fundus
Fimbria
Isthmus (uterine tube)
Cervical canal

Perimetrium
Vagina
Uterine tube
Internal os
Ampulla
Isthmus (uterus)
Endometrium
Cervix
Mesosalpinx
Mesovarium
Body of the uterus
External os

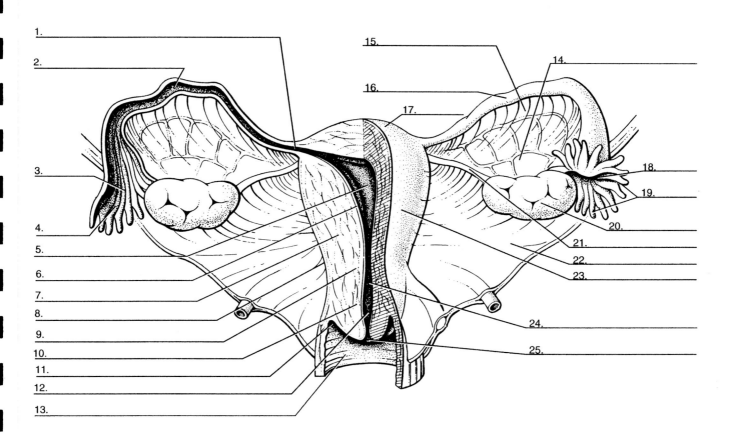

Figure 22.7 The female reproductive organs.

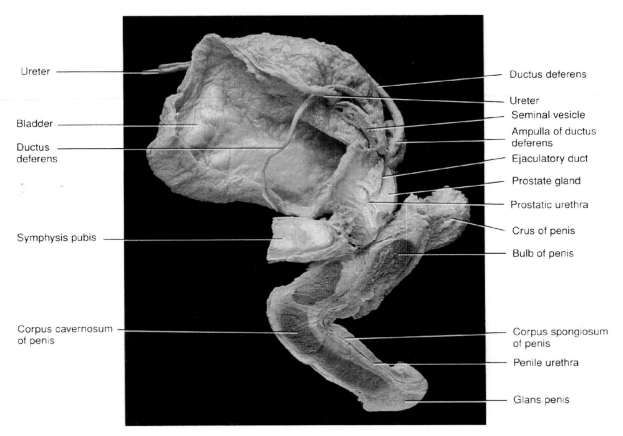

Ureter

Bladder

Ductus
deferens

Symphysis pubis

Corpus cavernosum
of penis

Ductus deferens

Ureter

Seminal vesicle

Ampulla of ductus
deferens

Ejaculatory duct

Prostate gland

Prostatic urethra

Crus of penis

Bulb of penis

Corpus spongiosum
of penis

Penile urethra

Glans penis

Plate 22. Longitudinal section of the penis, prostate gland, seminal veside, and bladder.

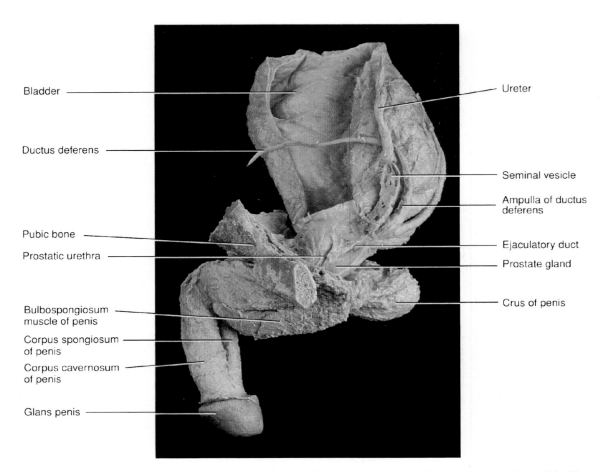

Bladder

Ductus deferens

Pubic bone

Prostatic urethra

Bulbospongiosum
muscle of penis

Corpus spongiosum
of penis

Corpus cavernosum
of penis

Glans penis

Ureter

Seminal vesicle

Ampulla of ductus
deferens

Ejaculatory duct

Prostate gland

Crus of penis

Plate 21. External anatomy of the penis, ductus deferens, seminal vesicle, prostate gland, and bladder.

POSTERIOR   ANTERIOR

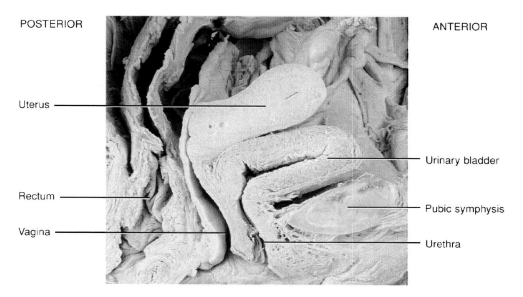

Uterus

Rectum

Vagina

Urinary bladder

Pubic symphysis

Urethra

Plate 19. Midsagittal section of the rectum, uterus, vagina, and bladder.

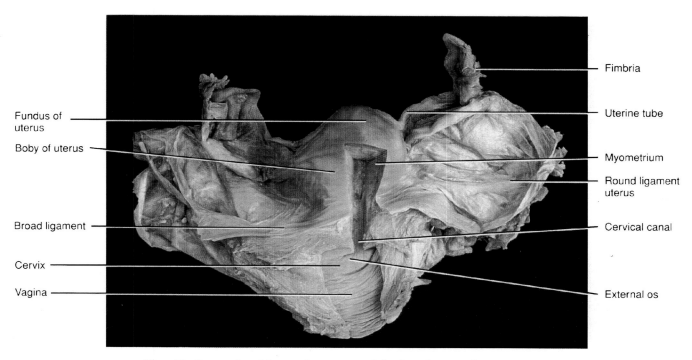

Fundus of
uterus

Boby of uterus

Broad ligament

Cervix

Vagina

Fimbria

Uterine tube

Myometrium

Round ligament
uterus

Cervical canal

External os

Plate 20. External and internal anatomy of the female reproductive organs.

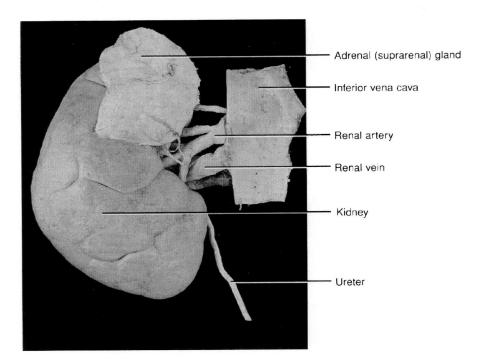

Adrenal (suprarenal) gland

Inferior vena cava

Renal artery

Renal vein

Kidney

Ureter

Plate 17. Anterior view of the right kidney with its associated structures and adrenal (suprarenal) gland.

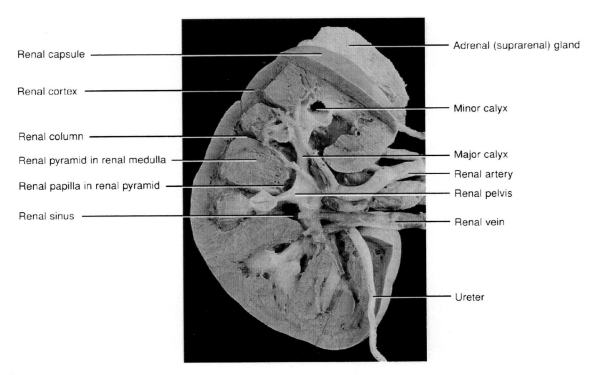

Renal capsule

Renal cortex

Renal column

Renal pyramid in renal medulla

Renal papilla in renal pyramid

Renal sinus

Adrenal (suprarenal) gland

Minor calyx

Major calyx

Renal artery

Renal pelvis

Renal vein

Ureter

Photograph of frontal section of right kidney (internal view)

Plate 18. Internal anatomy of the kidney with its associated structures and blood vessels.

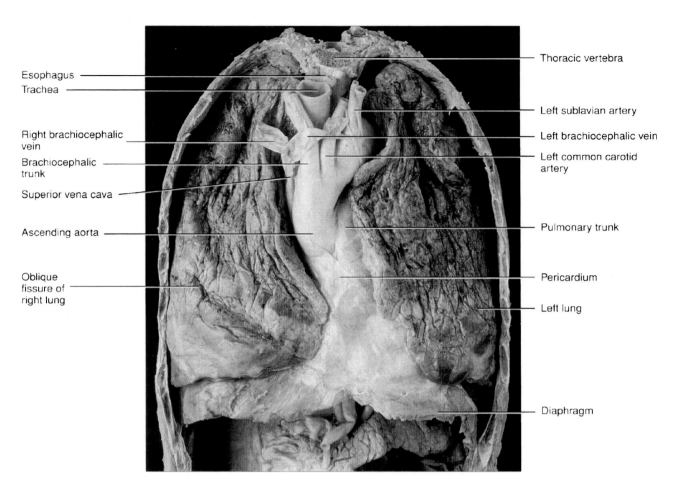

Esophagus

Trachea

Right brachiocephalic
vein

Brachiocephalic
trunk

Superior vena cava

Ascending aorta

Oblique
fissure of
right lung

Thoracic vertebra

Left sublavian artery

Left brachiocephalic vein

Left common carotid
artery

Pulmonary trunk

Pericardium

Left lung

Diaphragm

Plate 15. Anterior view of the thoracic cavity, its contents, and associated structures.

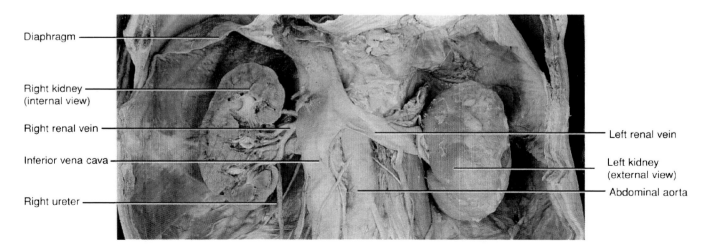

Diaphragm

Right kidney
(internal view)

Right renal vein

Inferior vena cava

Right ureter

Left renal vein

Left kidney
(external view)

Abdominal aorta

Plate 16. Anterior view of the internal and external anatomy of kidneys and associated structures.

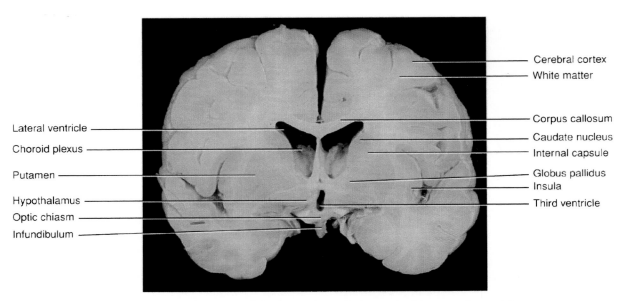

Cerebral cortex
White matter

Corpus callosum
Caudate nucleus
Internal capsule
Globus pallidus
Insula
Third ventricle

Lateral ventricle
Choroid plexus
Putamen
Hypothalamus
Optic chiasm
Infundibulum

Plate 13. Coronal section of the brain through the optic chiasm and ventricles.

Brachiocephalic veins

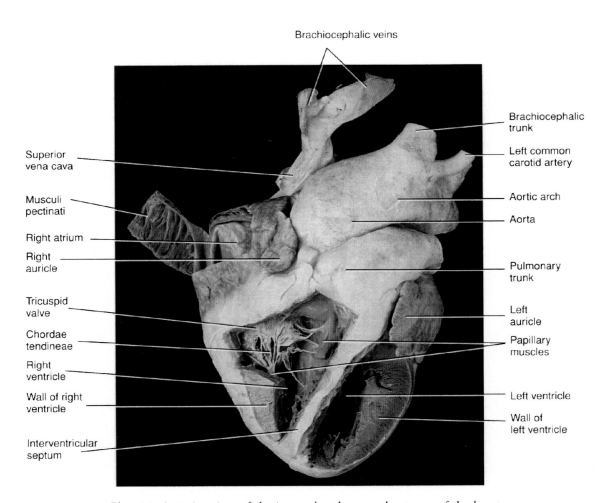

Brachiocephalic
trunk
Left common
carotid artery

Aortic arch
Aorta

Pulmonary
trunk

Left
auricle

Papillary
muscles

Left ventricle

Wall of
left ventricle

Superior
vena cava
Musculi
pectinati
Right atrium
Right
auricle
Tricuspid
valve
Chordae
tendineae
Right
ventricle
Wall of right
ventricle
Interventricular
septum

Plate 14. Anterior view of the internal and external antomy of the heart.

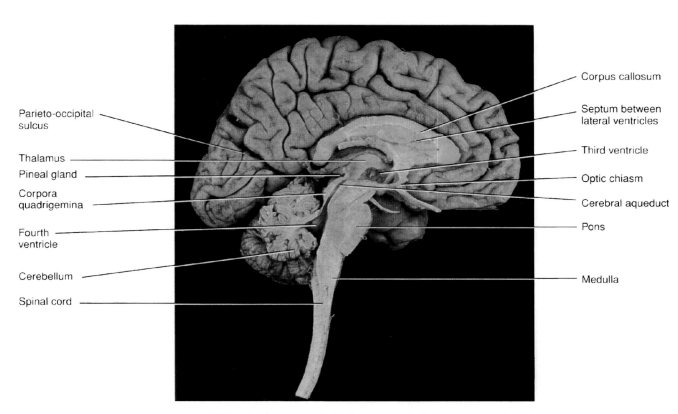

Parieto-occipital sulcus

Thalamus

Pineal gland

Corpora quadrigemina

Fourth ventricle

Cerebellum

Spinal cord

Corpus callosum

Septum between lateral ventricles

Third ventricle

Optic chiasm

Cerebral aqueduct

Pons

Medulla

Plate 11. Midsagittal section of the brain, cerebellum, and spinal cord.

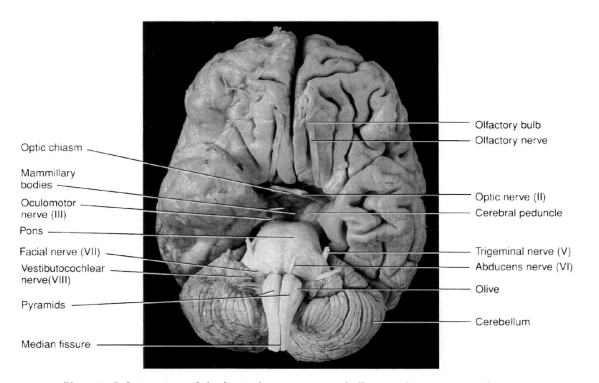

Optic chiasm

Mammillary bodies

Oculomotor nerve (III)

Pons

Facial nerve (VII)

Vestibutocochlear nerve(VIII)

Pyramids

Median fissure

Olfactory bulb

Olfactory nerve

Optic nerve (II)

Cerebral peduncle

Trigeminal nerve (V)

Abducens nerve (VI)

Olive

Cerebellum

Plate 12. Inferior view of the brain, brain stem, cerebellum, and certain cranial nerves.

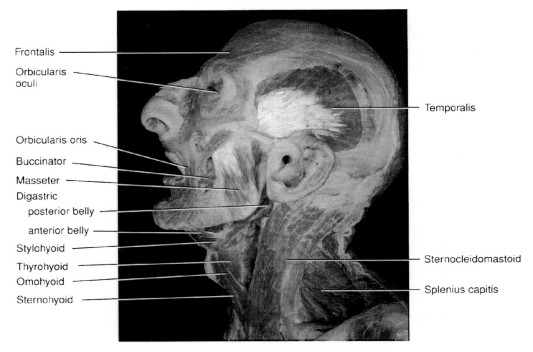

Plate 9. Superficial muscles of the skull, face, and neck.

Frontalis
Orbicularis oculi
Temporalis
Orbicularis oris
Buccinator
Masseter
Digastric
  posterior belly
  anterior belly
Stylohyoid
Thyrohyoid
Omohyoid
Sternohyoid
Sternocleidomastoid
Splenius capitis

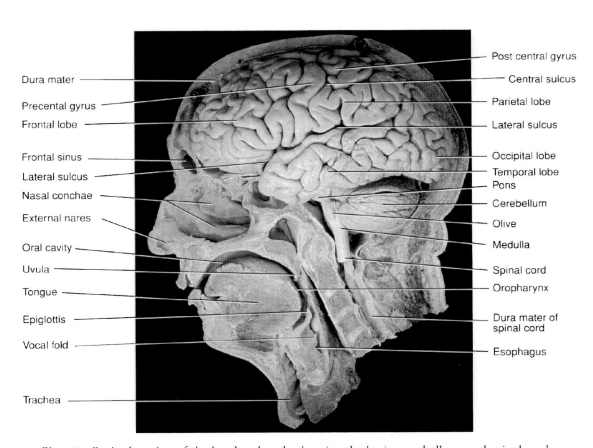

Plate 10. Sagittal section of the head and neck, showing the brain, cerebellum, and spinal cord.

Dura mater
Precental gyrus
Frontal lobe
Frontal sinus
Lateral sulcus
Nasal conchae
External nares
Oral cavity
Uvula
Tongue
Epiglottis
Vocal fold
Trachea

Post central gyrus
Central sulcus
Parietal lobe
Lateral sulcus
Occipital lobe
Temporal lobe
Pons
Cerebellum
Olive
Medulla
Spinal cord
Oropharynx
Dura mater of spinal cord
Esophagus

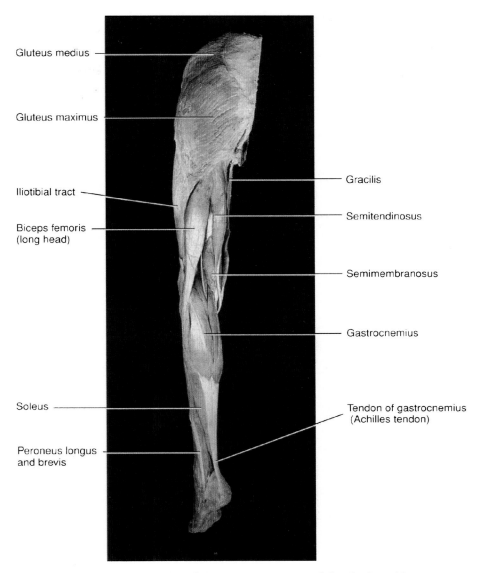

Gluteus medius

Gluteus maximus

Iliotibial tract

Biceps femoris
(long head)

Gracilis

Semitendinosus

Semimembranosus

Gastrocnemius

Soleus

Peroneus longus
and brevis

Tendon of gastrocnemius
(Achilles tendon)

Plate 8. Superficial muscles of the posterior aspect of the thigh and leg.

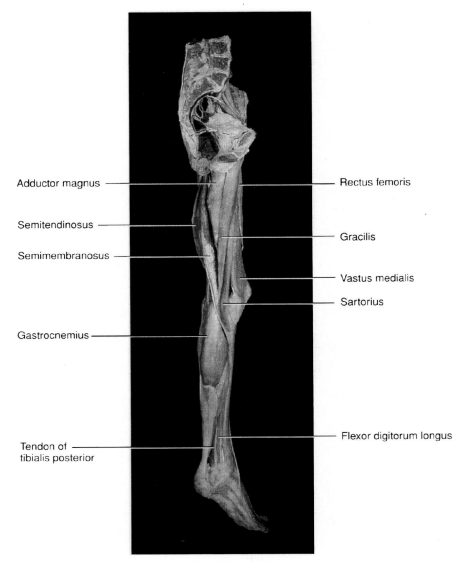

Adductor magnus

Semitendinosus

Semimembranosus

Gastrocnemius

Tendon of
tibialis posterior

Rectus femoris

Gracilis

Vastus medialis

Sartorius

Flexor digitorum longus

Plate 7. Superficial muscles of the medial aspect of the thigh and leg.

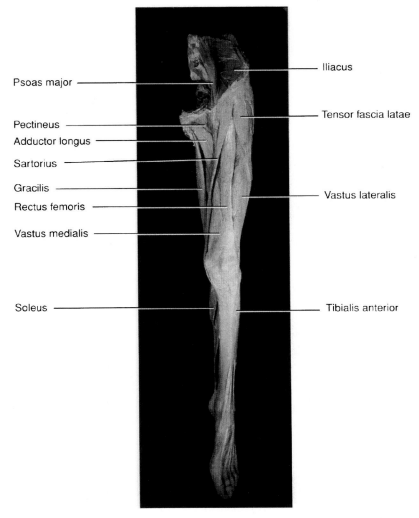

Psoas major ——————

Pectineus ——————
Adductor longus ——————
Sartorius ——————

Gracilis ——————
Rectus femoris ——————

Vastus medialis ——————

Soleus ——————

—————— Iliacus

—————— Tensor fascia latae

—————— Vastus lateralis

—————— Tibialis anterior

Plate 6. Superficial muscles of the anterior aspect of the thigh and leg.

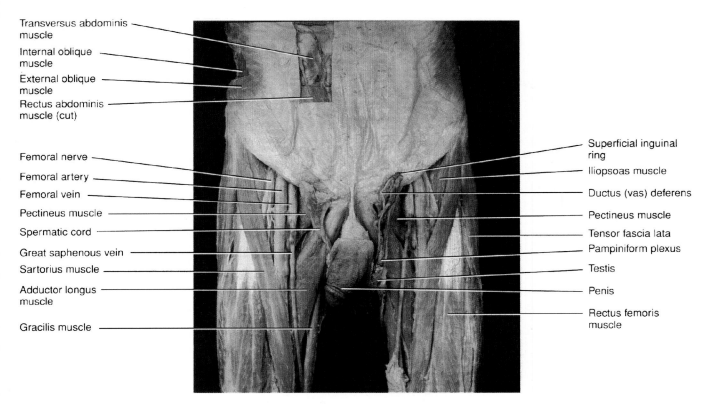

Transversus abdominis muscle

Internal oblique muscle

External oblique muscle

Rectus abdominis muscle (cut)

Femoral nerve

Femoral artery

Femoral vein

Pectineus muscle

Spermatic cord

Great saphenous vein

Sartorius muscle

Adductor longus muscle

Gracilis muscle

Superficial inguinal ring

Iliopsoas muscle

Ductus (vas) deferens

Pectineus muscle

Tensor fascia lata

Pampiniform plexus

Testis

Penis

Rectus femoris muscle

Plate 5. Anterior view of the body, showing muscles of the abdominal wall, thighs, blood vessels, and male reproductive organs.

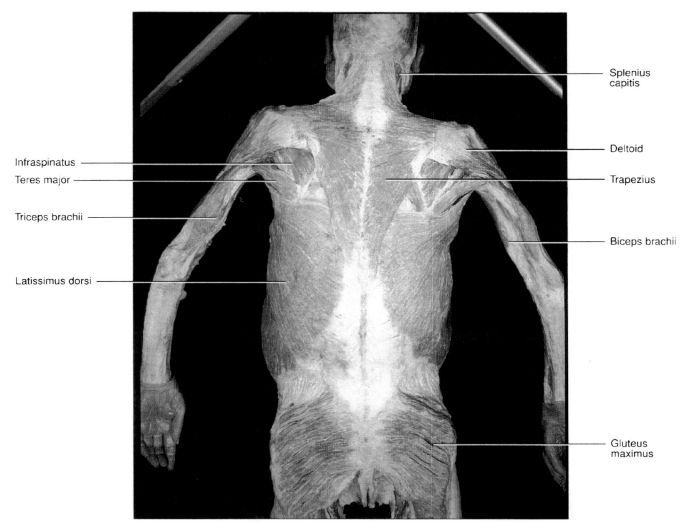

Splenius
capitis

Deltoid

Trapezius

Biceps brachii

Gluteus
maximus

Infraspinatus

Teres major

Triceps brachii

Latissimus dorsi

Plate 4. Posterior muscles of the neck, thorax, abdomen, and buttocks.

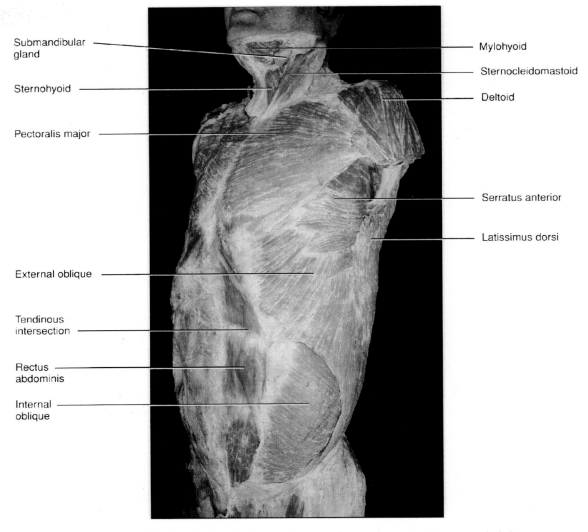

Submandibular gland

Sternohyoid

Pectoralis major

External oblique

Tendinous intersection

Rectus abdominis

Internal oblique

Mylohyoid

Sternocleidomastoid

Deltoid

Serratus anterior

Latissimus dorsi

Plate 3. Submandibular gland and the anterior muscles of the neck, thorax, and abdomen.

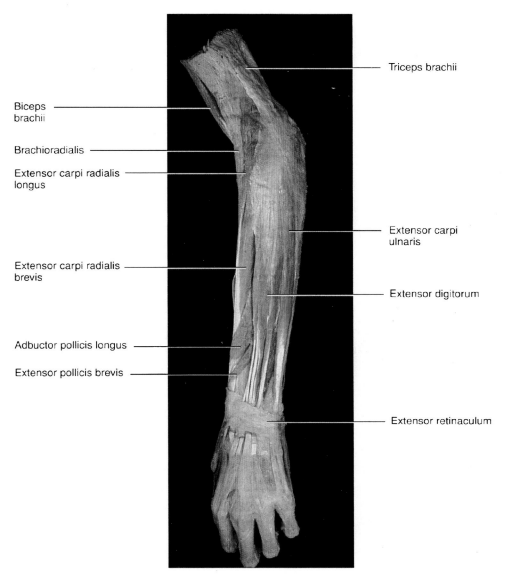

Triceps brachii

Biceps
brachii

Brachioradialis

Extensor carpi radialis
longus

Extensor carpi
ulnaris

Extensor carpi radialis
brevis

Extensor digitorum

Adbuctor pollicis longus

Extensor pollicis brevis

Extensor retinaculum

Plate 2. Superficial muscles of the posterior forearm.

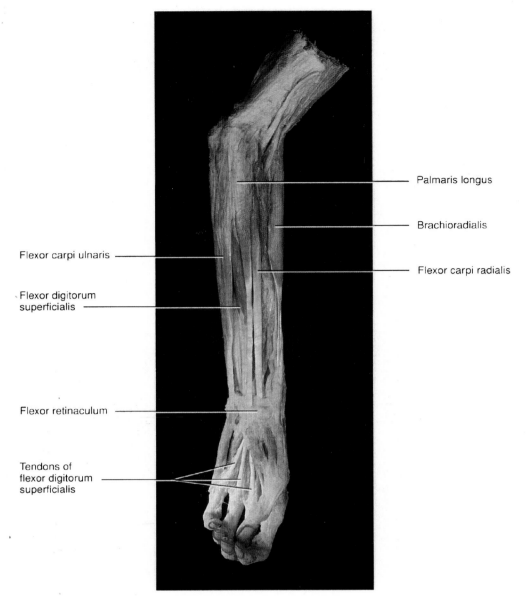

Palmaris longus

Brachioradialis

Flexor carpi radialis

Flexor carpi ulnaris

Flexor digitorum superficialis

Flexor retinaculum

Tendons of flexor digitorum superficialis

Plate 1. Superficial muscles of the anterior forearm.